Energy Security

NATO Science for Peace and Security Series

This Series presents the results of scientific meetings supported under the NATO Programme: Science for Peace and Security (SPS).

The NATO SPS Programme supports meetings in the following Key Priority areas: (1) Defence Against Terrorism; (2) Countering other Threats to Security and (3) NATO, Partner and Mediterranean Dialogue Country Priorities. The types of meeting supported are generally "Advanced Study Institutes" and "Advanced Research Workshops". The NATO SPS Series collects together the results of these meetings. The meetings are co-organized by scientists from NATO countries and scientists from NATO's "Partner" or "Mediterranean Dialogue" countries. The observations and recommendations made at the meetings, as well as the contents of the volumes in the Series, reflect those of participants and contributors only; they should not necessarily be regarded as reflecting NATO views or policy.

Advanced Study Institutes (ASI) are high-level tutorial courses to convey the latest developments in a subject to an advanced-level audience

Advanced Research Workshops (ARW) are expert meetings where an intense but informal exchange of views at the frontiers of a subject aims at identifying directions for future action

Following a transformation of the programme in 2006 the Series has been re-named and re-organised. Recent volumes on topics not related to security, which result from meetings supported under the programme earlier, may be found in the NATO Science Series.

The Series is published by IOS Press, Amsterdam, and Springer, Dordrecht, in conjunction with the NATO Emerging Security Challenges Division.

Sub-Series

A.	Chemistry and Biology	Springer
B.	Physics and Biophysics	Springer
C.	Environmental Security	Springer
D.	Information and Communication Security	IOS Press
E.	Human and Societal Dynamics	IOS Press

http://www.nato.int/science
http://www.springer.com
http://www.iospress.nl

Series C: Environmental Security

Energy Security

International and Local Issues, Theoretical Perspectives, and Critical Energy Infrastructures

edited by

Adrian Gheorghe
Old Dominion
University, Norfolk
Virginia, U.S.A.

and

Liviu Muresan
EURISC Foundation
Bucharest, Romania

 Springer

Published in cooperation with NATO Emerging Security Challenges Division

Proceedings of the NATO Advanced Research Workshop on
Energy Security in the Black Sea Area - Critical Infrastructure Protection
and System of Systems Engineering
Bucharest, Romania
19–23 October 2008

Library of Congress Control Number: 2011921812

ISBN 978-94-007-0721-4 (PB)
ISBN 978-94-007-0718-4 (HB)
ISBN 978-94-007-0719-1 (e-book)

Published by Springer,
P.O. Box 17, 3300 AA Dordrecht, The Netherlands.

www.springer.com

Printed on acid-free paper

CONTENTS

LIST OF CONTRIBUTORS

Abdallah, Tarek	Engineer Research and Development Center (ERDC), US Army Corps of Engineers, USA
Bores, Radu	EURISC Foundation, Bucharest, Romania
Bostan, Ion	Technical University of Moldova, Chisinau, Moldova
Bostan, Viorel	Technical University of Moldova, Chisinau, Moldova
Bosworth, Eric	Old Dominion University, Norfolk, VA USA
Caceu, Septimiu	EURISC Foundation, Bucharest, Romania
Celac, Sergiu	Romanian Institute for Sustainable Development, Bucharest, and EURISC Foundation, Romania
Degeratu, Claudiu	EURISC Foundation, Bucharest, Romania
Dicusară, Ion	Technical University of Moldova, Chisinau, Moldova
Ditmer, Renae D.	Center for Risk Management of Engineering Systems, University of Virginia, Charlottesville, VA, USA
Dulgheru, Valeriu	Technical University of Moldova, Chisinau, Moldova
Faas, Henryk	Institute for Energy, Joint Research Centre, Petten, the Netherlands
Fulli, Gianluca	Institute for Energy, Joint Research Centre, Petten, the Netherlands
Gheorghe, Adrian V.	Old Dominion University, Norfolk, VA USA
Gracceva, Francesco	Institute for Energy, Joint Research Centre, Petten, the Netherlands
Johnson, Melanie D.	Engineer Research and Development Center (ERDC), US Army Corps of Engineers, USA
Kanovits, Cristian	EURISC Foundation, Bucharest, Romania
Karvetski, Christopher W.	Center for Risk Management of Engineering Systems, University of Virginia, Charlottesville, VA, USA
Lambert, James H.	Center for Risk Management of Engineering Systems, University of Virginia, Charlottesville, VA, USA
Lenes, Leonela	EURISC Foundation, Bucharest, Romania
Linkov, Igor	Engineer Research and Development Center (ERDC), US Army Corps of Engineers, USA
Marcus, Solomon	Romanian Academy, Bucharest, Romania
Masera, Marcelo	Institute for the Protection and the Security of the Citizen, Joint Research Centre, Ispra, Italy
Minchev, Zlatogor	Institute for Parallel Processing-Bulgarian Academy of Sciences, C4I Department, Bulgaria

Minullin, Yaroslav	International Institute Applied Systems Analysis, Laxenburg, Austria
Mullen, Robert K.	Independent Consultant, USA
Muresan, Liviu	EURISC Foundation, Bucharest, Romania
Nerlich, Uwe	Center for European Security Strategy, Munich, Berlin, Germany
Prebensen, Chris	Norwegian Atlantic Treaty Association, Oslo, Norway
Refvem, Trygve	Norwegian Atlantic Treaty Association, Oslo, Norway
Schrattenholzer, Leo	International Institute Applied Systems Analysis, Laxenburg, Austria
Serbanescu, Dan	DG ENER D2 Unit Nuclear Safety, Decommissioning and Transport, Luxembourg
Shalamanov, Velizar	Institute for Parallel Processing-Bulgarian Academy of Sciences, C4I Department, Bulgaria
Sochireanu, Anatol	Technical University of Moldova, Chisinau, Moldova
Tomescu-Hatto, Odette	International Relations and Security Studies at *Sciences Po Paris*, France
Umbach, Frank	Center for European Security Strategy, Munich, Berlin, Germany
Umberger, Harold	Old Dominion University, Norfolk, VA USA
Vamanu, Dan V.	Horia Hulubei National Institute of Physics and Nuclear Engineering, Bucharest, Romania
Voeller, John	Senior Vice President, Black & Veatch, USA
Zio, Enrico	Ecole Centrale Paris-Supelec, Paris, France – Politecnico di Milano, Milano, Italy

ACKNOWLEDGMENTS

The editors would like to acknowledge NATO, Science for Peace and Security Program for the generous support and help in the organization of the event that resulted in this book. We also wish to thank to workshop participants and invited authors for their exceptional contributions to the book and peer-review of manuscripts. We are deeply grateful to Mrs. Kim Bullington Sibson, and Mr. Ersin Ancel for their excellent editorial assistance and management of the production of this book. The support from NATO – Supreme Allied Commander Transformation (SACT), Norfolk, Virginia, USA, through the personal attention from General James N. Mattis was highly relevant and encouraging to the workshop participants. NATO provided financial support for the workshop. Additional support was offered by the Old Dominion University, Norfolk, VA, USA, Batten College of Engineering and Technology through the professional involvement of Dean Oktay Baysal, EURISC Foundation, Transelectrica – Romania, University Politehnica Bucharest as well as ROSTREC project, financed by a grant in Romania from Norway through the Norwegian Cooperation Programme for Economic Growth and Sustainable Development. The patience and perseverance offered by Mrs. Wil Bruins, Springer editor of the book was highly appreciated.

INTRODUCTION

It is already acknowledged that, in terms of energy security the Black Sea region is important to Europe. So far, most of the debates on that subject have been focused on the vital supplies of oil and gas from Russia, the Caspian basin and the northern segment of the Middle East. Repeated disruptions of the transit of Russian gas through Ukraine, whatever the political or/and commercial motivation may have been, revealed some serious system vulnerabilities and highlighted the strategic nature of the pursuit of energy security in Europe today. Inevitably, and for very good reasons, a lot of attention has been given to the existing and planned pipeline routes going around or across the Black Sea. As it is indicated in the present book, much less attention has been given to the development of the Black Sea energy market in its own right and to the potential advantages of coping with some current and future energy issues in a multilateral regional format rather than through individual action at national level.

This book addresses, in a comprehensive manner, the current problematic of energy security, and goes beyond pipeline politics, without playing down their continued significance; it addresses some topical questions related to the sustainability and resilience of energy systems as applicable to the Black Sea region. More often than not, the topic of energy security is still being discussed, especially when it comes to hydrocarbons, in confrontational terms, as a zero-sum game of continental and inter-continental proportions. This does not necessarily have to be the case. The point of this book, which was generated by an NATO ARW (Advanced Research Workshop) event, which took place in Bucharest, Romania on October 19–23, 2008, is not to advocate a naive stance: let's all be friends, forget about geopolitics and see energy security as the perfect opportunity to prove the worth of the win-win approach. Ideally, such a change of attitudes may come about sometime in the future, but not just yet. For the time being, we shall have to operate within the confines of reality, which is not always pleasant. It is to be expected that the transition to a new pattern of energy production and consumption worldwide in the twenty-first century will be long and complicated, entailing continued competition for the control of energy resources and transit routes, increased vulnerability for some states as a result of political manipulation of energy markets, opportunistic arrangements and broken promises. The inherent inertia of the energy systems requiring huge investment over long periods of time will further compound the intricacies of this process.

The book brings the topic of energy security from various perspectives, and it is different from those studies published till now. The main chapters address the following aspects:

- diversity on energy security; an international perspective
- theoretical perspectives to energy security
- critical energy infrastructures: operational efficiency, security, and governance

A. Gheorghe and L. Muresan (eds.), *Energy Security: International and Local Issues,*
Theoretical Perspectives, and Critical Energy Infrastructures,
DOI 10.1007/978-94-007-0719-1_1, © Springer Science+Business Media B.V. 2011

Within this approach, concepts such as Security 3.0, Energy War Room, Tangibles and Intangibles, Energy plus Information, Infranomics are brought into discussion and adopted as additional new models and tools in order to deal with the current discourse of energy security.

It was agreed by the contributors to this work that, the planet we are living is under stress. The mankind is pushing the supportability of our "global shuttle" beyond the known limits. In most of the main issues we are about to exhaust the resources of water, fresh air, energy a.o but also, more challenging, of supportability and time. In this environment the man is producer and also consumer of instability/insecurity. What is he now looking for? The main international institutions as results of the Second World War are in a moment of a need of reinventing themselves. How relevant are they? How to avoid overlapping? How to overcome the institutional egos or the national priorities? All that is in a moment of a complex crisis, with effects on states, and international community. Which could be the answer to all that? Possible that the answer could be in the point where insecurity is meeting the citizen with his basic needs. A lot of sequential approaches human security, energy security, water security, food security a.o as well as hard and soft security. The present NATO ARW indicated a few novel aspects related to the concept of security, concluding that:

Security 1.0 War As Usual
Security 2.0 Cold War Era (No War and No Security)
Security 3.0 Distributed Security vs./or/and Global Sustainable Security.

Energy security issues, part of the new Security 3.0 paradigm, are usually related to tangible dangers, such as the increase in energy prices, increased world competition for energy resources faced with the rapid emergence of some huge economies like China and India, the lack of equitable distribution between major powers, sometimes at the expense of less developed countries. Dangers may appear in some controversial ways with regard to energy-related renewable technologies, energy policy related to unexpected changes in international relations, and at the role of decentralization. In all these respects, a long and a short – term perspective are both necessary.

Despite the unquestionable importance of all these issues, there is however a need to supplement them with a more theoretical perspective, to consider a broader context by going back to the historical development of the energy paradigm and, according to this development, to stress its interaction with the information and the computation paradigms. Moreover, we focused our attention on a universal paradigm strongly related to the security paradigm: the equilibrium paradigm. Indeed, equilibrium is a condition of security; it is a pattern goes across mechanics, thermodynamics, chemistry, biology, economics, sociology, linguistics, art etc. The mathematics of the idea of equilibrium gives the foundations, the link, and the common denominator of all versions of equilibrium, from Lagrange's analytical mechanics to the Nash equilibrium in the theory of strategic games. As was indicated by one of the authors of the present book, the mathematical theory of equilibrium is applicable all the way from quantum mechanics to psychiatry pointing out different degrees of instability.

Another source of innovation in the study of energy security comes from the challenges of physics. Topics such as "computation with physical constraints" and "wireless sensor networks" lead to a new, emergent topic: energy as a new computational resource. There is now an urgent need for a theory of computational energy complexity.

Energy security is one of the most discussed topics in examination of the issues of sustainability and most of these discussions cover the same ground with different perspectives on the same set of topics. Most of these discussions focus on what can be called direct or first order terms in the security equation. There is no question these terms are the most common or easily understood ones but they are not necessarily the most important or most easily disruptive to genuine security.

Much of the current view of security comes from an era where the major western nations were both the largest consumers and makers of the elements of energy from down-hole to light socket. Neither of these perspectives is accurate today and will become even less true over time as the emerging nations become not only the major consumers and makers but also innovators for reasons the early adopters did not have or choose to address. It is here that several critical components of energy security in numerous contexts become apparent.

This book brings into picture a new paradigm defined as energy <u>insecurities</u>. The focus on this shift of emphasis is the most critical topic for the next period. In looking at the many issues of pursuing certainty, the perspective of security contains a hidden problem. The problem is that security tends to focus on removing or neutralizing a list of known threats. The broader and more important perspective involves looking at the sea of circumstances to identify possible insecurities not yet cataloged.

Using the insecurity perspective and the undefined threat viewpoint is critical and it is interesting that this alternative viewpoint was introduced into U.S. government literature in 2006 by the US Department of Homeland Security produced National Critical Infrastructure R&D Plan Update and first appeared in the European Union's guiding research literature in Framework 7 in 2008. However, the bulk of the current operational efforts and industrial offerings are still focused on known security concerns. The following are just some of these higher order components of energy security.

In recent years, concerning the concept of energy security, we have consolidated the belief on the need for: (i) a systematic approach to the understanding of all the interrelated aspects that constitute the critical infrastructure topic; (ii) taking into consideration all tangibles and intangibles aspects, and all the concurrent concordance-discordance facets; and (iii) the rigorous structuring of the decision making matter in consideration of all stakeholders.

A new concept Infranomics, has been recently proposed as a crucial discipline for this century. Neither authorities, nor industrial or academic bodies could afford to ignore the advent of the convolution of opportunities and risks accompanying the implementation of the new generation of infrastructures e.g. energy infra-structures. The shape of our society will be determined by the characteristics of, and the services delivered through, those infrastructures. In this picture, communications, mobility, commerce, energy, health, finance, education, environmental impact, security, and up to international relations, each and all of them cannot be treated independently.

The overall message of this book is that our society will be as efficient and as secure as the infrastructures constituting it, would allow. The texture of interconnected infrastructures, each one of them a System-of-Systems by itself, is evolving into a Metasystem (a System-of Systems-of-Systems), which – for some features – will have to be considered as a whole (Gestalt notion). At that level of aggregation, where the Metasystem is contemplated from the society standpoint, new principles can be hypothesized: ASAIA (As Secure As Infrastructure Allow) and AEAIA (As Efficient As Infrastructure Achieve).

The new generation of infrastructures will be constructed upon the existing ones, however the services, architecture, business models and attributes of those newly developed systems will signify a qualitative leap forward, and not just more of the same.

Infranomics is the body of disciplines supporting the analysis and decision-making regarding the Metasystem. Infranomics is the set of theories, assumptions, models, methods, and associated scientific and technical tools required for studying the conception, design, development, implementation, operation, administration, maintenance, service supply, and resilience of the Metasystem. Because none of the currently existing disciplines provides a complete solution.

Within the economy of this book, Infranomics might manifest through inter alia the following acts:

- Analyzing how stakeholders interact, coordinate their functions, establish legal regimes and normative rules, set the economics of their services, handle normal and abnormal situations;
- Analyzing how infrastructural systems work and how they can fail (e.g. blackouts), their impact onto the resilience of society, taking into consideration all technical and organizational elements and external factors;
- Analyzing how infrastructures evolve with the introduction of new technologies, how vulnerable they are in different security scenarios, and how their adequacy and performance might degrade over time;
- Developing theories and methodologies for the modeling, simulation, assessment, critical analysis, and empirical investigation of infrastructures and their dependability;
- Assess infrastructures as a whole in the interaction among the political, legal and economic governance of societies, and mainly with respect to cross-border and multi-jurisdictional settings.

One of the key objectives of Infranomics should be to support decision making in view of achieving a set of goals, at the corporate, regional, national and international levels. For this aim, Infranomics should integrate the engineering, economic, political and social sciences, considering the inter-relationships among infrastructures and stakeholders, in a multi-national, multi-jurisdictional context.

The topic of energy security adopted in this book, generously initiated by NATO activities within the Science for Peace Program, intends to address this new interdisciplinary domain i.e. *energy security* in a novel way by simultaneously using new concept such as resilience, governance, critical energy infrastructures.

The Editors
Norfolk, Virginia, USA
November 2010

PART I: DIVERSITY ON ENERGY SECURITY:
AN INTERNATIONAL PERSPECTIVE

EUROPEAN SECURITY – A EUROPEAN PERSPECTIVE

HENRYK FAAS*, FRANCESCO GRACCEVA,
GIANLUCA FULLI[1]
*Institute for Energy, Joint Research Centre, Petten,
the Netherlands*

MARCELO MASERA
*Institute for the Protection and the Security of the Citizen,
Joint Research Centre, Ispra, Italy*

Abstract.   In spite of its critical role, energy security lacks a universally agreed definition, which given its complexity may well just be unrealistic. The concept is still used in a wide range of reports and documents, often without discussion of its dimensions and their significance. As a consequence, the literature is characterized by an almost overwhelming focus on securing supplies of primary energy sources and geopolitics. Clearly, views on energy security also differ widely between nations. The European Union's approach towards energy security is presented in this paper; it can be derived from several policy legislations and proposals that followed the European Commission's 2000 Green Paper "Towards a European Strategy for the Security of Energy Supply".

Introduction

The horrors of two world wars and a long history of violent conflict between European powers prompted a desire to bridge the divisions between European nations and move towards more integration. On May 9, 1950, the French Foreign minister Robert Schuman proposed to place the French and German production of coal and steel, essential military resources, under a common High Authority, "open to the participation of the other countries of Europe". By inextricably linking economies, so the hope was, the creation of a European Steel and Coal Community would make war "not only unthinkable but materially impossible". Today, European economies and specifically the energy systems are closely interlinked and cooperation is essential for securing our energy supply.

---

* To whom correspondence should be addressed: Henryk Faas, Institute For Energy, Joint Research Centre, Petten, The Netherlands e-mail: Henryk.FAAS@ec.europa.eu
[1] Disclaimer: The views expressed are purely those of the authors, and may not in any circumstances be regarded as stating an official position of the European Commission.

A. Gheorghe and L. Muresan (eds.), *Energy Security: International and Local Issues,
Theoretical Perspectives, and Critical Energy Infrastructures,*
DOI 10.1007/978-94-007-0719-1_2, © Springer Science+Business Media B.V. 2011

Energy is indispensable to our economies and our way of life. Since production and consumption for most energy carriers are separated by long distances and since energy cannot be easily substituted or stored, energy security is an essential concern for industrialized nations. It should not be forgotten, however, that energy poverty still afflicts over a third of the world's population, which lives without access to electricity.

The major challenge for both policymakers and policy-oriented research is to select economically rational strategies for energy security, i.e. strategies based on a careful consideration of their costs and benefits and the possible interactions between policies. This will require a robust, rigorous and transparent approach to assess future energy security challenges. The aim of this chapter is to present the key elements of such an assessment, and to highlight some of the main challenges for securing Europe's energy.

The Concept of Energy Security

In spite of its critical role, energy security lacks a universally agreed definition, which given its complexity may well just be unrealistic (Chester 2010). The concept is still used in a wide range of reports and documents, often without discussion of its dimensions and their significance. As a consequence, the literature is characterized by an almost overwhelming focus on securing supplies of primary energy sources and geopolitics. Clearly, views on energy security also differ widely between nations (Yergin 2006).

Any consistent assessment of energy security issues requires an explicit discussion of the concept and measures of energy insecurity. This vagueness of energy security lends itself to justify various policies or actions, and be "exploited in various ways by different interest groups" (Loeschel, Moslener et al. 2010). Depending on how the problem of energy security is framed (for example how many sectors are considered), the same energy system can be considered either secure or insecure (Winzer 2010). Starting from the general definition of security as "the condition of being protected from or not exposed to danger" (according to the Oxford English Dictionary) and from a review of several definitions, Winzer reduces energy security to the "absence of or protection of a system against energy supply related threats". There are two key elements in these definitions, one general and one specific: First, the condition of "not being exposed" to danger, or the "absence" of threats requires an assessment of the probability that critical events can occur, as well as their potential magnitude and impact. Second, the condition of being "protected" from threats, or of not being vulnerable to possible adverse events, can be synthesized to the concept of energy system resilience, i.e. the "capacity of an energy system to tolerate disturbance and to continue to deliver affordable energy services to consumers".

A corollary of this concept of energy security is the need to extend the restrictive concept of security of supply towards the more systemic concept of "energy services security" of the energy system as a whole (Tosato 2007, DBERR 2007, Jansen 2009). A system approach aims at including all significant elements of the system under study, emphasising the relations and synergies between them and the integral nature of the system. Energy systems include the complex, interrelated chains of commodities and processes linking the extraction of primary energy to the satisfaction of the demand for energy services. The extended concept of energy services security has much wider implications than securing the smooth behaviour of the global crude or natural gas markets. It encompasses all energy sectors and all steps in the energy chain, from the oil reservoir or coal mine to the passenger kilometres or the warm water demanded by the end user (Tosato 2009).

In the short term, the options for an economy to substitute energy for capital or labour are limited, which means that, for industrialized countries, energy possesses the role of a limiting factor for production as well as consumption. Problems due to energy supply disruptions can therefore rapidly and extensively affect the welfare of a country. The impact of possible adverse events on energy security is determined by the economic system, as well as by the timing of incidents. This is due to the wide variety of factors determining the capacity of the system to cope with those events, i.e. what is generally known as 'energy system resilience' (Lovins 1982).

In summary, even if many policies focus on a specific supply chain, a comprehensive assessment of energy security needs to take a systemic approach. A robust measure of energy security "should focus on what society cares about, i.e. the well-being of its members" [DBERR, 2007].

Quantitative Approaches to Energy Security

A key question for policy oriented research in energy security is how different policy options shape an energy system that is resilient enough to disruptive events to satisfy society's needs for energy services? This requires an assessment of the impacts of a policy on the energy system and an understanding of what is resilient 'enough'. All energy systems deliver some level of security for consumers (NERA 2002) and any increase in resilience comes at a price. Beyond a certain point, the costs of increased resilience exceed the expected benefits.

If the need for a policy intervention has been accepted, then both the nature and the extent of that intervention have to be carefully assessed, as any policy may have a considerable influence on the energy system. There is also a risk that market interventions are being justified by the need to increase energy security without proper assessment, in fact furthering different objectives. Policies on climate change and economic competitiveness, directed towards goals other than energy security, will have a strong impact on the structural characteristics of the energy system. Despite the synergies between those policies, it needs to be examined how they impact energy system resilience.

Two broad approaches to the quantitative assessment of energy security can be distinguished: The first is based on energy security indicators, sometimes combined with energy scenario analyses. Indicators are useful policy instruments because they enable a certain level of standardization. Composite indicators of high complexity lose the advantage of being transparent, while simple indicators can be too numerous or too narrow to be applicable in real situations. A more structural limit of this approach is that even the most sophisticated indicators are not measures of the resilience of the entire energy system. The second approach encompasses efforts to estimate costs and benefits of energy security. Some of these estimations are simply derived through ad hoc assessments. In other cases they are based on more complex macroeconomic, econometric or energy system models. An advantage of this approach is the possibility to carry out comprehensive, quantitative assessments of energy security and to assess mitigation strategies (Gallaghar, 2009).

European Energy Policy and Energy Security Implications

The European Union's approach towards energy security can be derived from several policy legislations and proposals that followed the European Commission's 2000 Green Paper "Towards a European Strategy for the Security of Energy Supply". This stated that "The EU's long-term strategy for energy supply security must be geared to ensuring [...] the uninterrupted physical availability of energy products on the markets, at a price which is affordable for all consumers, while respecting environmental concerns and looking towards sustainable development". In 2006, a second Green Paper titled "A European Strategy for Sustainable, Competitive and Secure Energy" of 2006 (COM(2006)105) defined security of supply as one of the three pillars of European energy policy, alongside competitiveness and sustainability.

A central element of the European energy policy is the will to fundamentally change its energy system to a low carbon economy, with the explicit goal of breaking "the cycle of increasing energy consumption, increasing imports, and increasing outflow of wealth created in the EU to pay energy producers" (COM(2008)781). Also, Member States agreed to drastically cut greenhouse gas (GHG) emissions until 2050 (Second Strategic Energy Review 2008). This requires a shift towards new technologies laid out in the Strategic Energy Technology Plan (SET Plan), including renewable electricity, second-generation biofuels, smart grids, electricity storage, transport sector electrification, and carbon capture and storage among others. Clearly, time frames in the lead up to 2020 are too short for fundamental infrastructure changes, but many of the foundations need to be laid, including financial and market structures. The IEA estimates that 2 trillion EUR are needed in energy infrastructure investment by 2030 (IEA 2009), which could be financed by returns from the Emission Trading Scheme or other support mechanisms.

In January 2008 the Commission launched the so called "20-20-20" Energy and Climate package (agreed by the European Council in March 2007), which requires the EU to reduce its GHG emissions by 20%, to increase the share of renewables in energy consumption to 20% and to save 20% of total primary energy consumption compared to baseline ('business as usual', i.e. no major changes), all by 2020. It is remarkable that all 27 Member States have committed themselves to a legally binding target of introducing 20% of renewable into their energy system by 2020, requiring national action plans that establish pathways for the development of renewable energy sources and creating cooperation mechanisms to help achieve the targets cost effectively (Directive 2009/28/EC). In the medium-long-term, the EU's 20-20-20 strategy calls for "an energy system with a diversity of non-fossil fuel supplies, flexible infrastructures and capacities for demand management [that] will be very different in energy security terms than today's system." In the short to medium term, effective provisions for preventing and dealing with supply crises must be made, in order to diminish the vulnerability to energy supply shocks (Figure 1).

Figure 1. Target for the share of renewables in energy consumption in 2020.

Key elements to reach the three underlying objectives of EU's new energy and environment policy are the idea of energy security as an issue of common EU concern and further integration of energy markets. "Solidarity between Member States is a basic feature of EU membership, and strategies to share and spread risk, and to make the best use of the combined weight of the EU in world affairs can be more effective than dispersed national actions" (Second Strategic Energy Review, COM(2008)781 final). While the energy mix is still the responsibility of the Member States, the Lisbon Treaty has established the role of the European Union, reflecting the increasing interdependence of national energy systems and the benefits of a more coordinated external energy policy. Notwithstanding the trust in market mechanisms and market integration as the most effective ways for increasing energy security, the explicit goal of reducing energy imports, expressed in the EU Energy Security and Solidarity Action Plan, still reflects a preventive attitude with "energy supply autonomy" as a useful strategy to reduce the exposure of the economy to international energy crises (Gnansounou 2008).

Yet striving for energy independence alone would be too narrow a view of energy security. In a global economy, to seek independence would be an exception to the wider free trade Western policy and could on balance be negative for consumers. The EU has thus developed advanced dialogues with important energy producers, transit and consumer countries, including Early Warning Mechanisms to deal with energy supply disruptions. Energy security is thus embedded in a context of international organizations, institutions or mechanisms, such as the Energy Community, the Energy Charter Treaty, the IAEA, IEA, UNECE, and the EU-Russian Federation Partnership.

Energy Security Challenges for Europe

To identify the main European energy security challenges, it is useful to briefly consider some current trends. World demand for energy is set to increase by around 40% between 2007 and 2030 (IEA 2009). China is now the second biggest economy, and non-OECD countries already account for more than half the world energy consumption. The EU's gross inland energy consumption reached 1806 Mtoe in 2007, corresponding to about 15% of the world's consumption. Europe's primary energy demand is projected to grow by just 0.2% per year by 2030 (IEA 2009). At the same time, Europe's indigenous hydrocarbons resources are shrinking as well as its overall energy production. As a result, the European Union imports more than half of its energy, a trend that is expected to continue through 2030. Russia is still the EU27 main energy partner, providing about 33% of imported oil, 42% of imported gas and 26% of imported coal. The EU is becoming increasingly exposed to the effects of price volatility and price rises on international energy markets. Energy security concerns in European Member States differ widely as a result of different energy national consumption patterns. Transnational energy transport is increasing and European infrastructure is highly interdependent. Some Member States depend fully on imports. For example, after the decommissioning of its nuclear reactor, Lithuania lost 80% of its electricity production capacity, which was largely replaced by (nuclear) electricity imports from Russia. Therefore, initiatives like the Baltic Energy Interconnection Plan and Mediterranean Energy Ring promote regional integration.

As a result of these trends, on the world stage Europe is still an important consumer, but increasingly sharing this role with other emerging players. Europe is and will continue in the medium term to be highly dependent on import from a small number of producer countries and long reaching transport corridors. If prices could adjust freely, security of supply could in principle be guaranteed (Helm

2002). But if market failures occur, a pure market approach does not produce a desirable outcome for society. To the extent that access to energy is a public good, public intervention (through subsidies, taxes or carbon pricing for example) can be justified.

Making energy production more sustainable through the introduction of renewable generation also presents energy security challenges. A reduction of hydrocarbon energy consumption will decrease imports, but also implies fundamental changes to the European electricity and wider energy system. In order to comply with the energy and climate change policy targets of the EU by 2020, the grids must be capable to host 'Renewable Energy Sources for Electricity' (RES-E) covering at least 30–35% of the EU electricity consumption compared with a 16% share recorded in 2006. To meet this objective, it is necessary to change the electricity infrastructure, primarily in order to cope with large amounts of variable generation from renewable energy sources such as wind and solar power generation. Adjectives such as 'super' and 'smart' are therefore more and more adopted when analysing future electricity grids to reflect features such as improved adequacy, flexibility, reliability and controllability (SET-Plan technology map (JRC 2009).

Flexible, coordinated and adequate electricity networks designed according to new architectural arrangements and embedding innovative technological solutions, are essential to address the risks of deterioration of reliability and security of supply. Information and Communication Technology (ICT) can contribute to increase the adequacy and robustness of the system, thus reducing the need for building new infrastructures, as well as augmenting its monitoring and governing.

While there is much debate on how Europe's ambitious targets can be reached, it is clear that one of the most important ways is to increase energy efficiency. Consumer behaviour does not yet support these ambitious goals, although average energy savings benefits for a household per year can be up to 1,000 EUR (COM (2008)772). Energy consumption is about 20% higher than for economic reasons necessary.

One important step to reduce energy consumption is to curb oil utilization, which remains the most important energy carrier. Transport in Europe is still almost exclusively dependant on oil, making demand highly inelastic. Consequently, replacing oil as a transport fuel is a focus of development and research. The capacity of EU policymakers to put in place efficient policies on transportation will play a key role. In the meantime, however, the classic energy security concerns associated with oil still apply and oil price volatility remains a major concern.

Specific European Challenges and Initiatives

In the following we highlight, by way of example, three specific challenges for energy security in Europe, and describe key policy initiatives designed to tackle them: the role of natural gas as a bridge energy carrier, the need for a remodelling of the European electricity grids and the problem of cybersecurity. The common characteristic of all these challenges is in their "network nature", implying that dealing with them relies on the reduction of the vulnerability of energy networks and increased resilience.

SECURITY OF GAS SUPPLY

Natural gas is likely to play an important role in the medium term as a bridge energy carrier on the road to a 'low carbon energy system'. It has considerably lower emissions than oil and coal and global gas supply is abundant, projected at 150 times current annual consumption levels (MIT 2010). However, for the European Union, in particular its Eastern Member States, securing natural gas supply is a key energy security concern.

Gas is a network bound energy carrier and for the whole of Europe, supply routes are quite evenly distributed: roughly one quarter comes from North Africa, the North Sea, the Middle East and Russia respectively (Figure 2). Yet, many Eastern European countries receive supplies almost exclusively from Russian gas fields and 80% of Russian supplies are transported via Ukraine. There are wide discrepancies in the role of natural gas in Member States' primary energy mix and in their ability to substitute it with other energy carriers. Since the Member States' ability to store gas also differs widely partly due to geological differences, trans-national pipelines are used to their maximum capacities during cold peak winter periods (Winter Outlook by Gas Infrastructure Europe). Problems became evident when in January 2009 a trade dispute between Russia and the Ukraine led to a 2 weeks interruption of 20% of Europe's gas supply, equivalent to 30% of imports. While gas transmission operators were able to restore supplies, the incident showed that at the core of the difficulties were inadequacies in channelling gas flows through pipelines along new routes, like reverse flow capabilities, but also some organizational weaknesses, for example different emergency measures called by Member States without sufficient coordination.

The European Commission identified two market failures in relation to security of gas supply (SEC/2009/0980 final): the insufficient integration (of physical networks and network management) and transparency of the internal gas market, and the issue of security of supply as a public good. As a consequence, the European Council and the European Parliament have called for a revision of the existing Directive (2004/67/EC) concerning measures to safeguard security of natural gas supply. Important elements of the new regulation are an improved information exchange, an assessment by individual Member States of risks to their security of gas supply and the development of national emergency plans.

A big unknown is the role of unconventional gas in Europe. For long, gas was extracted from permeable geologic formations. The development of advanced drilling and reservoirs stimulation methods have increased the production rate of the unconventional gas resources and made the production of shale gas economically viable. Shale gas extraction is seen as a potential game changer in both regional and, in longer term, world gas markets. Potential large shale gas resources exist in the United States, China and also in Europe. The long term global recoverable gas resources represent more than 850 tcm (trillion cubic meters), of which 45% are unconventional. In Europe, countries with large potential shale gas resources are Austria, France, Germany, Hungary, Poland and the UK. The first interdisciplinary shale gas initiative in Europe (GASH) was launched in 2009, focusing primarily on Denmark and Germany. According to estimates of Polish Ministry of Environment, Polish shale gas reserves alone could amount to 1.4–3 tcm, however no shale gas field has been documented yet. Despite its prospects, shale gas extraction has raised concerns about environmentally sound drilling and the protection of groundwater and surface water. Also, the land leasing regulations in Europe, as opposed to those in the United States, seem to be less favourable for shale gas development.

Figure 2. Existing gas network in Europe and Trans-European gas projects.

THE EUROPEAN ELECTRICITY GRID AND RENEWABLE ENERGY INTEGRATION

The European power system, one of the largest and most complex machines in the world, is aging, experiencing increasing congestion and undergoing challenging market liberalisation and renewable integration processes. Developing, and to a certain extent remodelling the European electricity grids will be a crucial step in the pursuit of the EU's competitiveness, sustainability and security of energy supply objectives for 2020 and beyond.

In order to better understand why and how this redesign will take place, European grids can be distinguished into transmission and distribution networks. These differ in terms of their function, structure, planning and operation philosophies. The pan-European transmission network hosts large-scale power plants, which constitutes the lion's share of generation in Europe (some 70–80% of installed capacity) and carries power over relatively long distances. It features higher voltages and a multi-terminal, so-called meshed, interconnected structure (composed of a few hundred thousand kilometers of wiring). The regional distribution networks embed lesser quantities of small-scale generation (roughly 20–30% of the installed capacity) and transfer power passively from the upstream transmission system to the final customers. They feature lower voltages and generally simpler radial structures (i.e. point-to-point connections, including several million km of lines).

Moving Europe to a low carbon, resource efficient and climate resilient economy also entails increasing our reliance upon renewable energy and what are known as 'Distributed Energy Resources'. These include small-sized power plants, such as certain renewable generation units (e.g. photovoltaic panels), as well as storage technologies and electric vehicles. The share of renewable electricity consumed in Europe (30–35% by 2020 according to EU targets) will increase further after 2020 and impact both transmission and distribution grids (with the integration of onshore/offshore wind and solar power playing a key role).

Both the European electricity system transmission and distribution grid will have to adapt, which is described as becoming 'super' (for the transmission grid) and 'smarter' (distribution grid):

The Super transmission grid: The best locations for the generation of renewable electricity are not uniformly distributed across the continent, and are often in places where connections to the electricity network are weak. To fully utilise these resources, the power grid must be enhanced to allow electricity to be transported to the main centres of demand and storage. A super grid can be defined as an electricity transmission system designed to transport large amounts of electricity from remote areas to consumption centres. A super grid could well be a high transfer capacity layer superimposed to the traditional transmission system.

Smart distribution grids: Due to the rising deployment of distributed energy resources, distribution networks will have to change their control properties and become more similar to the transmission network we have today: that is, they will need more 'active' control features. Distributed units will be fully integrated into the management of the electricity system, and collectively serve a role comparable to large conventional power stations. The electricity grid will enable this by becoming 'smarter' and more interconnected. This so-called smart grid requires hardware, software and data networks capable of delivering and responding to information quickly: the installation of smart meters could reduce energy consumption and make both generation and consumer demand more responsive and flexible.

The move towards renewed and redesigned power grids should be carefully monitored and studied system-wise and technology-wise. A number of large-scale power disruptions highlighted the risks attached to lack of coordination and foresight in the electricity systems operation and development. Modernized electric power systems can be increasingly vulnerable to different threats affecting security of energy supply and fundamental societal functions (Figure 3).

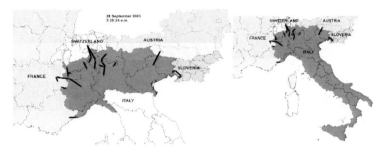

Figure 3. The electricity blackout in Italy on Sept 28, 2003.

The 2008 International Energy Agency Reference Scenario for Europe quotes investments in excess of EUR 1.5 trillion in the period from 2007 to 2030 in order to revamp the electrical system from generation (two thirds of the investment) to transmission and distribution (one third). In turn, distribution needs account for 75%, against 25% for transmission of the expected investment in electricity grids according to the SET Plan technology map (JRC 2009). These structural changes are expected to be enabled by a pervasive deployment of Information and Communication Technology (ICT) for upgraded monitoring, control, and protection functionalities. Improving monitoring and control of the networks through the deployment of metering, telecommunication and remote control technologies is also conducive to a more secure and reliable grid operation with an increased share of distributed energy resources. As an example, even if a large penetration of electric vehicles can strongly affect the distribution grids architecture and operation, it can also help in optimising power system management. Installation of smart meters coupled with demand side energy management measures may rationalise energy consumptions and make the load more responsive and flexible. The development and improvement of cost-effective and coordinated high-power energy storage systems will also play a vital role.

THE ROLE OF ICT IN THE NEW ENERGY SYSTEM AND CYBERSECURITY

Energy networks make extensive use of ICT means: their control, protection, measurement, management etc. at the company level are all ICT based. Not least, links with the customers are mediated by cyber means for the metering and billing of consumption. Moreover, the operation of the energy infrastructure requires the exchange of data over wide-area networks among operators and with authorities. The evolution of the energy networks in Europe will require more ICT inter-connections, crossing national borders and jurisdictions. The lack of common accepted cyber security standards and criteria, does not contribute to the confidence on the existence of proper protection of the ICT systems.

This cybersecurity risk was negligible until a decade ago, but it has had to be taken into consideration with the connection of energy systems to open communications networks like the Internet. The security of industrial ICT is so impending because most technologies have not been designed with cybersecurity in mind. Control protocols are old and vulnerable to attacks. Security countermeasures derived from the general purpose ICT field (such as firewalls and anti-viruses) are only partially applicable in energy systems. Those solutions rarely considered the real-time and other particular requirements of industrial ICT.

On the one hand, the security of those ICT functions is crucial for the efficient management of the installations, but increasingly it is the security of energy supply that is at stake. Malicious actions affecting the availability of the data or their integrity could have direct impact on the operations of the energy systems. One can easily imagine how the failure of a control system might cause the impairment of a technical installation, and possibly bringing it to a halt. A key aspect to consider is the use of similar technologies in many different installations. This might be the cause of common failures – for instance in case of some malware affecting that technology. In addition, ICT components age rapidly, much faster than the electrical and mechanical components of the energy infrastructure. This requires appropriate responses, which at times should be coordinated across systems (e.g. for solving problems of interoperability). In Europe there is still no common agreed approach to tackling these issues.

In the last years, some initiatives have tried to discuss the cybersecurity problems in energy infrastructures in Europe, most notably the expert group on energy supply within the working group organised by the European Committee for Standardization (CEN) on the "Protection and Security of the Citizen" (formally called CEN BT/WG 161).

PROTECTION OF EUROPEAN CRITICAL ENERGY INFRASTRUCTURE

The previous brief discussion on some key challenges regarding the European energy networks demonstrates how energy systems extensively deployed over EU territory are exposed to security threats. In particular, LNG terminals and pipelines, the power grid and substations, are soft targets. So far, energy infrastructures in Europe have not been subject to major attacks by terrorist groups, which may be due to terrorist targeting rather than lack of opportunity (Toft, Duero et al. 2010). Beyond the EU borders, attacks were more frequent, for example in 2008, when the BTC gas pipeline was attacked in Turkey. While so far, the only substantial interruptions to energy supplies have been accidental (e.g. the 2003 blackout affecting Italy) or the result of trade disputes (2006 and 2009 for gas supplies from Russia), an increasing complexity makes these networks vulnerable to malicious attacks with more severe impacts.

The Council Directive 2008/114/EC on the identification and designation of European critical infrastructures was designed to ascertain the vulnerabilities of

European relevance in the energy system (i.e. with incidents possibly affecting two or more Member States) and improve their protection. In synergy with this, the European Programme for Critical Infrastructure Protection (COM(2006)786) emphasized the need for Member States to manage their national critical infrastructures. It is based on an all hazards approach, not confined to terrorism, but also including criminal activities, natural hazards and other causes of accidents.

Concluding Remarks

Given the challenges to European energy security, a comprehensive and coherent approach to energy policy is essential. A major task, both for policymakers and policy-oriented research, is to select economically rational strategies for energy security and to assess the interactions of different energy strategies. This will require adopting a rigorous, robust and transparent approach to assess the future energy security challenges, not only to verify that the implementation of EU policies is not self-defeating, but also to identify new opportunities for synergy between policy domains.

References

Chester, L. (2010). "Conceptualising energy security and making explicit its polysemic nature." Energy Policy 38(2): 887-895.

COM (2006)105, "A European Strategy for Sustainable, Competitive and Secure Energy", Green Paper, Commission of the European Communities.

COM (2008) 781, "An EV Energy Security and Solidarity Action Plan", Commission of the European Communities, Brussels, 13.11.2008.

Department for Business, Enterprise and Regulatory Reform (2007). "Expected energy unserved". Contribution to the Energy Markets Outlook Report.

Gallagher, K. S. (2009). "Acting in Time on Energy Policy".

Gnansounou, E. (2008). "Assessing the energy vulnerability: Case of industrialised countries." Energy Policy 36(10): 3734-3744.

Helm, D. (2002). "Energy policy: security of supply, sustainability and competition." Energy Policy 30(3): 173-184.

IEA (2009). World Energy Outlook.

Jansen J.C, (2009), "Energy services security: concepts and metrics", ECN-E–09-080

JRC (2009). Technology Map of the European Strategic Energy Technology Plan (SET-Plan).

Loeschel, A., Moslener, U. et al. (2010). "Energy security-concepts and indicators." Energy Policy 38: 1607-1608.

Lovins, A. (1982). "Brittle Power: Energy Strategy for National Security".

NERA (2002). Security in gas and electricity markets. London.

Toft, P., Duero A., et al (2010). "Terrorist targeting and energy security." Energy Policy 38(8): 4411-4421.

Tosato G. (2007). "Energy security from a systems analysis point of view: introductory remarks", Proceedings of the 1st International Conference of the FP7 project REACCESS.

Winzer, C. (2010). "Defining Security of Supply". IAEE European Conference. Vilnius.

Yergin, D. (2006). "Ensuring Energy Security". Foreign Affairs 85(2).

THE EUROPEAN UNION AND THE WIDER BLACK SEA
REGION: INTERESTS, CHALLENGES AND OPPORTUNITIES

ODETTE TOMESCU-HATTO*
*International Relations and Security Studies at Sciences Po Paris,
France*

Abstract. The enlargements of NATO and EU, the involvement of the North Atlantic Alliance in Afghanistan, the increasing dependence of EU energy security on Russia and the focus on Caspian Sea as a potential alternative, have moved the center of gravity of European security towards South-Eastern Europe, the Caucasus, and Central Asia – three very sensitive areas in terms of *soft* and *hard security*. The paper discusses on the collective and individual security of EU members, which depends on a vast array of new threats and sources of instability that rise from inside and around the *Wider Black Sea Area.*

Introduction

The consecutive enlargements of NATO and EU, the involvement of the North Atlantic Alliance in Afghanistan, the increasing dependence of EU energy security on Russia and the focus on Caspian Sea as a potential alternative, have moved the center of gravity of European security towards South-Eastern Europe, the Caucasus, and Central Asia – three very sensitive areas in terms of *soft* and *hard security*. The collective and individual security of EU members depends on a vast array of new threats and sources of instability that rise from inside and around the *Wider Black Sea Region.*[1]

The increasing importance of the Wider Black Sea Region (WBSR) as the focal point of Euro-Atlantic and Euro-Asian security is encapsulated in the conceptualization of the region itself. The definition of the region is not only a matter of geography but is also related to politics, economy, security, and culture. In this area, Turkey, Russia, Ukraine, Eastern Balkans, and the South Caucasus come together. The security environment in the region is a product of diverse interests of littoral states and their neighbors. Some of these interests coincide

---

* To whom correspondence should be addressed: Odette Tomescu-Hatto, International Relations and Security Studies at Sciences Po Paris, France, e-mail: odette.hatto@sciences-po.org

[1] Ronald HATTO & Odette TOMESCU, "The EU and the Wider Black Sea Area: Challenges and Policy Options", GARNET *Policy Brief* No. 5, CERI – Sciences Po January, 2008.

A. Gheorghe and L. Muresan (eds.), *Energy Security: International and Local Issues,
Theoretical Perspectives, and Critical Energy Infrastructures,*
DOI 10.1007/978-94-007-0719-1_3, © Springer Science+Business Media B.V. 2011

with those of Euro-Atlantic structures while others reflect the security agenda of old and new regional powers. The Black Sea region is equally home to a patchwork of separatist conflicts in Abkhazia and South Ossetia in Georgia, Nagorno-Karabakh in Azerbaijan and Transdniestria in Moldova.[2]

The geopolitical assets of the region as well as all the challenges and opportunities that emanate from the area has motivated the German Marshall Fund of the United States to promote an Euro-Atlantic strategy towards the WBSR.[3] The strategy is meant to enhance cooperation between the various actors of the region including the international organizations and the littoral states and to address common challenges in a coherent and consistent manner: from weak states, poor governance, and frozen conflicts to issues such as organized crime and energy supply. The American initiative has been followed by the publication by the European Commission on April 2007 of the Black Sea Synergy a New regional Cooperation Initiative is concrete evidence that the European Union is beginning to think strategically about the Black Sea.[4]

This article analyzes the importance of the Wider Black Sea Region for European Security. To this end it begins by responding to the following question: what are the stakes for the EU? In the second section of this article, the strategies and policies that the various State actors have developed are analyzed for the past decade according to their interests in Black Sea region. Stability in this region depends largely on the interactions, convergences, and divergences between a wide range of international and regional actors (Russia, other bordering countries, NATO, EU, the United States, international organizations). The third part of the article underlines existing frameworks of cooperation that have been established in the Black Sea Region and evaluates their contribution to the democratization and stabilization of the region.

What Stakes for the EU?

The EU's interests in the region can be defined through five dimensions. The first dimension concerns the development of democracy, respect for human rights and good governance. One of the major interests of the European Union is to promote democracy in its own neighborhood. Over the past two decades, the force of Europeanization in Central and Eastern Europe have transformed a number of illiberal regimes into consolidated democracies, one of the greatest successes of the EU in terms of promoting democracy. The WBSR offers a mixed landscape of

---

[2] Svante E. CORNELL; Anna Jonsson, Niklas NILSSON, Per HAGGSTROM, "The Wider Black Sea Region. An Emerging Hub in European Security", Silk Road Paper, Central Asia-Caucasus Institute, John Hopkins University-SAIS, December 2006.

[3] Ronal ASMUS & Bruce JACKSON, The Black Sea and the Frontiers of Freedom: Towards a New Euro-Atlantic Strategy, Policy Review, June-July 2004, Hoover Institution available at: http://www.hoover.org/publications/policyreiew/3437816.html; ASMUS, Ronald, Next Steps in Forging a Euro-Atlantic Strategy for the Wider Black Sea Region, The German Marshall Fund of the United States, Washington DC, 2006.

[4] COM (2007) 160 final, Commission Européenne, Black Sea Synergy – A New Regional Cooperation Initiative, Bruxelles, 11.04.2007.

fragile liberal democracies (Bulgaria and Romania), illiberal democracies or pseudo-democracies (Ukraine and Turkey), weak states (Azerbaijan, Georgia, Moldova), and authoritarian regimes (Russia). The fall of the communist rule and the uncertain economic and political transitions that followed have impacted the quality of life of individuals as the region is plagued by poverty, corruption, and other forms of human insecurity. Besides, a number of Black Sea countries are multiethnic states and the cultural and religious differences are still hard to reconcile. Last but not least, the area lacks democracies in its immediate vicinity. In the East, Central Asia offers various forms of authoritarian governments and in the North a more and more assertive Russia dominates its former Soviet space. The democratic deficit of the WBSR suggests that the EU has several pragmatic reasons to duplicate the strategies used in the democratization of the Central and Eastern Europe and to export its values and standards in the WBSR.[5]

The frozen conflicts and the regional stability represent the second dimension of European interest. In geographic and cultural terms, the WBSR is not only an important crossroad of civilization, but also a bridge between Europe and more distant and troubled areas. This vast region of highly political and economic vulnerability has faced important challenges of *hard security* since the end of the Cold War. The conflicts in Transnistria, South Ossetia, Nagorno-Karabakh, and Chechnya are a part of today's EU neighborhood and increase the risks of violence and instability in the area. Thus it is not surprising that the *European Security Strategy* (December 2003), the *Communication from the European Commission – European Neighborhood Policy* Strategy Paper (COM 2004 – 373) and the *Eastern Partnership* (COM 2008 – 823) emphasize that the EU's task is to "make a particular contribution to stability and good governance in the immediate neighbourhood [and] to promote a ring of well governed countries to the East of the European Union".

A third aspect of European interest in the WBSR involves the fight against organized crime and terrorism. The prosperity of European space strongly contrasts with the areas of conflict and instability found in certain countries bordering the Black Sea but also in Central Asia and in the Middle East. The frozen conflicts encouraged the over-militarization of the area and increased the risks of terrorism. The process of globalization contributed to a growth in criminal phenomena including weapons and drug and human trafficking. Terrorist activities also benefit from organized crime and often use the money laundering networks of criminal groups. The weakness of the Caucasian divided states acts as a magnet for such criminal groups. The need to face these types of threats constitutes one of the EU's preoccupations in the region.

Another aspect of European interest in the WBSR involves the EU's quest for alternative sources of energy. Since 2006 the European Union and its member states have intensified their efforts meant to secure external energy supplies for the EU energy market. The Austrian EU presidency put energy security on the forefront

---

[5] Svante E. CORNELL, Anna JONSSON, Niklas NILSSON, Per HAGGSTROM, The Wider Black Sea Region. An Emerging Hub in European Security", *Silk Road Paper*, Central Asia-Caucasus Institute, John Hopkins University-SAIS, December 2006.

of the nation's agenda, the EU Commission issued a new *Green Paper on A European Strategy for Sustainable, Competitive and Secure Energy*[6] and an Energy Summit was organized on March 23, 2006. All this happened in the aftermath of the Russian-Ukrainian gas dispute of January 1–4, 2006. The Green Paper was the first document that drew the attention to the level of dependence on gas imports from sources outside the European Union. Furthermore, over 40% of European natural gas consumption is, at present, imported and the forecasts indicate that this level of dependence could rise to 70% in 2020.[7] EU-27 gas production in 2008 amounted to 190.3 bcm, whereas consumption levels were at 490.1 bcm.[8] Domestic production, thus, covers less than half of the EU's natural gas need. In 2008, Norway was the largest gas producer in Europe (99.2 bcm), followed by the UK (69.6 bcm) and the Netherlands (67.5 bcm).[9] In 2006, 62.7% of gas consumption in the EU was imported.[10] In spite of all of the EU's efforts to increase energy efficiency and encourage the use of renewables European dependence on imported energy will most likely to increase.

Natural gas supply to the EU is a sensitive issue as Europeans produce only 37% of their gas consumption. In 2008, the most important suppliers of crude oil and natural gas were Russia (33% of oil imports and 40% of gas imports) and Norway (16% and 23% respectively). Overall, in 2008 the EU-27 energy dependency was estimated at 53.8%. The most vulnerable countries being Germany (61.3%), Malta (54.6%) Ireland (52.1%), and Italy (51.4%).[11] Norway has proved to be one of the most reliable western gas suppliers for the EU. Algeria has also supplied gas to Eastern Europe since 1970 but nowadays Russia remains the main resort in terms of gas supply.

Last but not least the WBSR constitutes a test for the EU's ambitions to promote security and stability at its borders. The Black Sea region is part of the struggle over spheres of influence between Russia and the EU and to a certain extent between the EU and the US. The EU-Russia Summit (May 2005) ended with an agreement on the *Road Map for the Common Space and External Security*. Targeted to enhance the cooperation in their common neighborhood, the agreement makes no major progress in sensitive areas such as the frozen conflicts in the post-Soviet space and energy. If the EU wants to assert itself as a foreign policy actor, it needs to negotiate difficult matters of *hard security* with Moscow, without jeopardizing its geo-strategic interests. The WBSR is both a challenge and opportunity for the EU to act within its neighborhood and evaluate its instruments of cooperation, stabilization, and democratization, including the ENP and the bilateral Action Plans, which tend to be regarded as strictly soft security tools.

---

[6] Green Paper, *A European Strategy for Sustainable, Competitive and Secure Energy*, {SEC(2006) 317}, Brussels, 8.3.2006, COM(2006) 105 final.

[7] *Ibidem*.

[8] British Petroleum (BP), *Statistical Review of World Energy* 2009, www.bp.com/statisticalreview, pp. 27–29. During the same year the EU-27 consumed 17.9% of world oil production.

[9] *Ibidem*.

[10] Gerhard MANGOTT & Kirsten WESTPHAL, "The Relevance of the Wider Black Sea Region to EU and Russian Energy Issues" in Daniel HAMILTON & Gerhard MANGOTT (eds.), *The Wider Black Sea Region in the 21st Century*, Washington, Center for Transatlantic Relations, 2008, pp. 146–149.

[11] Europe's Energy Portal, http://www.energy.eu/#dependency.

The Communication of the European Commission *Black Sea Synergy, a New Regional Cooperation Initiative* encompasses the five dimensions described above and reflects the EU's concerns regarding the WBSR. Nevertheless, this document gives little attention to the way in which the EU is going to negotiate a number of *hard security* issues with the main regional powers: Turkey and Russia.

The Interests and Strategies of State Actors

In order to understand the complexity of the region and to emphasize different scenarios for the EU involvement in the area, we must examine the geo-strategic interest of the Black Sea State Actors. There are three groups of states interested in the WBSR today: (1) the Western powers like the EU and the United States, (2) Russia and Turkey which are moving closer and closer together, creating a block of Status Quo powers in the WBSR, and (3) the "Community of Hopeful" – the other Black Sea countries.[12] For the European Union, NATO or any other international organization, the challenge of promoting a security regime in the WBSR is compounded by the fact that the area is home to a range of countries with diverse and competing interests and security agendas.

Russia is incontestably the most important actor in the WBSR and is likely to represent a factor of regional instability. For Russia, the WBSR falls under its spheres of influence – the "near abroad" and Moscow has adopted two parallel strategies regarding the area. The first strategy is a defensive one and is based on the resistance to all types of "westernization" of the region. The second strategy is an active one and focuses on the use of energy weapons against the dependent Eastern and Western countries. In order to affirm its predominance in the WBSR, Russia uses the economic and political weakness of post-Soviet republics (Georgia, Moldova, and Ukraine) as well as its oil and energy resources. Through Gazprom, Moscow is establishing relations with its neighbors on the basis of energy domination. The gas crisis between Russia and Ukraine has proven that Moscow is ready to use the geopolitical dimension of energy to protect its *near abroad*. Moscow has recently paid close attention to the Caspian Area and the way in which gas is integrated into European and Asian markets.

Along with economic sanctions, Russia has tried on several occasions to influence the electoral processes in neighboring countries by openly supporting the pro-Russian candidates (Moldova and Ukraine). Furthermore, Moscow never contributed to the resolution of the frozen conflicts in the post-Soviet space. On the contrary, it did everything to preserve the *status quo*. By encouraging separatist Republics (Transnistria in Moldova; Abkhazia and South Ossetia in Georgia), Russia weakens the Moldovan and Georgian State and justifies the presence of Russian military bases on the territory of these states. The Russian strategy in the WBSR is part of a broader tactic to keep the former Soviet space under direct influence.

---

[12] Ognyan MINCHEV, *Major Interests and Strategies for the Black Sea Region. Framework Analytical Review*, Sofia, Institute for Regional and International Studies, 2006.

Turkey is the second important regional actor in the WBSR. Turkey's role as a regional power has been reinforced because of its active participation in the Gulf War (1990–1991). Starting in 2005, Turkey became an official candidate for EU integration and Ankara has placed its relations with Brussels on the forefront of its political agenda. In geo-strategic terms, Turkey could play an important role in any EU strategy in the Black Sea. Turkey controls the water supplies for the Middle East (Iraq, Iran, Syria) as well as the two straits—Bosporus, Dardanelles, and the Marmara Sea—which connects the Black Sea and the Aegean Sea. Turkish territory is an important route for the transport of hydrocarbons and energy from the Caspian Sea towards Europe. The pipeline Baku-Tbilissi-Ceyhan (BTC) was opened in May 2005, with the help of EU contributions. The BTC represents an alternative option to Russian resources. The Baku-Tbilissi-Erzurum gas pipeline (in service since 2006), makes the transport of the Azeri and Turkmen natural gas possible towards Europe and completes the East-West energy corridor. This "double energy corridor" brings the Caspian area closer to the European Union.

However, Turkish frustrations concerning the last American intervention in Iraq and the European hesitations regarding the further enlargement, seems to push Turkey far from its Western allies in a time when Europe and the US need Ankara to negotiate their relations with Russia, to secure the energy transport and to stabilize the WBSR. The increased cooperation between Turkey and Russia is undoubtedly one of the most important geopolitical events since the end of the USSR in 1991. The two Black Sea powers are struggling for power and influence in the WBSR but they are likely to cooperate in order to keep the West at a distance. In 1997, Ankara signed an agreement with Moscow concerning the construction of a gas pipeline which crosses the Black Sea – Bluestream. In spite of small divergences over some geo-strategic issues such as the NATO enlargement or Kosovo, Moscow, and Ankara have strengthened their economic and political ties and even find some points of convergence. The joint condemnation of the American presence in the Black Sea is another example of congruent points of view.

The Ukraine is another important actor in the WBSR. Its political culture and identity are strongly influenced by its Soviet past, and Kiev has often found it difficult to choose between the East and the West. This hesitation is even more visible at the level of foreign policy. Kiev has chosen a "multi-vector approach" that allows it to play several cards at once: participation in the Community of Independent States (CIS), adhesion to the NATO Partnership for the Peace (PfP, 1994), a role inside GUAM, and as a beneficiary of the ENP since 2005 and part of the Eastern Partnership (2008). Strongly dependent on Moscow, the Ukraine remains a Russian pivot point in the WBSR. In order to maintain its position of regional power and to facilitate Russian access to its fleet in Crimea, Moscow needs to keep Ukraine under its influence.[13] During the last elections (February 2010) Viktor Yanukovych, the leader of the Party of the Regions (PoR), won the second round of presidential elections in Ukraine with 48.9% of the votes, leaving

---

[13] Alexandra GOUJON « Les nouveaux voisins de l'Union Européenne. Stratégies identitaires et politiques en Ukraine, Biélorussie et Moldavie », *Les Etudes du CERI*, CERI - Sciences Po, Paris, No.109, septembre 2004.

behind Yulia Timoshenko of the Blok Yulia Timoshenko (BYUT) with a difference of only 3.49%. The OSCE declared "professional, transparent and honest voting and counting" and certified the Ukrainian elections as meeting "most OSCE and Council of Europe commitments". According to certain analysts Yanukovych's victory is seen as a warning both for domestic developments in the country and for Ukraine's further relations with the EU.[14]

Among the six Black Sea States, Georgia is certainly the one presenting the most important challenges in terms of *soft* and *hard security*. Georgia is a weak state located at a crossroads between Europe and the Caucasus; it represents Russia's *near abroad* and is one of the EU's new neighbors. The frozen conflicts in Abkhazia and South Ossetia are challenging the integrity of the Georgian States. The areas of conflict have become a haven for illegal traffic and smuggled goods. As is the case for other former Soviet states, Georgia cannot solve its internal problems without Russian interference. The position of Moscow in the "resolution" of the two frozen conflicts has always been ambiguous. On one hand, Moscow declared its support for the consolidation of the Georgian State, while on the other it opened the door to Ossetian separatism and the integration of South Ossetia with the Russian Federation. In order to counterbalance Russian influence, Georgia is seeking recognition from the West (through international organizations, the EU, NATO and the United States).[15] The Georgian War (August 2008) has proved has not changed the balance of power in Eurasia. It simply announced that the balance of power had already shifted. The United States has been absorbed in its wars in Iraq and Afghanistan, as well as potential conflict with Iran and a destabilizing situation in Pakistan. This has opened a window of opportunity for Russians to reassert even more their spheres of influences in the post-soviet space. Georgia represents a neuralgic point in the Caucasus and draws the attention of big regional and international powers. With regards to energy transport, Georgia's location, between the Caspian Sea and the Black Sea, the Southern Caucasus and the Northern Caucasus, made it an ideal place to pass the Baku-Tbilissi-Ceyhan pipeline (BTC) and the South Caucasus Pipeline (SCP).

After struggling with economic and political reforms for several years, Bulgaria and Romania are the only Black Sea States that have successfully integrated into NATO (2004) and the EU (2007). The two countries have shared a common past as Soviet satellites and members of the Warsaw Pact during the Cold War. Additionally, they followed similar paths of transformation and democratization during the 1990s. As for their importance as Black Sea States, the security environment of Bulgaria and Romania is very complex. Since EU enlargement, both countries seem to operate as a buffer zone between an unstable East and the wealthy European space. Bulgaria and Romania are close to areas of instability (Kosovo) and Transnistria (North-East) and therefore they are often perceived as countries of transit for illegal goods, merchandises, and migration. At the same time, their geo-strategic position has motivated the United States to redeploy its

---

[14] Sabine FISHER, "Has EU lost Ukraine?" *EUISS Analysis*, European Union Institute for Security Studies, February 2010.

[15] Tracey, GERMAN, « Abkhazie et Ossétie du Sud : le choc des intérêts russes et géorgiens », Russie. Nei. *Visions* No. 11, IFRI, juin 2006.

military equipment and its armed forces from Western Europe to Bulgaria and Romania. Both countries are important routes for energy and oil transit. The "Progress" pipeline connecting Russia and Turkey crosses the south-east of Bulgaria and Romania. The gas pipeline "Nabucco" will also pass through Bulgaria, Romania, and Hungary before being connected to Austria. The Nabucco project is included in the EU Trans-European Energy Network program and a feasibility study for the Nabucco pipeline has been performed under an EU project grant.

Despite their common past and many other similarities, Bulgaria and Romania do not have the same interests in the WBSR. Romania tends to be more active in the WBSR than its neighbor. Bucharest's interest in the Black Sea area has been confirmed on several occasions; President Traian Basescu placed the Black Sea area on the forefront of his political Agenda. Romania inaugurated the first session of the Black Sea Forum for Partnership and Dialogue (BSF). For Romania, the WBSR provides an opportunity to participate at the side of the big Western powers and take revenge for years of international isolation and Soviet submission. Yet the Russian factor plays an important role in the formulation of Romanian foreign policy towards the Black Sea. Bucharest is constantly concerned with the situation in neighboring Moldova and denounced the Russian support of the separatist regime in Transnistria. The presence of NATO and the US in the Black Sea region is perceived by Bucharest as one of the most valuable solutions to the Russian dominance in the region.

As for Bulgaria, its interests and position regarding the WBSR were underlined in the Bulgarian Foreign Policy Strategy 2001–2005. Three main lines can be identified: (1) the development of the relations between Bulgaria and Russian Federation, (2) the development of bilateral relations with Moldova and the Ukraine; (3) intensification of the dialog with the Caucasian Republics. However, Bulgaria is less dynamic than Romania in the WBSR. This passive attitude could be explained by historical affinities with Russia and by an important Turkish minority population. In other words, Sofia is subjected to the pressures of the "status quo powers". The attitudes towards Russia vary. According to a study published in November 2007 by the European Council on Foreign Relations, Bulgaria is qualified as a "friendly pragmatist" in its relations with Russia while Romania remains a "frosty pragmatist" (together with Czech Republic, Estonia, and Latvia) meaning that Bucharest focuses on business interests but also speak out against Russian behavior on human rights or other issues.[16]

The United States is the most recent player in the WBSR. US interests in the WBSR are both idealistic and pragmatic. Among its idealistic notions, its support of the democratization of countries in the region is notable. The Rose and Orange Revolutions in Georgia and Ukraine have been perceived in Washington as a signal of the revival of democratic societal forces. These tiny democratic transformations have paved the way for debate amongst American decision-makers. Several analysts believe that a Euro-Atlantic strategy towards the Black Sea region that merges

---

[16] Mark LEONARD & Nicu POPESCU, "A Power Audit of EU-Russia Relations", European Council on Foreign Relations, November, 2007.

European Security and Defense Policy (ESDP) and NATO instruments together is the only way to securely anchor the Black Sea countries to the West.

In more realistic terms, the US engagement in the Black Sea region could be seen as part of wider foreign policy tendencies that focus on US national interest. The terrorist attacks of September 11th altered the broad outlines of American Foreign policy and brought specific challenges such as counterterrorism and the fight against the proliferation of weapons of mass destruction (WMD) to the top of the political agenda. In this respect, the WBSR is an appealing field of action for US (with its proximity to Iran, Iraq, and Afghanistan). Second, energy security has become a major issue of concern not only for the EU but also for the US. The Black Sea is already an important route for the transit of oil and gas from the Middle East and Central Asia to Europe, from ports on Russia's Black Sea coast through the Turkish straits and into the Mediterranean. Third, a growing US presence in the WBSR is likely to contribute to the consolidation of the American supremacy at the top of the international order and counteract the dominance of other powers in the region.[17] However, the change in leadership in Washington in 2008 suggests that Eastern Europe does not represent necessarily a priority for the Obama Administration. The tergiversations regarding the shield anti-missiles in the Czech Republic and Poland (September 2009) and the negotiations between Moscow and Washington with the Strategic Arms Control Treaty (START) on which rest Russian-American relations have strengthened the idea that the New Europe (including the Black Sea area) comes after other urgent American issues.[18]

Current Frameworks of Cooperation

The existing frameworks of cooperation in the WBSR have been established using various definitions of the area and the principal external actors, including the Euro-Atlantic institutions, have employed flexible approaches to the area.

The Black Sea Economic Cooperation Organization (BSEC) represents the most inclusive framework of collaboration in the area. The BSEC was created in 1992 as a result of a Turkish initiative and includes 12 states (six Black Sea states, plus Albania, Armenia, Azerbaijan, Greece, Moldova, and Serbia) and 11 observers (Germany, Belarus, Croatia, the United States, France, Israel, Italy, Poland, Czech Republic, Slovakia, and Tunisia). During its 18 years of existence, the BSEC has undergone several transformations and has established a number of institutions such as the Parliamentary Assembly of BSEC (PABSEC), a Permanent Secretariat (PERMIS), a Development Bank (BSTDB), and an International Center for Black Sea Studies (ICBSS).[19]

---

[17] HATTO, RONALD & Odette TOMESCU, Les Etats-Unis et la « Nouvelle Europe ». La stratégie américaine en Europe centrale et orientale, Paris, éditions, CERI-Autrement, 2007.

[18] Daniel HAMILTON et Nikolas FOSTER, « The Obama Administration and Europe », in *The Obama Moment. European and American Perspectives,* Alvaro de VASCONCELOS et Marcin ZABOROWSKI (dir.), European Union Institute for Security Studies, 2009.

[19] Mustafa AYDIN, "Europe's Next Shore: the Black Sea Region after EU Enlargement", *Occasional Paper No. 53,* Institut de Sécurité de l'Union européenne, 2004.

Since its creation, the BSEC has been mainly a *soft power* actor intended to strengthen economic cooperation between its members. However, growing political instabilities have forced the organization to redefine its objectives. During its 10th Summit (Istanbul, June 2002), the Council of Ministers called on the BSEC to reinforce the security and the stability in the WBSR, officially including cooperation in the field of *hard security* in the BSEC's agenda. However, the role of the BSEC in the field of *hard security* is confronted with several obstacles: (1) each state has its own preoccupations in terms of *hard security* and the national security strategies of countries in the region do not always converge; (2) the BSEC is both too small and too wide of an international organization to address issues of *hard security*; the big Western powers with credible defense capabilities are not part of BSEC and the diversity of its members would make coordination difficult; (3) the member states of BSEC belong to various political and security organizations and this makes policy coordination too complex; (4) the BSEC is characterized by the weak political engagement of its members. Most of the time, member states prioritize other instruments of foreign policy and they look to other multilateral forums of cooperation to achieve their goals; (5) BSEC members are hesitant to create a framework of regional security in which the big regional powers (Russia, Turkey) would play an essential role; (6) the BESC has not yet developed concrete mechanisms that would allow it to act in the field of *hard security*; (7) there is a major lack of interface and coordination between the BSEC and other international organizations such as the OSCE and NATO.[20]

The European Union began its involvement in the WBSR in 1990 when the Union launched and participated in a number of regional projects (Baku Initiative, INOGATE, PETrAS, SYNERGY Program, TRACECA). In the context of the SYNERGY program, in 1995 the EU established the Black Sea Regional Energy Center (BSREC). The *Baku Initiative* launched in 2004 was also designed to promote cooperation and legal harmonization in the field of energy and include all Black Sea states and the Caspian Sea countries (including Turkmenistan and Iran 'if political conditions permit'). However, the majority of EU programs and projects have been negotiated on a bilateral basis, which is the case for the European Neighborhood Policy and its consequent Action Plans.

The Black Sea Synergy – A New Regional Cooperation Initiative launched in April 2007 constitutes the first EU attempt to treat the Black Sea Area as a region. The purpose of Black Sea Synergy is to reorganize existing regional policies and enhance the existing regional cooperation and dialog. Among sectors and issues mentioned in the document: promotion of democracy and respect of human rights, fight against trafficking and organized crime, frozen conflicts, energy, transport, environmental problems, trade, research, and education. Nevertheless, the Strategy does not adopt a clear and efficient political position on the regional agenda and appears to be simply a declaratory initiative that does not engage in action.

During the past few years the EU has also strengthened its relations with the BSEC. The co-operation between the BSEC and the EU started to develop in 1997

---

[20] CELAC, Sergiu & Panagoita MANOLI, "Towards a New Model of Comprehensive Regionalism in the Black Sea Area", *Southeast European and Black Sea Studies*, Vol. 6, No. 2, June, 2006.

when the European Commission submitted to the EU Council a Communication on the Regional Cooperation in the Black Sea Area relating to the state of affairs and perspectives of EU action that encouraged further development (Doc.COM (97), 597 Final Brussels, 14 November 1997). Eight years later, the Ministers of Foreign Affairs of the BSEC Member States adopted in Chisinau (28 October 2005) a declaration that included the main sectoral objectives that could be addressed by the BSEC and the EU together: (1) development of infrastructure, including transport and energy; (2) trade and economic activities, including cross-border cooperation; (3) environmental protection; (4) institutional and social sectors; and (5) science and technology. Nevertheless, as underlined by several analysts from the International Center for Black Sea Studies, the framework of interaction between the EU and the BSEC is currently developed only at a conceptual level; there is no concrete definition of the structure, nor the content of their interaction.

The *Eastern Partnership* is the latest EU initiative towards the countries from the Black Sea Region. The commission's proposal for a new Eastern Partnership represents a step towards a change in the EU's relations with Armenia, Azerbaijan, Belarus, Georgia, Moldova, and the Ukraine. This ambitious partnership foresees a substantial upgrading of the level of political engagement, including the prospect of a new generation of Association Agreements, far-reaching integration into the EU economy, easier travel to EU for citizens providing that security requirements are met, enhanced energy security arrangements benefitting all concerned, and increased financial assistance.

NATO became a constant actor in the WBSR since the Alliance decided to increase its membership to include two Black Sea states (Bulgaria and Romania). NATO's Concept of Defense against Terrorism, published in 2002, underlines the interest of the Alliance in the WBSR and its desire to contribute to the stability and security of this area. The use of Black Sea airspace by NATO and the US during their missions in Afghanistan and Iraq reinforced the bonds between the Black Sea countries and the Atlantic Alliance. In order to reinforce the military capacities of the countries in the area, NATO inaugurated the Individual Partnership Action Plan (IPAP) in 2002.[21] Georgia and Azerbaijan were the first two countries to sign the IPAP in 2004 and 2005, followed by Armenia, Kazakhstan and Moldova. However, the presence of NATO in the WBSR has created tensions between the Alliance and some of its members (Turkey) and non-members (Russia). Russia and Turkey expressed their disagreement regarding the presence of NATO in the WBSR and warned against the over-militarization of the area. NATO's Open Door Policy and the relations developed with a new generation of partners and aspirants from South Caucasus suggest that Bulgaria and Romania are not the final frontier of the Alliance.

Another important framework of regional co-operation in the WBSR is GUAM (Georgia, Ukraine, Azerbaijan, Moldova). GUAM was created in 1996, by the presidents of Georgia, Ukraine and Azerbaijan with the assistance of the United States. It was meant to weaken the military and political control of Russia

---

[21] ZULEAN, Marian, "Reforming the Security Sectors in South Eastern Europe: Lessons Learned and their Relevance for a Wider Black Sea Area Policy", *The Quarterly Journal*, Vol. 3, No. 4, 2004, p. 96.

in the area. Moldova joined the organization in 1997 and Uzbekistan in 1999. The latter left the organization in 2005. At a meeting in Kiev in May 2006, GUAM was formally institutionalized and renamed the "Organization for Democracy and Economic Development (ODED-GUAM)". The last GUAM Summit (Black Sea Resort in Batumi July 2008) focused on three questions: the democratization of the region, the co-operation with NATO and the EU, and new approaches to solve *frozen conflicts,* including new strategies to attract international involvement.

In order to increase the military level of integration and cooperation in the Black Sea, Turkey launched the *Black Sea Force* (BLACKSEAFOR) in April 2001. The BLACKSEAFOR is a naval initiative of Bulgaria, Georgia, Romania, Russia, Turkey, and the Ukraine. The initiative involves rescue naval operations and assistance to civil ships.

The *Black Sea Forum for Dialogue and Partnership* represents the latest regional initiative in the WBSR driven by active involvement and interaction between stakeholders interested in the future of the Black Sea region. The goals of the Forum are multiple: to reinforce mutual trust, to facilitate synergy between the various regional initiatives, to support the pragmatic regional projects, and to share information and past experiences between the countries of the area and other partners.

An analysis of the existing frameworks of cooperation in the WBSR shows how it is impossible for the different institutional actors to unite their forces and to formulate a cohesive and global strategy towards the WBSR. The complexity of the area has made the coordination between the various actors and the fight for spheres of influence impossible and prevails inside some of the institutional frameworks to the detriment of coordination and cooperation.

Conclusion: The Way Ahead

The challenges and opportunities offered by the region itself, the contrasted interests of new and old actors, and the limits of existing frameworks of cooperation suggest that wider involvement of the West in the WBSR is needed. Today, the EU represents one of the most powerful regional organizations in the region, but the multiple challenges call for a joint Euro-Atlantic strategy that would bring together the *soft and hard powers* of NATO, the US and the EU.

The need for such a strategy has created serious attention from both sides of the Atlantic. Nevertheless, in Washington and in some European capitals, decision-makers and analysts have started to believe that it will be difficult to decouple the idea of a Euro-Atlantic strategy from the prospect of a further wave of enlargement of NATO and the EU. In fact, how can one envision a Euro-Atlantic strategy without offering a perspective of accession (no matter how vague) to the Black Sea States? From a long-term perspective of the enlargement of European and North Atlantic structures, debates have moved to the issue of whether it is NATO or the EU who should lead the effort to anchor the fragile democracies around the Black Sea to the West. This debate is not new since Westerners have already been confronted by similar dilemmas for the enlargement of the Euro-Atlantic structures to Central and Eastern Europe.

In 2006 there were three concurrent scenarios regarding the prospect of a Euro-Atlantic strategy towards the WBSR: (1) NATO first? (2) the EU first? (3) Should both approaches occur simultaneously?[22]

NATO Bucharest Summit Declaration (April 2008) welcomed Ukraine's and Georgia's Euro-Atlantic aspirations for membership in NATO and the Alliance made clear that it supports these countries' applications for Membership Action Plan (MAP) but no further promises were reached. The war in Georgia and the tensions between EU-Russia and NATO-Russia suggested that the NATO's first scenario is not yet a realistic perspective. Moreover, the concept of WBSR is new for NATO. The role and activities of the North Atlantic Alliance in the area are important but their political impacts are rather limited. In spite of the increasingly institutionalized relations with countries in the region, the Alliance continues to negotiate on a bilateral basis. Until now, no long-term, global approach to the region has been discussed inside the NATO circles. The complexity of frozen conflicts spread throughout the WBSR and its focus on more urgent matters, such as the Balkans and Afghanistan, have kept NATO from developing a coherent regional strategy towards the Black Sea area. Then, as it was already mentioned NATO's interest in the WBSR, the WBSR is expected to radicalize Russian positions *vis-à-vis* countries in the region and perpetuate old rivalries between Moscow and the West. NATO's image in Moscow is that of an "aggressive" organization that serves American interests.

The "EU first" scenario seems to be more realistic. In Javier Solana's (the former EU high Representative for the CFSP) optimistic words, "the EU could provide an institutional framework larger than the region itself, providing security and critical mass for regional cooperation... Finally, it has transformational power directly applicable to conflict situations because on its focus on common values, such as human rights, the rule of law and the market economy." Geographically speaking, the EU is already largely involved in the region through a number of programs and bilateral agreements. The German Presidency introduced an extensive package of a reformed *European Neighborhood Policy* (ENP +) that brought together existing Action Plans of the EU with the non-EU Black Sea countries and the Swedish-Polish cooperation in the field of European foreign policy ended up with the adoption of the *Eastern Partnership*. However, in spite of all the attributes it possesses and all the programs and technical aid it deploys, it is still premature to assess the transformative power of the EU and its capacity to anchor the Black Sea countries to the West.

There is no definitive evidence in terms of how far the European Security and Defense Policy (ESDP) capabilities could contribute to the stabilization of the WBSR and how far the previous EU experiences in crisis and post-conflict management may be duplicated in the Black Sea Area. The European Security Strategy (ESS), adopted on December 2003, focuses on the building of a "secure

---

[22] RUMER, Eugene & SIMON, Jeffrey, "An Euro-Atlantic Strategy for the Black Sea Region", Institute for National Strategic Studies, National Defense University, Washington, January 2006, p. 10.

Europe in a better world." Effective multilateralism through cooperation with other international organizations is the core principle of the ESS. The "prevention" and not the "preemption" stands as the main code of the European Security Strategy. Then the EU should reassess its CFSP and ESDP instruments in order to act effectively in the area. This requires reconsidering its *soft* and *hard power* tools to adapt them to more complicated regional tasks. Then Brussels should reconsider its relations with Turkey and Russia—two important Black Sea States that would be impossible to ignore in the elaboration of any strategy in the region. The questionable europeanization/westernization of the region remains therefore in the hands of a complex web of actors (littoral states, international organizations, regional powers).

The EU is not the only player with interests in the WBSR. This article intentionally addressed the piece of the puzzle for each actor in the region. In this concise analysis, it is neither proposed to anticipate the developments in the debate surrounding the EU initiative towards the Black Sea region, nor to predict the future of the region. In fact, for the time being it is too early to assume that the EU, NATO, and US initiatives would exceed a declaratory level and emerge into real policy actions. However, it is advisable – and this was the primary goal – to begin reflecting on the importance of the Wider Black Sea Region as a concept and as a goal for future policy actions not only in Brussels and Washington but also in the capitals of Black Sea countries. Indeed, the reasons for stronger Western involvement in the WBSR are to be found in its potential to act as a *bridge vs. border* in Euro-Atlantic and Euro-Asian Dynamics.

About the Author: Odette Tomescu-Hatto holds a Ph.D. in European Studies from The Institut d'Etudes Politiques de Paris and a Master of Science in International Relations from Université de Montréal. She is a Senior Lecturer in International relations and Security Studies at *Sciences Po Paris*. She is the author of many studies on security and democratization in Central and Eastern Europe, including *Politique et Société dans la Roumanie contemporaine* (with Alexandra Ionescu, 2004), *Promoting Human Security: Ethical, Normative and Educational Frameworks in Eastern Europe* (with Shahrbanou Tadjbakhsh, UNESCO, 2007), *Les Etats-Unis et la "nouvelle Europe". La Stratégie américaine en Europe centrale et orientale* CERI-Autrement, 2007 (with Ronald HATTO).

ISSUES OF ENERGY SECURITY IN THE BLACK SEA REGION

SERGIU CELAC*

Romanian Institute for Sustainable Development, Bucharest, Romania

Abstract. In terms of energy security, the Black Sea region is important to Europe. So far, most of the debates on that subject have been focused on the vital supplies of oil and gas from Russia, the Caspian Basin, and the northern segment of the Middle East. Repeated disruptions of the transit of Russian gas through Ukraine, whatever the political or/and commercial motivation may have been, has revealed some serious system vulnerabilities and has highlighted the strategic nature of the pursuit of energy security in Europe today. Inevitably and for very good reasons, a lot of attention has been given to the existing and planned pipeline routes going around or across the Black Sea. Much less attention has been given to the development of the Black Sea energy market in its own right and to the potential advantages of coping with some current and future energy issues in a multilateral regional format rather than through individual action at national level. This paper proposes to go beyond pipeline politics, without down playing their continued significance, and to concentrate instead on some topical questions related to the sustainability and resilience of energy systems as applicable to the Black Sea region.

Topics in Debate

More often than not, the topic of energy security is still being discussed, especially when it comes to hydrocarbons, in confrontational terms, as a zero-sum game of continental and inter-continental proportions. This does not necessarily have to be the case. The point of this paper is not to advocate a naïve, do-gooder stance: let's all be friends, forget about geopolitics and see energy security as the perfect opportunity to prove the worth of the win-win approach. Ideally, such a change of attitudes may come about sometime in the future, but not just yet. For the time being, we shall have to operate within the confines of reality, which is not always pleasant. It is to be expected that the transition to a new pattern of energy production and consumption worldwide in the twenty-first century will be long and complicated, entailing continued competition for the control of energy resources and transit routes, increased vulnerability for some states as a result of political manipulation

* To whom correspondence should be addressed: Sergiu Celac, Romanian Institute for Sustainable Development, Bucharest, Romania, e-mail: Sergiu.celac@ncsd.ro

A. Gheorghe and L. Muresan (eds.), *Energy Security: International and Local Issues, Theoretical Perspectives, and Critical Energy Infrastructures,*
DOI 10.1007/978-94-007-0719-1_4, © Springer Science+Business Media B.V. 2011

of energy markets, opportunistic arrangements and broken promises. The inherent inertia of the energy systems requiring huge investment over long periods of time will further compound the intricacies of this process.

And yet some signs of positive change can be glimpsed even at this early stage. There are at least three developments that are beginning to shape mindsets and impact policy making, even though in an oblique, indirect manner: (i) a growing universal awareness of the fact that global climate change is a real and present danger, that it is, at least in part, the result of human activities related to energy production and consumption, and that something must and can be done about it; (ii) the demonstrably realistic promises of technological breakthroughs in energy production and distribution, and incremental advances in energy efficiency; (iii) the emerging alliance between business interests and environmental responsibility as clean energy from renewable or alternative sources begins to make clear commercial sense. Regardless of hidden political agendas and strategic designs, these three engines of change will most likely generate in the medium and long run a stream of irrepressible incentives for more transparent and market-based energy policies.

Since energy is the least internationally regulated sector of the world trade, it is to be expected that more meaningful efforts will be undertaken in the future to produce comprehensive agreements striving to strike a rational balance between the respective interests of all major actors, meaning security of supply for the consumers, security of demand for producers and security of steady revenue from costly investments in infrastructure for transit countries.[1] Such a balanced approach is particularly relevant for the Black Sea region, where one finds all three types of countries, some of them playing a dual or even triple role along the hydrocarbons supply chain.

Indeed, the Black Sea region displays a high degree of diversity in terms of the size and physical power of the countries involved, levels of socio-economic development, systems of governance, maturity of democratic institutions, sophistic- ation of business culture and financial structures, and human development indicators. Moreover, some countries belong to different, though not necessarily antagonistic, political-military alliances or other forms of association such as EU membership, others aspire for candidate status, while still others may have different plans. The WTO rules do not apply to all Black Sea countries and neither does the Energy Charter Treaty. Consequently, it is hardly realistic to strive for comprehensive regional integration in a conventional sense at least in the short and medium term. Nevertheless, diversity should not be seen as an insurmountable obstacle to wide ranging cooperation but rather an incentive for creative thinking and pragmatic action based on clearly identified mutual interests and flexible accommodation among various political cultures. What can otherwise be regarded as a liability for regional cohesiveness may yet turn out to be an asset for devising a new pattern of constructive regionalism?

When the Black Sea Economic Cooperation (BSEC) initiative was launched by Turkey in 1992, high hopes were expressed about the prospects of regional

---

[1] David G. VICTOR & Linda YUEH, "The New Energy Order", *Foreign Affairs*, volume 89, no. 1, January–February 2010.

development. It was seriously expected that what coal and steel had done for the stability and prosperity of Western Europe in the aftermath of World War II, energy cooperation could accomplish in the Black Sea area after the end of the Cold War. The statistics of sectoral BSEC meetings at ministerial and working group levels show that the largest number by far were devoted to energy issues.[2] The results, however, have been meagre, to put it mildly.[3] Some project ideas may have been unrealistically overambitious, adequate funding was not readily available and, above all, the political will to move ahead in a regional format was not there. Geopolitical rivalries, protracted regional conflicts and the use of energy as leverage for political ends also played an inhibitive role.

This being said, the individual economic performance of most of the Black Sea countries over the past decade was impressive, averaging as a region one of the highest rates of GDP growth in the world. Considering that the post-communist states had to undergo a difficult and often painful transition to pluralistic democracy and functional market economy, this was no mean achievement. The current crisis dealt a severe blow to all the countries of the region, revealing a number of structural vulnerabilities and affecting also the energy sector. No matter when and how the consequences of the crisis are going to be overcome, it has become obvious that the prevailing pattern of growth that relied on uneconomical exploitation of resources with little regard for environmental impacts will have to change. A strong body of evidence supports the growing consensus of the expert community that the way out of this predicament goes through consistent implementation of the principles and goals of sustainable development.

The primary responsibility for articulating a coherent vision of what the Black Sea region should look like 10 years from now rests with the regional actors themselves. If they prove unable or unwilling to seize the considerable opportunities for regional cooperation, it should come as no surprise if the geopolitical forces now at play inside and outside the region continue to pursue their respective, and not necessarily convergent, agendas.[4]

The United States discovered early on the potential merits of a regional approach to the complex situation in the Black Sea, first in support of the energy security of its European allies, then it terms of the region's strategic value for the ongoing operations in Afghanistan and Iraq. Diplomatic demarches were supplemented by a considerable amount of learned contributions essentially advocating a joint US-EU approach to energy security with particular reference to the Black Sea region.[5] The current American administration reiterated the earlier US commitments to the

---

[2] *The BSEC at Fifteen: Key Documents 1992–2007*, International Centre for Black Sea Studies, Athens, 2007.

[3] For a critical analysis of BSEC's performance in the economic sphere, see Panayota MANOLI, "Limited Integration: Transnational Exchanges and Demands in the BSEC Area", *Agora without Frontiers*, 10 (4) 2005, Athens (via www.idec.gr/iier/new/tomos10).

[4] Daniel HAMILTON & Gerhard MANGOTT, Eds., *The Wider Black Sea Region in the 21st Century: Strategic, Economic and Energy Perspectives*, Center for Transatlantic Relations, Washington, D.C., 2008.

[5] Ronald D. ASMUS *et al.*, *A New Euro-Atlantic Strategy for the Black Sea Region*, The German Marshall Fund of the United States, Washington, D.C., 2004; Jeffrey MANKOFF, *Eurasian Energy*, Council on Foreign Relations, Special Report No. 43, New York, February 2009.

region but also made it clear that the US intended "to play a supporting, not leading, role in Europe's energy security and the development of Caspian oil and gas".[6] The three main components of the US-Eurasian energy strategy, covering also the Black Sea region, as described in that policy statement are: (i) to encourage the development of new oil and gas resources; (ii) to support Europe in its quest for energy security; and (iii) to help the Caucasus and Central Asian producer countries find new routes to market for their oil and gas. A more recent study[7] specifically recommends increased transfer of US technology and know-how to the countries of Central and Eastern Europe for the mapping and development of non-conventional gas resources, carbon capture and sequestration, renewable energy sources, improved energy efficiency and introduction of smart grids.

After the accession of Romania and Bulgaria in 2007, the European Union has actually become a Black Sea power and no longer counts as an outsider. After some initial hesitation and extensive consultations, the European Commission adopted a Communication entitled *Black Sea Synergy – A New Regional Cooperation Initiative*[8] and followed up a year later with a report on the first year of implementation.[9] The policy paper gives prominence to regional cooperation in the field of energy emphasizing its sustainable and ecological dimension, improved energy efficiency and construction of new energy infrastructure. It also commits the EU to explore the feasibility of an overall legal framework with clear, transparent, and non-discriminatory rules for the interaction of energy producer, transit and consumer countries. A subsequent EU initiative on Eastern Partnership is also relevant for the Black Sea region since it supplements the multilateral approach of the *Synergy* with targeted actions in a bilateral format. The combined effect of these two European policy initiatives is likely to be considerable in the medium run, provided they evolve into a real regional strategy with adequate financial backing.[10]

Taking a longer view, the Black Sea region is going to be quite a different place 20 years from now. Some analysts even predict that not only Turkey but also the Ukraine and even Russia may well become EU members or closely associated with it by 2030.[11] Significantly, this vision of the future seems to be shared by an influential Russian think tank.[12]

---

[6] Ambassador Richard MORNINGSTAR, Special Envoy for Eurasian Energy, *2010 Outlook for Eurasian Energy*, speech at the Center for American Progress, Washington, D.C., US Department of State, 28 January 2010.

[7] Keith C. SMITH, *Russia-Europe Energy Relations: Implications for U.S. Policy*, Center for Strategic and International Studies, Washington, D.C., February 2010.

[8] Commission of the European Communities, COM(2007) 160 final, Brussels, 11 April 2007.

[9] COM(2008) 391 final, Brussels, 19 June 2008.

[10] Michael EMERSON, *The EU's New Black Sea Policy: What kind of Regionalism is This?*, Centre for European Policy Studies, Working Document No.297, Brussels, July 2008.

[11] Parag KHANNA, "A Postmodern Middle Ages", *Spiegel Online* (via www.spiegel.de/international/europe), 23 July 2009.

[12] I. YURGENS, E. GONTMACHER, N. MASLENNIKOV, L. GRIGORIEV, *Rossia XXI-go veka: obraz zhelaemogo zavtra*, Report of the Institute of Contemporary Development, Econ-Inform, Moscow, 2010 (via www.riocenter.ru/files/ Obraz_gel_zavtra.pdf).

In the meantime, whatever the European Union does, or fails to do, in the Black Sea region is bound to carry increasing weight. The EU package of directives on energy and climate change sets ambitious targets for 20% reduction of carbon dioxide emissions, 20% improvement of energy efficiency and 20% share of renewables in the overall production of electricity by 2020. Naturally, the non-EU Black Sea countries are not bound by those commitments, but it is reasonable to expect that the developments within the EU will have a considerable impact throughout the region in terms of progressively harmonized standards and operating procedures leading to improved transparency and accountability. Quite interesting in that respect is the recent suggestion coming from a German think tank[13] to introduce a new type of benchmarking for the performance of the EU member states relying on the 'naming and praising' concept. By emphasizing a carrots-rather-than-sticks approach and giving preference to rewards over penalties without renouncing the tested method of conditionality this new concept may render good service to the future EU undertakings in the Black Sea area.

There is no single magic solution to the twin challenges of growing energy demand and the need to mitigate the nefarious consequences of man-made climate change. Successful transition to cleaner, greener post-carbon economy will require a wider choice from a more diversified range of primary energy sources, including continued, though gradually diminishing, direct burning of fossil fuels. Eventually, they will have to be replaced by alternative, renewable, and advanced nuclear sources of energy in a process that should be steadily reinforced through conservation measures. This will take time, effort, and considerable investments.

Technology is coming to the rescue. At no time since the advent of nuclear power have we seen such a massive invasion of the energy market by innovative and eco-efficient methods to generate heat and power at increasingly competitive cost. Wind farms and solar panels have become a familiar part of the landscape. A recent technological breakthrough has enabled the United States to mass produce natural gas from unconventional sources (shale and tight gas) resulting in a downward price adjustment in the world LNG market and the postponement of strategic investments in high-cost upstream development.[14]

The countries of the Black Sea region are, inescapably, part of this process. Most of them have to face a set of daunting problems that have been largely overcome in other parts of Europe: overdependence on a single source of primary energy, aging power production facilities and transportation networks, poor cross-border interconnection of pipelines and electricity grids, low energy efficiency per unit of product, and large pockets of energy poverty. The panoply of policies and pro-active measures that are currently being considered can be roughly divided into two categories: (i) upgrading the national systems of energy production,

---

[13] Joachim FRITZ-VANNAHME, Armando GARCIA SCHMIDT, Dominik HIERLEMANN & Robert VEHRKAMP, *Spotlight Europe,* a Bertelsmann Foundation publication, (via www.bertelsmann-stiftung.de/spotlight), 11 February 2010.

[14] For an exciting presentation of the realistic prospects for further imminent breakthroughs in energy, see Bill GATES, "Innovating to Zero", speech at TED – Ideas worth spreading (via www.ted.com/speaker/bill_gates.html), February 2010; for recent developments on the gas scene *q.v.* also Anders ASLUND, "Gazprom is the Essence of the Energy Curse", *The Moscow Times*, 24 February 2010.

transport, distribution and use; and (ii) promoting international cooperation primarily in those areas where a regional approach is politically feasible and provides added value compared to individual country efforts. This paper intends to focus on the second category and offer some indicative suggestions on the way forward in areas that are seldom, if at all, being considered in the framework of the existing Black Sea regional organisations and initiatives. It deliberately leaves aside the politically sensitive questions of energy sector reform, upstream development of hydrocarbon resources and pipeline routes, which have been extensively discussed in recent writings on energy security.

Renewable Resources and Smart Grids

Some of Europe's largest wind power compounds are currently being developed along the Black Sea littoral. Energy production from solar thermal and photovoltaic installations is also set to take off.[15] The problem is that, in the entire Black Sea area, the national power grids in their present form are poorly equipped to cope with massive but inherently intermittent inputs of additional electricity. Wind doesn't blow and sun doesn't shine all the time. The danger of major disruptions, including serious blackouts, once those new facilities come on line is real, and it can give a bad name to the very notion of renewable energy. Such difficulties were encountered by some West European countries in the early days of wind power development, but they were able to manage thanks to their sophisticated and internationally interconnected power grids. The antiquated systems in the Black Sea countries cannot deliver such flexibility.

A solution is needed, and urgently. Such a solution exists and it has been successfully tested on a scale large enough to warrant positive conclusions. The advances in digital technology make it possible to implement an intelligent system of transmission and distribution bringing the producers and consumers together in a symbiotic, interactive relationship. The system would be built on micro-grids and smart grids to be linked up into an interconnected continental super grid that would guarantee security of power supply at all times. According to recent studies[16], smart grid technologies increase system efficiency to 60% (compared to the current world average of 33%) and also reduce electricity consumption by as much as 30% if combined with smart metering. The implementation of such systems is advancing rapidly. By 2014, global cumulative spending on smart grids is likely to exceed US$33 billion up from $12 billion in 2008. The estimated investment for introducing the new system amounts to less than 5 euros a year per household, which is largely compensated by savings on the energy bill.[17]

---

[15] A comprehensive country-by-country review of the potential sources of renewable energy in the Black Sea region (wind, biomass, solar, geothermal) can be found in *Greening the Black Sea Synergy*, WWF-World Wide Fund for Nature and Heinrich Boell Foundation, Brussels, June 2008, pp. 63–72.

[16] *Energy Predictions 2010*, Deloitte Energy and Resources (via www.deloitte.com/energy), London, 2009.

[17] "Smart grids enable renewables at 'moderate' cost", press report by *EurActiv* on the joint study released by the European Renewable Energy Council (EREC) and Greenpeace on 4 February 2010 (via www.euractiv.com/en/energy, 5 February 2010).

In view of the specific circumstances in the Black Sea area and considering the anticipated consequences of climate change, serious consideration of the feasibility of a regional smart grid system makes eminently good sense. Individual countries will have to develop such systems anyway. By pooling national financial resources, resorting to imaginative ways to bring in foreign investment and providing attractive incentives for public-private partnerships the Black Sea countries can reap the advantages of the economy of scale, improve the cost-effectiveness of investment and serve the interests of their citizens in a very tangible sense.

Energy Poverty and Smart Metering

Energy poverty is defined as the condition of a household when its energy costs are greater than 10% of its disposable income. In most Black Sea countries that percentage is often greatly exceeded, even though energy prices at the consumer's end continue to be subsidised, in some cases heavily. Many households in this vulnerable group have to resort to extreme measures such as voluntarily disconnecting themselves from the grid or/and district heating schemes or using improvised and dangerous substitutes. Energy theft is common. Collection rates on energy bills for both commercial and household use are rarely published and whatever statistics are available show a distressing picture. For good reason energy poverty is considered to be a significant social problem.

There is no simple answer to that question. It will require a complex set of measures ranging from improved legislation on social and community services, better regulation of the energy market, combined heat and power generation, enhanced energy efficiency in buildings, and direct support where necessary.[18] One of the important components of the solution is offered, once again, by technological progress. Italy was the first European country to implement a smart metering system to help households keep track of their electricity consumption, resulting in improved collection rates, reduction of service interruptions by more than half, lower meter management costs and significant savings for the customers.[19]

That experience is highly relevant for the Black Sea region. It will involve adequate agreements on setting regional rather than national standards for metering systems, which are now quite diversified. Even more importantly, it will require a thorough shake-up of the intricate web of state aid measures and cross-subsidies that are still prevailing in the energy systems of many of the region's countries.

---

[18] "EU urged to prioritize tackling energy poverty", *EurActiv*, 6 January 2010 (via www.euractiv. com/en/energy).

[19] "Italy reaps first-mover benefits of smart meters: Enel", *EurActiv*, 2 February 2010 (via www.euractiv.com/en/climate-environment).

Waste to Energy

Waste, especially municipal waste collection, selection, recycling and disposal are another common problem for the Black Sea countries. For many decades, waste used to be simply dumped in dedicated sites (not proper landfills) with no regard for the environmental impact or the sheer nuisance it caused to the neighboring communities. Those mountains of accumulated, and often toxic, waste have to be dealt with as a matter of urgency. Even EU member states like Bulgaria, Greece, and Romania face infringement procedures and hefty fines for non-compliance with European regulations within the established deadlines. Most municipalities find it hard to cope with the problem on their own. Again, a comprehensive set of legislative, regulatory and administrative measures has to be put in motion in each of the region's countries.

Part of the answer lies in the application of modern technologies. State-of-the-art, ecologically safe incinerator *cum* cogeneration units are perfectly adequate to take care of large urban agglomerations, but they are simply too big to cover the needs of middle-sized or smaller communities. This entails shipping large amounts of waste from the surrounding area over considerable distances. One recent example involves the planning for a modern, fully eco-efficient waste incinerator plant using Swedish technology in the Romanian harbor city of Tulcea on the Danube, close to the borders of Ukraine and the Republic of Moldova. Preliminary calculations showed that it would be much cheaper to send in waste by barges from the adjoining areas in the two neighbor countries in order to enable the plant to operate at full capacity, while providing the communities in the nearby areas across the borders with additional inputs of power at affordable prices.

Before it becomes possible to develop a regionally integrated waste management, such cross-boundary projects, even on a local scale, should be encouraged, taking advantage of geographical proximity in order to deliver improved energy services to the population while safely eliminating waste.

Carbon Capture and Sequestration (Storage)

In all Black Sea countries coal-fired power plants still account for a sizeable portion of the energy mix. For more or less historically justified reasons of national self-sufficiency and system reliability those plans are frequently burning low-grade domestically produced coal. More than half of the plants are outdated; they use 30–40 years old technology and are a major source of emissions of carbon dioxide and other pollutants. Burning coal emits 1 tonne of carbon dioxide for every megawatt of electricity generated compared to roughly 0.4 tonne in the case of natural gas. Plans are under way in all the Black Sea countries to shift from coal to gas as a medium-term solution toward reducing dependence on fossil fuels. Until the new plants come on line, a bridging solution is needed.

Great expectations are placed in the new technology of carbon capture and storage, which allows the coal-fired plants to pump their emissions into depleted gas deposits or other natural underground cavities or porous rock formations. Up to 15 demonstrative coal-fuelled plants using such technologies are planned to become operational in the EU by 2015.[20] Only one of those is being developed east of the now defunct Cold War divide (namely in Poland) although the achievement of conformity with EU and international emissions standards is much more of a problem in that part of Europe.

It stands to reason that international efforts to reduce emissions of greenhouse gases should concentrate on those regions where there is more room for substantial improvements through cost-effective measures. The Black Sea is such a region and carbon sequestration is such a measure.

---

[20] "EU Energy", *Agence Europe*, 19 September 2007.

CONSIDERATIONS ON SUSTAINABLE DEVELOPMENT AND ENERGY SECURITY

JOHN VOELLER*
Senior Vice President, Black & Veatch

Abstract.   The current views and methods associated with sustainability in energy in all steps of the life cycle tend to focus on direct elements and the first tier cascade of related impacts or potential problems. There is little examination into secondary, tertiary, or even higher order terms in the overall sustainability model. As has often been the case in the past in other endeavors, we are often too ready to suggest the higher order terms of an equation are of less impact and can be removed for clarity and handling ease. This paper will discuss the higher order effects of energy security where the KISS principle strips out critical knowledge we must incorporate to have for a more encompassing view of real vulnerability.

Premise

Energy security is one of the most discussed topics in examination of the issues of sustainability and most of these discussions cover the same ground with different perspectives on the same set of topics. Most of these discussions focus on what I will call direct or first order terms in the security equation. There is no question these terms are the most common or easily understood ones but they are not necessarily the most important or most easily disruptive to genuine security.

Much of the current view of security comes from an era where the major western nations were both the largest consumers and makers of the elements of energy from down-hole to light socket. Neither of these perspectives is accurate today and will become even less true over time as the emerging nations become not only the major consumers and makers but also innovators for reasons the early adopters did not have or did not choose to address. It is here that several critical components of energy security in numerous contexts become apparent.

The examples provided in this paper are termed energy <u>insecurities</u>. The focus on this shift of emphasis is the most critical topic of this paper's examination. In looking at the many issues of pursuing certainty, the perspective of security contains a hidden problem. The problem is that security tends to focus on removing

---
* To whom correspondence should be addressed: John Voeller, Senior Vice President, Black & Veatch, e-mail: VoellerJG@bv.com

A. Gheorghe and L. Muresan (eds.), *Energy Security: International and Local Issues,*
Theoretical Perspectives, and Critical Energy Infrastructures,
DOI 10.1007/978-94-007-0719-1_5, © Springer Science+Business Media B.V. 2011

or neutralizing a list of known threats. The broader and more important perspective involves looking at the sea of circumstances to identify possible insecurities not yet cataloged.

Without this shift in perspectives, one is always playing catch-up and is assured of a period of damage until the insecurity becomes a defined security with prescribed solutions one can teach or sell. Our recent history includes having someone with the stature of Alan Greenspan admit that he could never understand Consolidated Debt Obligations (CDO) but allowed them to be created and sold. The same is true for concentration on computer viruses using a dictionary of past problems instead of addressing zero-day attack mechanisms which have caused the majority of real damage in the past decade. Using the insecurity perspective and the undefined threat viewpoint is critical and it is interesting that this alternative viewpoint was introduced into U.S. government literature in 2006 in the US Department of Homeland Security produced National Critical Infrastructure R&D Plan Update and first appeared in the European Union's guiding research literature in Framework 7 in 2008. However, the bulk of the current operational efforts and industrial offerings are still focused on known security concerns. The following are just some of these higher order components of energy security.

Specialty Materials

Late in 2009, Chinese organizations gained control of seventeen of the major providers of rare earth materials in the world. This involves the majority of known supplies and processors. In the past, this would not have been a major issue but today it impacts every cell phone, every compact electric motor, every high-density magnetic transducer, many chemical and biological sensors, batteries in applications from phones to cars, creep prevention turbine blades, the bulbs that are supposed to replace incandescent ones, most photovoltaic solar cells, the hafnium that allowed the multi-core chips to be developed, and so much more.

The Chinese did not do this to corner the market but rather they saw that as they westernized, they would need the majority of it for themselves. The problem was that even the largest firms did not see this coming, For example, GE explained at a recent meeting at the National Academies of Science that this caught them unaware and they had to do emergency work for a year to deal with it on just one element. They have since discovered other issues as well with additional elements where access is limited or lost.

This is not a market shift or consequence of the economic disruption but rather something that was predicted by the author as early as 1999 and presented as a likely future by the National Intelligence Council in 2001. Independent of the direct implications this situation creates, the higher order terms related to supply chain, spare parts access and cost, possible disruption, or loss of operations for long periods or at strategic times.

CATALYSTS

Hidden in many different facets of manufacturing and production in a wide range of areas of energy are physical and biological catalysts that enable reactions or enhance processes in unique ways. As with rare earths, we have found that in many cases, there are few and often only one material that accomplishes what we need in a commercially viable or safe. In some cases the catalysts participate to the point of consumption and in others they simply enable. In many cases, the amount of these available and the current production capacity to provide these is limited and often not under friendly control. The loss of access to these for even a few days can shut down a facility and, for several weeks, can destroy a company. The use of catalysts is well known in petrochemical processing but more invisible in specialty alloys, coatings and reaction chemistry. The geopolitical redistribution of wealth and consumption will have major impacts on this area and things from new assets to spare parts for existing ones will be impacted.

It is interesting to note that in the 2007 analysis done by the Director of National Intelligence on critical technologies to watch in the future, bio-catalysis was deemed one of the most significant. Unlike inorganic chemical catalysts and additives, bio-catalytic agents can be active and even exhibit learned behavior in their actions. This also means they can be crippled, disabled or perhaps taught to do other things than what we want.

The impacts of catalysts on all things energy is enormous and the intellectual property, the methods, and materials have always been controlled but the loss of access, volume, process, or IP can created lasting impacts. Conversely, lack of intense and aggressive management of catalysts in future processes and materials could mean the difference between a breakthrough and a one-hit wonder in the energy realm. With the many new forms of energy creation, production and use that make up sustainable energy futures, insecurities lurk in many of the new paths to the future.

LUBRICANTS

The idea that lubricants would ever be a security issue might seem silly but it is a very unique element of everything that moves or rotates. In some cases, there are many alternatives but in an increasing number, the alternatives are few as special formulations are needed to allow performance or behavior not possible with the more common lubricants. For example, in higher pressure control systems of large prime movers, conventional lubricants gave way to ester-based fluids in which several increasingly scarce substances are used. This affects every steam or gas turbine or turbo-compressor in use. It will also affect the next generation of both horizontal and vertical wind turbines and a new generation of nano-slurry sizers needed to make nano-materials in bulk at low cost.

ALLOYS

The additives common to specialty allows used in the different areas of the energy business include the usual suspects of chromium, titanium, molybdenum, tungsten, etc. However, even these well known elements are becoming a security issue as both the consumption and access create unexpected consequences on moving modern energy systems ahead whether new or maintained.

The conventional wisdom is that these are enhancing agents that can be created without but the lack of access to chromium in high chrome piping for power plants would lower the operating temperature, efficiency, and plant output significantly and reduce their economic value as well as increase their risk profile.

In the most sophisticated new materials we foresee in the next decade, the composition of materials will be very different with the incorporation of not only nanotechnology and molecular manipulation to create synthetic materials, but also opportunity to create "mutals" or metals that self-regulate their behavior in response to their loadings or conditions. A number of the methods and materials needed to make this possible are part of intellectual property not being created or controlled by those who are most dependent on success in these areas. Though obvious once said, the strategic implication of global shifts in key intellectual property control in this topic area is a building insecurity.

Meeting Standards

Standards are often assumed to have binary lives where they are either relevant when created or becoming irrelevant when the issues they address are no longer of concern. We have spent the entire Industrial Revolution developing vast collections of standards around the world and using them to not only allow interchangeability and efficiency but also to establish "legally accepted practice" to show what is considered sufficient in so many things we do, make, and use. We have also used them as a form of trade control or trade barrier, a trend we will see grow large again in the near future.

With the massive changes in who produces goods, the material and process content, the availability and choices of materials and resources, the press for lower carbon footprint, energy use and water use, and many other factors, the standards of the past 100 years must be examined proactively to expedite the changes forced by new circumstance and to prevent slowing progress due to lack of such changes. In something as unique as the MilSpecs used in the US and adopted in many ways around the world, we have dramatic impacts we must consider and in all the engineering and technical societies, we have similar efforts we must undertake.

This is not simply a technical issue. The speed of innovation and introduction of new materials and methods ahead and without benefit of supportive consensus-based standards creates legal exposures on a wide range of levels. The current speed of creation and adoption of standards has become so bureaucratic that this entire domain needs re-engineering to meet the needs of overhauling what we have assumed were static standards. This not a matter of simply rewriting these standards but many will require new methods, materials and delivery research, product liability research, and longevity research that all are part of a legitimate sustainability profile. Lack of critical standards related to major changes in materials, methods and delivery is another form of energy insecurity.

Distributed Power

The many variations of distributed power from the Smart Grid to the power fabric concepts being considered in South America all rely on a wide range of methods and protocols to accomplish connectivity and control along multiple channels for the different stakeholders. Beyond the obvious security issues of cyber intrusion and denial of service, there are actions possible in such environments that would be extremely difficult to detect without very special social and behavioral learning and monitoring. Such actions will cause economic loss, loss of consumer confidence, and thwarting of systems unless they can be prevented. The issue of sustainability in this case is one of dealing with the massive investment needed to build out and support such systems and the delay of return on that investment induced by extension of the return timeline due to added intervening costs and lower adoption levels.

The elements of distributed power also introduce new forms of service, supply, continuity, and emergency management and within each are new insecurities not present in current master energy systems. This is not a matter of breaking the big system down into little pieces but involves new forms of cascaded insecurities and more entry ways through which a threat can be executed by several orders of magnitude beyond the current largely analog bulk delivery system.

It is important to note that the massive proliferation of computing, communication, and savant devices in the Smart Grid or its successors introduce massive increases in cyber and data vulnerability and similarly massive increases in the variations of disruption one can generate with very simple techniques. With the very recent circumvention of the DEP protection system for Windows 7, the primary gross protection method of all Windows systems has been compromised. This makes the overall implementation of any distribution energy delivery system dramatically more vulnerable at multiple levels and in devices that cannot support protection systems but can be used as a gateway for attack.

Next Generation Communications

Perhaps the most interesting example of dependence, and therefore concern, is not an energy device but an increasing critical component of its future. Today it manifests itself in cell phones. Ten years ago, the average cell phone used fifteen of the elements of the Periodic Table. Today, a smart phone uses over sixty elements and the iPhone even more. This means the previously indicated issue on rare earths has a direct potential impact on this device.

However, one would ask, why is this energy security-related. The answer is that the exact same technologies are intrinsic to the machine-to-machine wireless control systems exploding across all industries. In energy, the build-out of the Smart Grid or its successor must have the ease of secure wireless to connect pieces and parts of distributed systems that will not be connected to the broader grid. In the US, this is still being considered but in the emerging nations, it is an intrinsic element. In energy discovery and production, water, power and transportation, the M2M element is a game changer but only if we can sustain it.

The last sentence may sound dramatic but the loss of access to critical elements and materials in the context of many of the plans for next generation energy and supportive utility systems can be severely limited by our inability to deliver the levels of M2M that are needed. The use of wired alternatives is not economically viable and the sophistication of the endpoints would be little different in a wired circumstance in either case. From the display screens in emergency operations and control centers to the individual anomaly sensor in an advanced perimeter security system, the next generation communications systems independent of the cyber component are a massive source of new insecurities of several kinds.

Summary

There are many more areas where the issue of sustainability can be damaged or prevented by factors not normally considered in the past or not present in prior eras of the world economy. Without comprehensive examination of the total array of possible impacting agents on sustainability to the energy space, it will be easy for us to mislead ourselves that we have crafted a strong solution set. The current perspectives on Smart Grid for example may rank as being as short-sighted in this regard as the first several generations of biofuels were as far as being green and low carbon footprint. In both cases in engineering parlance, the control volume around the area of examination was drawn too small and did not incorporate the second and third order terms that were obvious once a properly inclusive boundary was drawn. We have dozens of major examples across many of the alternative energy ideas being purveyed today with more to come if we do not look at the insecurities fully.

In a meeting at the National Academies of Science held by the Government, University and Industry Research Roundtable (GUIRR part of the NAS), a series of presentations not intended to be specific to energy were given on a number of these issues. The result was an initiative to pursue these concerns in 2010 as a national one. Several follow-on works have been triggered by this meeting whose proceedings are on the GUIRR site. However, in reviewing the issues, four of the insecurities presented here were clearly present.

It has been six decades since the world experienced the strong and broad impacts of trade wars where almost every aspect of access was disrupted or prevented by artificial means. As the emerging nations of new consumers and producers reach for their place in the global spotlight and do what is necessary to satisfy their citizens' expectations, there will be renewed opportunities for global trade wars of a type not seen since the ones that helped start WWII. We must imagine sustainability and energy security in this context and consider how we address the challenges above with the added burden that these political constraints would inject and do so for years to come.

Two simple habits pounded into every engineer in the early stages of their education involve writing down all assumptions and drawing a proper control volume around the problem before attempting to analyze it. Doing so properly ensures we will encounter the full collection of terms in an overall energy situation. An addition habit is to examine all terms fully before discarding them which is the discipline of seeking the insecurities that may not be obvious or intuitive. If we continue to over-simplify our view of energy security at these different levels, we will materially damage energy sustainability in irretrievable ways.

LESSONS LEARNED FOR REGIONAL AND GLOBAL ENERGY SECURITY

YAROSLAV MINULLIN, LEO SCHRATTENHOLZER[*]
International Institute Applied Systems Analysis, Laxenburg, Austria

Abstract. Energy supply security is analyzed from the perspective of strategic goals that enhance the long-term resilience of the global energy system in terms amenable to energy modeling. Identified indicators, which are argued to measure long-term energy security, are explored in view of the authors' experience in field of international and regional energy modeling.

Introduction

One major thrust for the production of the volume on Science and Technology for Homeland Security was the "need for a coordinated scientific and technological response to terrorism" (Voeller, 2007). As a political response to terrorism, the Homeland Security Presidential Directive/Hspd-7 established a national policy for Federal departments and agencies to identify and prioritize the United States critical infrastructure and key resources and to protect them from terrorist attacks (Bush, 2003). Voeller (2007, *op. cit.*) argues that the Presidential Directive does not emphasize the "need to mobilize the nation's skills in science and technology" as strongly as the operational concerns. This apparent lack of emphasis was the stimulus for the study by the National Research Council (NRC) *Making the Nation Safer: The Role of Science and Technology in Countering Terrorism* (NRC, 2002). In response to the research priorities identified by the NRC report, the present volume was proposed to create a major new reference resource. Picking up one of the aims proposed for this handbook, this overview article addresses "the international dimensions of homeland security", in view of energy security.

The background of the authors of this article is in the field of systems analysis. Together, they gathered some 40 years of work experience at the International Institute for Applied Systems Analysis (IIASA), a non-governmental global institute for scientific research (IIASA, 2008). Adhering to the scientific method requires working with terms and concepts that are rigorously defined. In this regard, it appears

---

[*] To whom correspondence should be addressed: Yaroslav Minullin, International Institute Applied Systems Analysis, Laxenburg, Austri, e-mail: minulin@gmail.com

A. Gheorghe and L. Muresan (eds.), *Energy Security: International and Local Issues, Theoretical Perspectives, and Critical Energy Infrastructures*,
DOI 10.1007/978-94-007-0719-1_6, © Springer Science+Business Media B.V. 2011

unfortunate that the term "terrorism" has not been unambiguously defined (that is, in absolute rather than relative terms) even politically, let alone scientifically. We therefore refrain from using this term in the description of our analysis following below. In our opinion, the term "deliberate attacks" is a more precise – and therefore better – descriptor of what is commonly referred to as "terrorist attacks".

There appears to be no universally accepted single definition of systems analysis. One of the reasons is that the term has a specific meaning in some fields. Even IIASA itself does not promote a unique interpretation of its name on its website, but it is obvious that interdisciplinary research of interactions of systems is a key characteristic of systems analysis. A particularly important aspect of studying interaction is to avoid the pitfall of neglecting crucial interdependencies between systems. For the general question of national security, any systems analytical approach must therefore consider relevant international aspects in the form of outside reactions to national policies. (EIA, DOE, 2009), (Energy Charter Secretariat, 2007).

The second major pillar on which we want to base our concept is a general and fundamental tenet of systems analysis, according to which systems vulnerability is proportional to systems efficiency.[1] To the extent that increasing efficiency of a system goes along with decreasing redundancy, the assertion appears immediately plausible because in a system with many redundancies, the possibilities for substitution (of a malfunctioning subsystem) are greater than in a system with fewer redundancies. Furthermore, if we understand resilience as the opposite of vulnerability, we see that an important systems analytical issue to study is the interplay (trade-off) between resilience and efficiency, that is, to find a joint optimum (maximum) of the conflicting objectives resilience and efficiency.[2]

As to energy systems, vulnerability and resilience are related to security in an obvious way. Energy security can be enhanced by increasing the resilience of the energy (infrastructure) system. For the purpose of analysis, we believe that it is useful to also distinguish between resilience in the short term and resilience in the long term. While short-term resilience is related to natural disasters and deliberate attacks, long-term resilience is more a fundamental structural feature of any system.

In the sequel, we cover two topics. i) basic design features of secure energy systems, (modeling of energy security and competition by presenting the design and some illustrative results of a model of gas-market competition), and ii) policy implications of the analysis presented.

---

[1] This general insight has been formulated and substantiated in many specific cases. See, for instance, Criado *et al.* (2006), who analyzed different efficiency, vulnerability and cost functions and found "that these magnitudes display strong correlations".

[2] Accordingly, one of the first major themes of research undertaken at the International Institute for Applied Systems Analysis (IIASA) was a collaborative effort on resilience by IIASA's Ecology and Energy Projects (Holling 1973; Häfele 1976).

Cooperative Energy Security: Resilient Energy Systems

In systems analytical terms, one important manifestation of resilience is a stable equilibrium of a system (Häfele, 1976). To illustrate this mathematically for the case of a one-dimensional dynamic system:

$$Y = \dot{x}(t),$$

a stationary state x_0 (defined by $\dot{x}(t_0) = 0$) is a stable equilibrium of the system if and only if:

$$\ddot{x} < 0$$

In an interval around x_0, the dynamics of the system move it towards the equilibrium point.

Applying the concept to an energy system means that the system can be called resilient if it is designed in a way that the dynamic forces determining the evolvement of the system lead it back to the original equilibrium state after a minor (accidental) shock has displaced it away from the equilibrium point.[3] An example of such a shock would be a deliberate hostile attack on a particular part of energy infrastructure. Addressing such a shock would be addressing the short-term resilience of an energy supply system.

Let us now turn to long-term energy system resilience and, consequently, on resilience as a fundamental structural feature of the energy system (Holling 1973).

Doing so requires looking at the long-term vulnerabilities of a national energy system. One important – maybe the most important – long-term security threat to any national energy system is the failure of international suppliers of energy to deliver according to signed agreements or established market principles. Looking at examples of the past, we find that supply disruptions often were justified with disagreements among the partners (consuming, producing, and transit countries) and thus political rather than technical. From this we conclude that long-term resilience of energy supply requires energy importers, exporters, and transit countries to share a strong common interest, which stabilizes any equilibrium and which treats 'shocks' as a stability problem that must be solved jointly.

As a prerequisite of such a system design is therefore symmetry in the sense that the system must serve the interests of all parties involved. Thus if security of supply is the main criterion for the long-term resilience of an energy system from the importer's perspective, security of demand must be seen as the symmetric criterion from the perspective of an exporter (Criado et al., 2006). Both criteria

---

[3] Perhaps one of the most well-known applications of the equilibrium concept is the model of equilibrium price, which results from the intersection of the demand curve and the supply curve. According to this model, for instance, a downward movement ('shock') of the demand curve is followed by an adjustment (reduction) of supply of the item in question. The shock and the adjustment together lead to a new price equilibrium. Alas, recent experiences (in particular in the years 2007 and 2008) have shown that in practice, and contrary to what the model suggests, prices can "run away" from levels that were considered stable.

taken together gave rise to cooperative energy security. We cover the two criteria one by one in the following two subsections.

Recent work in this area (Gheorghe and Vamanu, 2009) generalizes the concept of resiliency in relation to critical infrastructures (e.g. energy infrastructures) by introducing cooperative systems models for quantitative vulnerability assessment for interdependent complex structures. They describe the resiliency of systems by highlighting the coexistence of two meta-indicator sets defined as tangibles (investments, available resources, etc.), and intangibles (geopolitics, cultural aspects, etc.).

SECURITY OF SUPPLY

For a long time, energy security was more or less tacitly assumed to refer to the security of supply only. Accordingly, national energy security was defined as "adequate and reliable energy supply by reasonable prices to avoid damages of fundamental national goals and principles" (Yergin, 1973). Even the World Energy Council had only supply in mind when it defined energy security as the "[s]ecurity of citizens, economics, society and nations against damages and for sustainable fuels and energy supply". Similarly, the International Energy Agency sees energy security as the "availability of energy, sufficient (by volume) and available (by price)". In the case of the IEA, the lack of consideration of energy demand security is of course understandable on the grounds that its mission is defined as "energy policy advisor to 27 member countries in their effort to ensure reliable, affordable and clean energy for their citizens" (IEA, 2008b).

Recently, energy supply security was analyzed also from the perspective of strategic goals that enhance the long-term resilience of the global energy system in terms amenable to energy modeling. Schrattenholzer (2008) identified two indicators that are argued to measure long-term energy security. These are the resource-to-production (R/P) ratio of mineral primary-energy resources and equity, which are defined as follows.

The R/P ratio is defined, for any given year and any given resource, the amount of the resource left for consumption ('in the ground'), divided by the annual consumption in that given year. Equity was defined as the ratio of average GDP per capita in today's developing regions and today's industrialized world regions.[4] Most global long-term scenarios include the data that allow these indicators to be calculated. Important examples are the scenarios published by IPCC's Special Report on Emission Scenarios (IPCC, 2000).

SECURITY OF DEMAND

As we have argued above, symmetry is a necessary requirement for stable equilibria. Since the notion of energy demand security is a comparatively recent concept, we

---

[4] In 1990, this ratio was approximately 6%.

begin this discussion by deriving specific aspects of demand security from their 'mirror images' on the supply side.

Vulnerability of supply to natural disasters and deliberate attacks. Natural disasters and deliberate attacks on energy supply infrastructure are hazards to suppliers as well as to consumers and thus symmetric in principle. Recognizing this suggests that producers and consumers have a natural interest in jointly addressing the risks posed by this hazards.

Use of energy as a weapon by suppliers. This demand-side hazard figures strongly in the public coverage of the issue of energy security. Often the symmetric hazard of consumers using the same situation as a weapon against suppliers is not included in the discussion. Considering the fact, however, that energy deliveries, for instance those of natural gas, require sizable up-front investments, the possibility that consumers use sunk cost as leverage is an obvious hazard for suppliers.

Energy prices. Energy prices so high as to be felt as threatening the security of energy supply has, for suppliers, the mirror hazard of energy prices so low as to fail covering costs.

So far, we have mainly argued that a systems analytical approach suggests that the security of energy supply and the security of energy demand should be considered together. Now we turn to the question: *To which degree are security of supply and security of demand different concepts?*

To the extent that security is the absence of risk and risk is '*the probability of an unwanted event*', the concepts are the same — only the unwanted events are different! If consumers and producers follow identical concepts, it is easier for them to practice active energy security in an institutionalized dialogue. Moreover, as our analysis suggests, disruptions (of supply or demand) can be avoided by timely planning. This is another argument for embarking on a comprehensive dialogue, supported by analysis and research, between consumer, producer, and transit countries. Following this argument, one would, e.g., aim at minimizing the joint probabilities of all sides' unwanted events.

Also, the actors in such a global energy security management are likely to orient their assessment of the (subjective) probabilities involved in this exercise according to scenarios of future developments. Thus, energy projections have always played a major role in long-term national security considerations but also in short- and medium-term policy decisions of governments and international organizations such as the International Energy Agency (IEA) and the European Union.

Modeling Energy Security

AN ILLUSTRATIVE EXAMPLE: NATURAL GAS

In order to illustrate how the concept described above can be applied to the formal modeling of real-world issues, we turn to one of the most interesting examples in the wider area of energy security, the international natural gas markets (Victor, 2006).

Before summarizing the model, we want to note that in our opinion, what is usually referred to as natural-gas market in our opinion lacks important features of more conventional markets. The main reason for this opinion is the lack of a global referencing point for the price formation (as in case of oil), and the undeveloped trade on well-established markets. Due to its – relative to the other fossil energy carriers coal and oil – environmental friendliness, the share of natural gas in the primary-energy mix is on the increase, with some countries supplying 40–60% of their primary-energy needs with the "blue fuel". Although this "success story" began as early as in the 1970s, not much has been done in terms of establishing an institutional framework for efficient and reliable gas trade.

In pursuit of a suitable arrangement, in the beginning of 'gas era', consumers and producers tried to hedge the risks of both parties by negotiating long-term contracts (LTCs) which – as a rule – serve to protect producers' interests by introducing a minimum price and an indexing scheme (in some cases there is an additional condition called "take-or-pay", which increases the security of demand, but compromises the flexibility of the consumer) and consumers' interests by guaranteeing a certain delivery pattern throughout each year at predictable prices. The equivalent of *market clearance* occurs during re-negotiation phases, which can take years. These re-negotiations lead to contract amendments regulating the pricing mechanism. By adjusting the coefficients in the 'formula', round by round each party approaches an equilibrium price.

Today global gas trade is still following the formula "if there are no good relations between supplier and consumer, there is no gas trade". This condition immediately extends the matter from an economic layer to a political, if not – given the strategic interests of each party – geopolitical layer. The quoted formula wants to express that in most cases, physical gas deliveries under import-export deals require good bilateral relations between two or three countries. LNG (liquefied natural gas) imports into the USA[5] are a minor counterexample at best as only some 50% of these imports were delivered under short-term LNG contracts.[6] Likewise, regional and local distribution especially in Eurasia is also characterized by non-market price formation for end users.

These peculiarities of gas trade can be explained – among others – by the facts that (1) consumption and production centers are concentrated, (2) the infrastructure for natural-gas production, transportation, and distribution and is very inflexible (due to the long economic service life of the equipment involved) and cost-intensive (3) there are few alternative means of gas delivery, which are, again, inflexible, and, finally, (4) there are few major gas producers, and these are geographically scattered.

In any case, these types of international relations have an obvious bearing on national energy security. Our formal analysis of these relations builds on the notion of equilibrium in the form of agreed-upon prices and volumes (Klaassan et al., 2002), (Kryazhimsky et al., 2005).

---

[5] The USA is often believed to be a pioneer in the liberalization of natural-gas trade, and LNG has been attributed a role of a "dissolver" of long-term trading agreements.

[6] Moreover, the share of LNG imports in the US consumption is very small – only 3.7% in 2007.

In practice, such equilibrium suffers several drawbacks. First, the negotiation position of a consumer depends on a set of geopolitical factors and the state of bilateral relations. Second, there is very little or no competition between sellers at all (primarily due to non-flexibility of the import infrastructure), which provides more leverage to the supplier. Thus, the equilibrium is defined on a very narrow optimization interval, which is sensitive to externalities and influenced by many intangible factors.

On the other hand, such long-term arrangement provides two very important ingredients to cooperative energy security: it guarantees the long-term demand for the supplier, which is a key condition to invest into natural-gas production and transmission infrastructure; and it allows the consumer to do national energy planning at given volumes and prices.

In public discussions in Western consuming countries, this mutual dependence of a producer and a consumer is often misperceived as a burden of consumer only. It is obvious, however, that both parties should be interested to act and plan jointly, thus improving energy security of both. Although we recognize at the same time that such long-term interdependence, which extends to the energy sectors of both parties, can also reduce the flexibility of each of them, and therefore can be associated with the disadvantages of long-term form of gas contracts.

Responding to the perceived shortcomings of long-term markets, the established up-to-date practice of short-term gas deals had the primary goal of reducing the burden of mutual dependence – the heritage of long-term contracts. Whereas past local gas markets were formed from scratch, new spot-markets were formed by analogy to oil markets. The main instruments in this spot-market trade are financial derivatives, covered by physical deliveries under long- or short-term contracts. Among the advantages of such a scheme one could highlight the room for competition, transparency (prices and quantities are reported publically) and indifference with regard to suppliers.

However, while resolving almost all the supply-related drawbacks of the long-term contracts, markets (in their pure form) bring one big disadvantage with respect to security of demand, which is near zero. We can characterize such markets as *delivering energy security to consumer and supplier at the cost of efficiency and mutual dependence,* and we have encountered another manifestation of the tenet of systems analysis mentioned in the introduction, according to which efficiency (market efficiency from the perspective of suppliers) is proportional to vulnerability (of suppliers). Following the logic of cooperative energy security, it is thus natural to expect that the deterioration of suppliers' security will eventually reflect on consumers.

The arguments for the latter consideration go back to the issue of the high investment intensity of gas infrastructure. Payback times, usually equivalent to between 10 and 15 years of a pipeline operating at maximum capacity, pose rather strict necessary conditions even for consideration of an export project.

We summarize main characteristics of the two market types in the following table.

TABLE 1. Main characteristics of international gas trade models

Long-term, bilateral	Free-market
• Few producers, few consumers • Price is indexed by alternative fuels • "Shock-absorbing" price formation • Two-way reliability (prices and quantities) • Infrastructure development: small risk • Foreseeable thus dependable for consumers, producers, investors • Requires to maintain 'good' long-term political relations; creates mutual dependence	• More producers, more consumers • Competitive short-term price formation (seasonality pattern) • High price volatility • Infrastructure development: mostly operational (storage), risky for transmission lines • Less dependable because more erratic • Indifferent with regard to suppliers or consumers

This comparison makes it obvious that both market models have advantages and disadvantages. Therefore a harmonious solution of cooperative energy security should include features of both of them.

ECONOMIC ASPECTS AND RATIONALITY

In previous sections, we identified that competition can help mitigating a number of unwanted features of gas trade – not to mention that fostering competitive energy markets is one of the key elements in modern energy policies of almost all countries. Competition and energy security are thus the key ingredients of the model that we shall summarize here.

The model is named GASCOM (Gas Market Competition) and belongs to the family of gaming models. The key idea behind the model was to combine competition with existing methods of evaluating the economic efficiency of energy export projects. GASCOM covers gas trade from an evaluation of a transport corridor on a national level to precise supply schedules and corresponding cash flows.

One of the key characteristics of the gaming approach is that it provides an insight to all admissible strategies (in this particular example: of supply volumes and the timing of market penetration) of all agents. The model thus confirms with our understanding of systems analysis as described in the introduction because it includes all relevant feedbacks and interactions. The result is, for *all* agents,[7] an optimized supply schedule and an optimized time for entering the market.

Here are two interpretations of this model solution:

- In the case of long-term contracts the model replicates rounds of negotiations between the agents,[8] who iteratively update their knowledge about their competitors and adjust their own strategy accordingly. Thus, having collected all information about responses of others to their strategy, agents are capable of

---

[7] This feature is in contrast to models in which the NPV (net present value) of *one* (and only one) is maximized.

[8] In the example discussed here, is the agents are one importer and many exporters.

defining an 'optimal' time of entering the market, optimality being defined by the minimum time passed from the point of decision making to payback (this option of optimality criterion is closer to maximization of internal rate of return, IRR, of the project rather than traditionally applied maximization of the net present value, NPV). Having determined the time of market entry, agents engage in the second, distinct phase of negotiations, because – naturally – this represents another game, when agents control their supply to the market. Eventually (in most cases) all agents will reach a point (the Nash equilibrium), where varying the timing or supply schedule won't improve their own benefit. Thus, for long-term contracts case, a gaming model such as GASCOM, permits each agent to reveal the potential demand of the importer as well as the potential supply – a strategic advantage.

- In case of mid- and short-term contracts, the model imitates the market price formation (in our example, with regard to the futures with the delivery in 1 year), where market fundamentals and project's economical characteristics identify its competitive advantage. Technically, the solution is similar to the case of long-term contracts, but the essence of results is different. First, the equilibrium solution involves much higher market risks due to additional impact on price caused by agents' supply strategies. Second, the solution also presumes coordination between agents: no long-term plans will be valid in case there is some irrational behavior for one of the agents (i.e. deviation from equilibrium in pursuit of strategic interests).

Before presenting illustrative results we would like to mention that most arguments in this section apply for cases when there is a need for the construction of new upstream infrastructure. With growing world demand for natural gas and a trend for diversification of supply sources, this issue is most relevant. Another consideration is that once a gas transmission pipeline has paid back, the risk profile of this infrastructure changes. This, in fact, defines the turning point with regard to providing cooperative energy security: it is most crucial for the payback period; after return of investments, the supplier might gradually engage in free trade.

The three figures below illustrate the processes described above. They illustrate results from a recent case study in which GASCOM was used to analyze the perspectives and the potential of the emerging China gas market as well as of a set of proposed export projects from Russia, Kazakhstan, Turkmenistan and LNG in the Pacific Basin (Minullin, 2008a). In Figure 1, we present the discounted cash flows before optimization, that is, the cash flows as a consequence of realizing the projects as announced by the representative companies. Explanations inserted into the graphics and the shapes of the curves demonstrate that the model reflects all important peculiarities of a transmission project. Not only that without optimization one of the projects does not pay off before 2050, most of them also have a relatively low IRR (internal rate of return). The underlying reason for this is that all agents endeavor to take a strategic position on a newly emerging market and therefore all of them enter it as quickly as they can. Since demand is limited, achievable prices are insufficient for the candidate projects to be economically attractive.

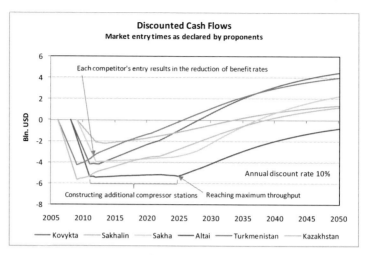

Figure 1. Discounted cash flow with timing as announced by the promoters.

Figure 2 displays the cash flows for the same set of candidate projects after GASCOM optimization. Now all the projects pay off in shorter time frame, and they have a higher IRR.

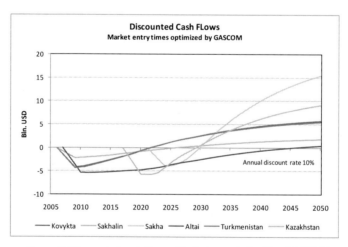

Figure 2. Discounted cash flow with timing optimized by GASCOM.

To give a broader picture, Figure 3 presents the supply schedule by the agents and how it compares to total demand in the market after optimization with the GASCOM model.

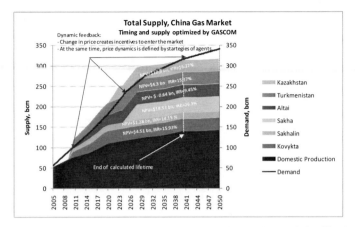

Figure 3. Total supply in the market, timing and supply schedule optimized by GASCOM.

From the past studies with the GASCOM model we would like to highlight two – as we believe – important conclusions.

In case of traditional gas trade, based on long-term contractual agreements, the suppliers have an incentive to cooperate with each other with the aim to mitigate market risks and to achieve the best financial results for all the parties.

In the case of market trade, we have observed that the suppliers acting at a market with limited representation (i.e. not so many suppliers and even less consumers – which is the case of today's natural gas markets) tend to keep it in deficit, thus increasing instantaneous profits – which, given no possibility to build on long-term planning is a more profitable strategy for them.

EXPANDING THE SCOPE: GEOPOLITICS AND THE INCLUSION OF TRANSIT COUNTRIES INTO THE ANALYSIS

So far we based our discussion of energy security on price and volume factors. However, there are a number of concerns, which make it necessary to enhance the existing analysis tools to be closely related to the real world.

- There has been a growing number of instances in which strategic aspects have dominated over traditionally-considered technical and economical feasibility in the process of decision-making in the field of international energy deals.
- Additional concerns are being raised by the growing complexity of the world energy infrastructure, which strongly exposes inter-dependability of its layers and sensitivity to short and long-term threats.
- The interruption of the export flows of Russian gas in January 2009 suggest that role of transit regions in providing the security of supply had been underestimated by research. It is also important that bypassing traditional transit regions will have a significant socio-economic impact on them, which, in turn, will affect cooperative energy security.

GASCOM is being enhanced to take these concerns into account.

By monitoring the unwanted events (e.g. a supply disruption for a critical period of time) the new system is will be capable of answering questions such as the following:

- Which setup and evolution of the import-export network guarantees the minimum vulnerability in case the given unwanted event occurs?
- How will alternative supply routes impact the cooperative security, that is, to which degree will it serve the suppliers' demand security, the consumers' supply security and the socio-economic objectives of the transmitters?

In our opinion, these examples of modeling energy security are distinct from previous approaches to energy systems and capable of dealing with the nexus of energy system and its economic-environmental solutions in a dynamic geopolitical global perspective.

Policy Implications

Viewing the resilience of a national energy system as a problem of homeland security can go a long way, but as we have argued in this article, going the whole way to international long-term energy security requires a systems-analytical approach. One of the most important ingredients of such an approach is to include symmetry, most importantly the symmetry of demand security and supply security. We have illustrated this conviction with examples of popular arguments, which we analyzed from the perspective of symmetry. The resulting recommended strategy we call cooperative energy security.

The second major focus of this article was the presentation of GASCOM, a concrete model of gas trade, which is one of the most important issues in long-term international energy supply and demand. At the core of that model is the notion of Nash Equilibrium, which is completely symmetric by its very formulation.

Applying the model we analyzed two trade modes, a long-term, bilateral mode and a free-market mode. From our analysis of the advantages and disadvantages of the two modes we conclude that the 'optimal' mode of the model is fully consistent with the strategy of cooperative security: long-term contracts until the end of the payback period, free trade afterwards. In addition to its security aspect, we think that cooperation is needed in any gas trade model due to limited nature of gas 'markets'.

How to implement symmetry of global energy security in the real world? Generally speaking, consumer-producer dialogues that recognize this kind of symmetry and that enter the dialogue in a spirit of joint problem solving, would appear as a prerequisite. Modest steps in this direction were undertaken jointly by IIASA's Dynamic Systems (DYN) and Environmentally Compatible Energy (ECS) Programs in 2004–2006. In the spirit of systems analysis, a forum was created, on which academia, industry, and policy making from consumer and producer countries regularly exchanged their views about energy security – in this case of natural gas in particular – in a scientific and neutral environment.

Activities of this forum were the basis of contributions, by the authors of this chapter and their colleagues, to the Energy Modeling Forum's study on "Prices and Trade in a Globalizing Natural Gas Market" (EMF-23), to the Civil G-8 activities preparing for the St. Petersburg G8 Summit and to industrial applications.

In order to better understand the phenomena illustrated in this chapter, IIASA's DYN Program has embarked on the Fragility of Critical Infrastructures (FCI) initiative. The purpose of FCI is to view critical infrastructures in the context of systems analysis that is assessing not only physical properties but also operational, regulatory and behavioral aspects of network nodes and agents involved.

In the meanwhile, the forum has transformed into the so-called *WIEN (World Independent Energy Network) Group*, moderated by the Institute of Energy and Finance, Moscow (Grigoriev et al., 2008). WIEN is an informal network of independent experts and acts as an assembly of individuals with academic, governmental, and industrial backgrounds who are interested in specific issues which are being addressed by the Network.

Also the thoughts presented in this chapter were inspired by the discussions on that forum and in the WIEN Group. Nonetheless, the authors are solely responsible for the contents presented here.

References

Bush, GW, 2003, Homeland Security Presidential Directive/HSPD-7, The White House.

Criado, R, Hernández-Bermejo B, Marco-Blanco J, and Romance M, 2006, Asymptotic estimates for efficiency, vulnerability and cost for random networks, Journal of Computational and Applied Mathematics, Volume 204, Issue 1, 1 July 2007, Pages 166–171.

EIA DOE, 2009: U.S. Natural Gas Imports and Exports: 2007, Special Report.

Energy Charter Secretariat, 2007: Putting a Price on Energy: International Pricing Mechanisms for Oil and Gas.

Energy Modeling Forum, EMF-23, Prices and Trade in a Globalizing Natural Gas Market, Stanford University, 2007. http://www.stanford.edu/group/EMF/projects/emf23/emf23.pdf.

Gheorghe, A., Vamanu D., 2009, Mining intelligence data in the benefit of critical infrastructures security: vulnerability modelling, simulation and assessment, system of systems engineering, International Journal System of Systems Engineering, Vol. 1, Nos ½.

Grigoriev, L et al., 2008, World Independent Energy Network, Draft Mission Statement.

Häfele, W, 1976, 'Resilience of Energy Systems', in: Häfele, W et al., Second Status Report of the IIASA Project on Energy Systems, Research Report 76-1, International Institute of Applied Systems Analysis, Laxenburg, Austria.

Holling, CS, 1973, Resilience and stability of ecological systems, Annual Review of Ecology and Systematics 4: 1-23; reprinted as Research Report 73-3, International Institute of Applied Systems Analysis, Laxenburg, Austria.

IEA, 2008b, http://www.iea.org/about/index.asp.

IIASA, 2008, http://www.iiasa.ac.at/.

IPCC, 2000, Nakicenovic N, Alcamo J, Davis G, de Vries B, Fenhann J, Gaffin S, Gregory K, Gruebler A, Jung TY, Kram T, La Rovere EL, Michaelis L, Mori S, Morita T, Pepper W, Pitcher H, Price L, Riahi K, Roehrl RA, Rogner H.-H, Sankovski A, Schlesinger M, Shukla P, Smith S, Swart R, van Rooijen S, Victor N, Dadi Z. 2000. Special Report on Emissions

Scenarios (SRES), A Special Report of Working Group III of the Intergovernmental Panel on Climate Change, Cambridge University Press, Cambridge, UK.

Klaassen, G., Kryazhimsky, A., Minullin, Ya., Nikonov, O., On a Game of Gas Pipeline Projects Competition, 2002, International Congress Of Mathematicians, Game Theory and Applications Satellite Conference (ICM2002GTA), Proceedings volume, pp. 327–334, Qingdao publishing house, China.

Kryazhimsky, A., Minullin, Ya., Schrattenholzer, L., Global Long-term Energy-Economy-Environment Scenarios with an Emphasis on Russia, 2005, Perspectives in Energy Journal, Vol. 9, pp. 119–137.

Minullin, Ya., "Queuing to China's Gas Market", The Oil of Russia Journal, #5, May 2008a (in Russian).

Minullin Ya., "Whose Pipeline will go East?", The Oil of Russia Journal, #3, March 2008b (in Russian).

NRC, National Research Council Committee on Science and Technology for Countering Terrorism, 2002, Making the Nation Safer: The Role of Science and Technology in Countering Terrorism, (Washington, D.C.: National Academies Press).

Schrattenholzer L, 2008,"Scenarios of Energy Demand and Supply until 2100: Implications for Energy Security", in: Hans Günter Brauch, Úrsula Oswald Spring, John Grin, Czeslaw Mesjasz, Patricia Kameri-Mbote, Navnita Chadha Behera, Béchir Chourou, Heinz Krummenacher (Eds.), Facing Global Environmental Change: Environmental, Human, Energy, Food, Health and Water Security Concepts, Hexagon Series on Human and Environmental Security and Peace, vol. 4, Berlin – Heidelberg – New York: Springer-Verlag, in print.

Victor, D., G., Jaffe, A., M., Hayes, M., H., Natural Gas and Geopolitics: From 1970 to 2040, Cambridge University Press, 2006.

Voeller, J. G., 2007, Wiley, Handbook of Science and Technology for Homeland Security, A Guide for Authors.

Yergin, D, 1973, "The Energy Crisis: Time for Action" Time Magazine.

PART II: THEORETICAL PERSPECTIVES TO ENERGY SECURITY

ENLARGING THE PERSPECTIVE: ENERGY SECURITY
VIA EQUILIBRIUM, INFORMATION, AND COMPUTATION

SOLOMON MARCUS*

Romanian Academy, Bucharest, Romania

Abstract. Despite the unquestionable importance of issues such as increase energy prices, increased world competition for energy, etc. there is however a need to supplement them with a more theoretical perspective, to consider a broader context by going back to the historical development of the energy paradigm and, according to this development, to stress its interaction with the information and the computation paradigms.

Main Motivation

Energy security issues are usually related to tangible dangers, such as the increase in energy prices, increased world competition for energy resources faced with the rapid emergence of some huge economies like China and India, the lack of equitable distribution between major powers, sometimes at the expense of less developed countries. Dangers may appear in some controversial ways with regard to energy-related renewable technologies, energy policy related to unexpected changes in international relations, and at the role of decentralization. In all these respects, a long and a short term perspective are both necessary.

Despite the unquestionable importance of all these issues, there is however a need to supplement them with a more theoretical perspective, to consider a broader context by going back to the historical development of the energy paradigm and, according to this development, to stress its interaction with the information and the computation paradigms. Moreover, we focus our attention on a universal paradigm strongly related to the security paradigm: the equilibrium paradigm. Indeed, equilibrium is a condition of security; it is a pattern goes across mechanics, thermo-dynamics, chemistry, biology, economics, sociology, linguistics, art. The mathematics of the idea of equilibrium give the foundations, the link, and the common denominator of all versions of equilibrium, from Lagrange's analytical mechanics to the Nash equilibrium in the theory of strategic games. In a pioneering paper, Marston Morse ("Equilibria in nature – stable and unstable", *Proceedings of the American*

---

* To whom correspondence should be addressed: Solomon Marcus, Romanian Academy, Bucharest, Romania, e-mail: solomon.marcus@imar.ro

A. Gheorghe and L. Muresan (eds.), *Energy Security: International and Local Issues,*
Theoretical Perspectives, and Critical Energy Infrastructures,
DOI 10.1007/978-94-007-0719-1_7, © Springer Science+Business Media B.V. 2011

Philosophical Society 93 (1949) 3, 222–225) warns us that "the mathematical theory of equilibrium is applicable all the way from quantum mechanics to psychiatry" and he points out different degrees of instability. His ideas created rich development in the mathematics in the last 60 years, and had a considerable impact on all other fields of research; but we are far from taking the whole benefit from the mathematics of the equilibrium.

Another source of innovation in the study of energy security comes from the challenges of physics. Topics such as "computation with physical constraints" and "wireless sensor networks" lead to a new, emergent topic: energy as a new computational resource. There is now an urgent need for a theory of computational energy complexity. This idea will be considered in the last part of this article.

From Mechanics to Economics, in the Nineteenth Century

In the nineteenth century, some ideas from Newtonian mechanics and from Hamilton's analytical mechanics were transferred into the field of economics. Lagrange's variational calculus was used by Hamilton, in the nineteenth century, in order to obtain a definition and a classification of the equilibrium states of a mechanical system. In 1874, L. Walras (*Elements d'economie politique pure.* Corbaz, Lausanne) observes the existence of a constant relation between the quantities of goods to be sold and their price. Starting from this view and inspired by Lagrange's method in defining the mechanical equilibrium, Walras formulates the relation between goods and their prices as an equilibrium problem. The value of a good is given by the energy incorporated in it: (I. Fisher, "Mathematical investigations of the theory of values and prices" *Transactions of the Connecticut Academy of Arts and Sciences*, 9, 1892, 111–126). This author connected the marginal utility with the force, the utility with the energy and the disutility with the mechanical work. A general analogy between field theory and utility theory was proposed by F.Y. Edgeworth (*Mathematical Psychics. An Essay on the Application of Mathematics to the Moral Sciences*, Paul, London, 1881), taking as a term of reference the fields of forces occurring in Maxwell's approach, so useful in the analysis of equilibrium states of the bodies interacting at distance.

Energy, Entropy, Information in the Nineteenth Century

The energy paradigm emerged in the nineteenth century, within the development of thermodynamics. Clausius (1850) introduces the notion of entropy by the formula $dS = (dQ)/T$, where dS is the increment of entropy, dQ is the increment of heat and T is the absolute temperature. Starting from the atomic constitution of matter, Boltzmann redefined entropy S of a macrosystem M with reference to the probability distribution of the states of M. So S is the product between the Boltzmann constant K and the logarithm of the number of states of M. Helmholtz evaluates the internal energy U of a thermodynamic system by the relation $U = F + TS$, where F is the free energy, while S and T are as before. In an isolated system, U is constant, while T and S are increasing, because F is decreasing. When $F = 0$, the maximal value of S is the ratio between U and the maximal value of T; the thermodynamic order O is given by the difference between the maximal value of S and its real value. One can interpret O as *thermodynamic information*.

Energy Explains More Than Matter

With Planck's quantum vision in physics and with Einstein's equivalence relation between energy and mass, energy gains, with respect to the explanatory power, its competition with the matter paradigm. Besides Planck's and Einstein's results, the emergence of the energy paradigm includes also the events related to the year 1927: Heisenberg's uncertainty principle and Bohr's complementary principle.

Equilibrium: From Thermodynamics to Economics

Let's go back to thermodynamics and its relation to economics. There is a kind of isomorphism between them. Energy means both force and utility. The analogy between the way Walras analyzes the equilibrium of a market and Lagrange's method of analyzing the mechanical equilibrium was discussed by A.G. Pikler ("Optimum allocation in econometrics and physics", *Weltwirtschaftliches Archiv* 66, 1951, 97–132) and L. Amoroso ("Théorie mathématique de l'équilibre économique", *Econometrica* 18, 64–80). In their research concerning budget theory, H.T. Davis (*The theory of econometrics*, Principia, Bloomington, 1942) and J.H.C. Lisman ("Econometrics and thermodynamics: a remark on Davis' theory of budgets", *Econometrica* 12, 1949, 59–62) developed an analogy between entropy and utility of currency, implying a further analogy, between income and the provided heat, between savings and internal energy, between expenses and the mechanical work done out of the system. Identification of entropy with utility and of income with heat is also proposed by A.G Pikler ("Utility theories in field physics and mathematical economics", *British Journal for the Philosophy of Science* 5, 1954, 47-58; 303–318). Other analogies are proposed by J. Bryant ("A thermodynamic approach to economics", *Energy Economics* 4, 1982, 36–50).

Samuelson's Itinerary: From Economics to Physics

The force of analogy is visible in P.A. Samuelson's works, for example in *Foundations of Economic Analysis* (Harvard University Press, Cambridge, Mass, 1948) and in "Maximum principles in analytical economics" (*American Economic Review* 62, 1972, 249–262). In the former, some economic relations are expressed as a variant of Le Chatelier's principle, stating essentially that if some constraints are imposed to a system, then the system will react trying to neutralize them. Comparing, for instance, isothermic changes with adiabatic changes, the value of the partial derivative of the volume with respect to the pressure is not higher in the first case than in the second case. Samuelson compares this relation with that of one occurring in a production process with two entries, quantities and prices, when some modifications are operated, some of them keeping the quantities constant, while the other keeps the prices constant. If the products have the prices p and P and quantities q and Q, we distinguish two situations. In the first one, we keep Q constant; then, increasing p implies (as Samuelson shows) that q will decrease, so the partial derivative of q with respect to p cannot be positive. In a second situation, we keep P constant. Increasing p, q will decrease to a value smaller than in the preceding case, so the partial derivative of q with respect to p, when P is fixed, is not larger than the partial derivative of q with respect to p, when Q is fixed. This is just the analogue of Le Chatelier's principle. But the economic principle appeared independently and only in a further step we observe that its physical analogue is Le Chatelier's principle.

Equilibrium States of Isolated, Closed and Open Systems

In isolated systems, the thermodynamic equilibrium is associated with maximum entropy. In closed systems (where there are exchanges of energy with the exterior, but no exchange of matter), the equilibrium is obtained when the free energy is minimal. According to Nicholas Georgescu Roegen, economic systems are open systems based on low entropy. In his *Entropy Law and The Economic Progress* (Harvard University Press, Cambridge, Mass., 1971) he states that the economic processes are governed by the second law of thermodynamics, telling us that usable free energy tends to disperse or become lost in the form of bound energy.

Ilya Prigogine (*Thermodynamics of Irreversible Processes*. Interscience, New York, 1961; *Self-Organization in Non-Equilibrium Systems*. Wiley, New York, 1977; *From Being to Becoming*. Freeman, New York, 1980) introduces the notion of dissipative structures, having a basic role in the study of thermodynamic systems far from equilibrium. With Prigogine, self-organizing systems become very important, as a third step in a chain whose first two steps are mechanical systems and entropic systems. In 1990, Prigogine became interested in the behavior of matter under nonequilibrium conditions. His *End of Certainty* (The Free Press, New York, 1997) points out that the explanatory power of determinism is smaller and smaller. It is symptomatic that a similar title belongs to another famous book: Mathematics: *The Loss of Certainty*, by Morris Kline (Oxford University Press, New York, 1980).

Equilibrium at the Level of the Human Brain: Stephane Lupasco

Lupasco proposes a new way to look at energy, involving a new logic, which is no longer binary, as the classical one, but triadic (*Le principe d'antagonisme et la logique de l'énergie*. Hermann, Paris, 1951; *L'énergie et la matière vivante*. Julliard, Paris, 1962; *La tragédie de l'énergie*.Casterman, Paris, 1970; *L'énergie et la matière psychique*. Julliard, Paris, 1974). The binary distinction order-disorder known from thermodynamics is replaced by a ternary one: three types of organization of matter, related to physics, biology and psychics. For Lupasco, the source of the systemic interactions is the conflict between antagonist energies. A system, be it physical, biologic, psychic, or social, remains alive as soon as it involves the coexistence of some forces of attraction and of some rejection forces between its elements or between its subsystems. Any force acts against another force, one of them gains what the other loses. This means that one of them is actualized, while the other is potentialized. But no of these processes (irrespective the nature of the forces: mechanical, cinetic, chemical, electrical, thermic, gravitational, electrostatic, or nuclear) can be totally realized. So, there is no total potentialization and no total actualization. Here, another fundamental distinction is proposed by Lupasco: that between *homogenization* and *heterogenization*. Beyond their variety, the conflicts between them are always between forces of homogenization and forces of heterogenization. The trend towards homogenization concerns the macrophysical level and it is related to the second principle of thermodynamics, while the trend towards heterogenization is characteristic for the living systems and it corresponds to their local decreasing of entropy.

According to these principles, there are three types of systems: physical systems, where homogeneity is actualized, while heterogeneity is potentialized; living systems, where the converse takes place: homogeneity is potentialized, while heterogeneity is actualized; and neuro-psychic systems, where there is a balance between homogeneity and heterogeneity: both are semipotentialized and semiactualized. This is what Lupasco calls the *state T*, the included third or the included middle, rejected by the principle of excluded middle of classical logic. But let us observe that, at the level of the human brain, both too much homogeneity and too much heterogeneity lead to states of illness; the state of health is characterized by the equilibrium between them, i.e., by the state T.

The Energy-Information Connection

The first connection between energy and information was provided by thermodynamics, via thermodynamic entropy. Then, Harry Nyquist ("Certain factors affecting telegraph speed" *Bell System Technical Journal* 3, 1924, p. 324) is concerned with telegraph signals and discusses quantifying "intelligence" and the "line speed" at which it can be transmitted by a communication system, proposing the relation $W = K \log m$, where W is the speed of transmission of "intelligence", m is the number of different voltage levels to choose from at each time step and K is a constant. The next step belongs to Ralph Hartley ("Transmission of information", *Bell System Technical Journal* 7, 1928, p. 535); for him, information is a measurable quantity, expressing the receiver's capacity to distinguish that one sequence of symbols had been intended by the sender rather than any other. This amount of information carried by a signal s he quantified as $m(s) = \log (1/p(s))$, where $p(s)$ is the probability of appearance of s. All the states of the system have the same probability of appearance. The natural unit of information was the decimal digit, later called the *hartley*. A similar unit of logarithm in base ten of the probability, the *ban,* has as its derived unit the *deciban* (one tenth of a ban) and was introduced by Alan Turing in 1940, within the framework of his approach to the breaking of the German World War II enigma cyphers.

The next step belongs to Claude Shannon ("A mathematical theory of communication", *Bell System Technical Journal* 27, 1948, 379–423 and 623–656); it is the transfer of entropy and information from thermodynamics into the field of communication. Shannon articulates five components of communication: the message, the source, the code, the channel, and the noise. He states that if the entropy of the source is not larger than the capacity of the channel, then there is a codification of the message providing a degree of accuracy as high as desired, in the transmission of the message. Norbert Wiener (p. 11 in his *Cybernetics or Control and Communication in the Animal and the Machine*, The MIT Press, Cambridge, Mass., 1948) completes this picture: the quantity of information of a system is a measure of its degree of organization, while the entropy of the system is a measure of its degree of disorder. But, on the other hand, the itinerary from thermodynamics to information can be reversed, as Leon Brillouin *(Science and Information Theory*, Academic Press, New York, 1962) argues, describing thermodynamics as a special case of information theory. Information is always obtained by production of entropy, so information is *negentropy*. Concomitantly, R. Ingarden and K. Urbanik ("Information without probability", *Colloq. Math.* 9, 1962) proposed a way to define information with no reference to probability, but Patrick Suppes (Probability and information, *The Behavioral and Brain Science* 15, 1983, 457–458) stresses the probabilistic nature of information.

The philosophical meaning of Hartley-Shannon information is in agreement with Karl Popper's proposal in *Logik der Forschung* (1934, p. 569): "The information provided by a statement S is the quantity of indetermination concerning the state of the universe, eliminated by S."

In thermodynamics, the link between energy and entropy is given by the second principle of thermodynamics: when one increases, the other decreases. Octav Onicescu ("Energie informationnelle", *C. Rendus de l'Académie des Sciences, Paris* 263, 1965, 22, 841–842) pointed out the existence of an information analogue of the second principle of thermodynamics, by defining the concept of information energy of a system whose states have definite probabilities of appearance. Its value is given (in analogy with the expression of cinetic energy) by the sum of the squares of the probabilities of the states. Onicescu shows that the information entropy increases (decreases) when the information energy decreases (increases).

Equilibrium and Security in Game Theory

The equilibrium in a strategic game means that all plans chosen by the agents can be executed, i.e., they are mutually consistent. The security level of the possible outcomes of a game is the value of the least favourable outcome. A Nash equilibrium is a collection of strategies by the players of a game such that no player can improve the outcome by only changing the strategy. This notion is central to John Nash's work, for which he got the Nobel Prize in economics in 1994. Many problems related to energy can be placed within the framework of game theory. Energy security becomes, in this way, a problem of security level in a strategic game.

The Physical Side of Computing: Richard Feynman

The first Conference on Physics and Computation was organized at MIT, in 1981, with Richard P. Feynman (Nobel Prize in physics in 1965) as the protagonist. At that moment, quantum theory had matured, computer science was up and running, computers had been used extensively for physical computation, while the understood links between physics and computation were already great: Maxwell's Daemon (relation between irreversibility in computation and in thermodynamics), universal reversible computation and equivalence to general computation, realization of (classical) Turing machine under quantum formalism. One year later, Feynman published his keynote speech at the MIT Conference in *International Journal of Theoretical Physics* (21 (1982)6/7, 467–488), as an article with the title "Simulating physics with computers". Feynman copes with questions such as: What kind of computer are we going to use to simulate physics? What kind of physics are we going to imitate? As a first difficulty it is seen the fact that "the speed of light would depend slightly on the direction, and there might be other anisotropies in physics that we could detect experimentally". Another difficulty: "the laws of nature are reversible, while the computer rules are not". The next sections are concerned with simulating time, simulating probability, quantum computers versus universal quantum simulators, while the next question is: Can quantum systems be probabilistically simulated by a classical computer?

"The Unity of Science Through Computation"

An important workshop on physics and computation took place in Vienna, August 25–28, 2008, organized by Cristian S. Calude and Jose Felix Costa.

The title above is the slogan proposed by Felix Costa; his main reference is to S. Barry Cooper and Piergiorgio Odifreddi, "Incomputability in nature" (in S. Barry Cooper and S. Goncharov (eds.) *Computability and Models*, Kluwer Academic, 2003, 137–160).

From the beginning, they express a pessimistic view on computational theory; it will never be as deep as mathematical analysis, with its 400 years of history. The word *never* appears to be very strange in this context.

Their first thesis is Galileo's principle of natural computation: "Computers exist in Nature when we abstract the physical entities". Their fourth thesis is the so-called Gandalf's Principle of Hidden Universe Computation: "Up to Turing power, all computations of the Universe are describable by suitable *programs,* which correspond to the prescription by finite means of some rational parameters of the system or some computable real numbers. From Turing power up we have computations that are not describable by finite means: *computation without a program*".

Some objects around us may be performing hypercomputation (in the sense of Sir Roger Penrose (*Shadows of the Mind*, Oxford University Press, 1994): we observe them, but we will never be able to simulate their behavior on a computer.

Martin Davis' ("The myth of hypercomputation", in Christof Teuscher, ed.: *Alan Turing: Life and Legacy of a Great Thinker*. Springer Verlag, Berlin et al., 2004, 195–212) fight against hypercomputation is, according to Cooper and Costa, related to his dream to build a hypermachine. In fact, some itineraries in computer science lead to hypermachines.

Wireless Sensor Networks and the Need of a Theory of Energy Complexity

According to Feng Zhao ("Technical perspective. The physical side of computing" *Communications of A.C.M.*, July, 2008), wireless sensor networks (or sensornets) are a new computing platform that blends computation, sensing, and communication with a physical environment. New programming models and new management tools are required by this different class of computers. The whole attitude towards computation has to change, because these new computers challenge "the design of the next generation internet, that not only connects people together, but also connects people with the physical environment". This new computing platform is characterized by severe constraints in resources, mainly in energy, unreliable hardware and communication links and the need to respond to time-critical events. So, a sensornet must gather and act on sensor data in a timely manner. In view of limited battery capacity, a sensor node limits the amount of onboard memory. The data collection and dissemination must be handled in an energy-efficient manner. Zhao considers that a fundamental computer science question arising from sensornet is the role of energy in its functioning. Traditionally, the main problems of computational complexity are those of space complexity and time complexity of algorithms. Now, we have in front of us the emergent task to build a theory of energy complexity in computing.

It is to be expected that research in this direction will raise specific questions on energy security.

FACES OF RESILIENCE

ADRIAN V. GHEORGHE*
Old Dominion University, Norfolk, VA USA

DAN V. VAMANU
Horia Hulubei National Institute of Physics and Nuclear Engineering, Bucharest, Romania

Abstract. A set of physical analogies emphasizing cooperative behaviour in systems are explored and were integrated into so called 'serious energy games', to contribute in a better understanding, monitoring and good governance of some of the complex security issues confronting the evolvement of contemporary energy systems and, most particularly, Europe's. It is believed that such mild mind-stretching exercises might be supportive in the necessary attempt, by the many stakeholders and actors involved, to bring balance and harmony to prevailing policies the first commandment of which seems to be "Go *change* yourself – in a *resilient* fashion!" An extrapolation of several findings to other than Energy systems tempted to implement drastic changes based on substantial substitutions – hardware, software and mindware – under relatively short time horizons cannot be ruled out.

Introduction

The 20-20-20 *Energy and Climate package* adopted by the European Council in March 2007 (v.e.g. EC Environment, 2010), requiring the EU region to reduce, by the year 2020, its greenhouse gas emissions by 20%, increase the share of renewables in energy consumption to 20% and save 20% of total primary energy consumption is perceived by many analysts and voices from the public as *the* most outstretched ambition the planning thinktank of the United Europe had ever set forth for the first half of this century, perhaps next only to an unanimously agreed Continental Constitution and Government. While the goal and cost figures may look far-fetched for those who have lived the fever of the wishful technology substitution triggered by the Oil Crisis of the 1970s – and killed subsequently by a mere drop in oil prices, it is a certain notional clash simmering ominously beyond the confident phrases of the reference policy documents that raise concerns, especially in the academic circles that enjoy the unfair luxury of staying in a

---

* To whom correspondence should be addressed: Adrian V. Gheorghe, Old Dominion University, Norfolk, VA USA, e-mail: adriangheorghe9145@gmail.com

A. Gheorghe and L. Muresan (eds.), *Energy Security: International and Local Issues, Theoretical Perspectives, and Critical Energy Infrastructures,*
DOI 10.1007/978-94-007-0719-1_8, © Springer Science+Business Media B.V. 2011

distance from painful decisions and responsibilities. For what the official European policy wants is – *drastically change* the Energy System to, hopefully, adapt it to the twenty-first century challenges, yet make sure that the shifting realities thus resulting would somehow maintain a *permanently resilient* System, this being seen as the very essence of a sought *Energy Security*. The mere fact that the 27 MS Ayed the documents in regard with the goal on renewables cannot and should not preclude a continued reflection on the feasibility of the overall project, and of the many specific tasks at hand. On the contrary, thinking hard of how one *can* do what, to many, may seem hardly doable within the given time horizon is believed to be a good, correct attitude (SDA Security and Defense Agenda, 2010).

These being said, we are driven back to the *wording*: energy system/services security, safety, risks, adaptability, vulnerability, resilience and others, that are time and again pondered, nuanced, interchanged, fusioned or split apart in a quest for clear, unequivocal messages calling for action.

In this discussion we gloss over *vulnerability*, as a road to perhaps get a better grip on what *resilience* may imply, when thinking of a system under the heavy pressure of unplanned challenges and planned reformation. Epistemologically, we take a pragmatic way in sorting out *terms*: take up the term; make clear your assumptions on it; embody these in a model of the processes of interest; test model's diagnostic and prognostic powers; and, on that, let the people decide what the term is good for.

Vulnerability – The Key In

The bare fact that feeding 'system vulnerability *modeling*' into one popular web search engine has produced 6,000,000 results, whereas calling for 'system vulnerability *definition*' had only resulted in 4,270,000 findings may be construed as just another proof that people – these authors included – indulge in employing things to a much higher degree that they understand them. This apologetic observation will not, however, discourage one to persist in this attitude, that is only consistent with human mind's proclivity for intuitive, inductive, heuristic, trial-and-error, get-it-by-doing-wise learning protocols.

A cursory scanning over the said 6 million web entries would then indelibly reveal that vulnerability 'modeling' is merely a shy way to look, actually, for a vulnerability *metrics*. This may be another reflex of our intellect, according to which *measuring* an esoteric thing would immediately and happily make it exoteric and, thereby, usable.

The term *Quantitative Vulnerability Assessment* (QVA) is, most probably, a result of a warranted analogy with *Quantitative Risk Assessment* (QRA) – the latter coined during the closing quarter of the past century and having made quite a career in the community of risk and safety managers emerging worldwide ever since. QVA is about expressing its subject – vulnerability – in numbers, in a scientifically defendable and practically meaningful way. Unlike QRA, QVA has to face the embarassing difficulty of not having available an agreed 'closed formula' for vulnerability, whereas for risk one *does* have a number of these, one widely accepted saying that the *risk* of a disruptive event equates the *probability* of the event's occurring, times the *measure of the event consequence* powered to a subjective *consequence perception exponent*.

The QRA task is to take a well-substantiated *noun* (risk) to a number. The QVA task is rather to take an *adjective* (i.e. 'vulnerable'), reflective of a virtuality ('prone/open to…') to a number. And *that*, it is believed, is what makes QVA more 'qualitative' than the QRA – in the sense that capturing vulnerability may require more than a single correlation: it may take a *process*, reflective of the targeted system *behavior*, and echoing in a *pattern* that, in turn, may be expressed in equations and numbers. And, with all respectful consideration for the prevailing, historically-established academic opinion that generic analogies may not be trustful in comparing processes too different in nature, one contends that digging into the wealth of analogies that Physics provide may prove worthwhile.

This act of faith comes from the observation that, specifically, Statistical Physics, although scoring its success stories in apparently mudane fields such as the magnets, the alloys, the elastic and thermal properties of materials etc., still remains extraordinarily generic and comprehensive because it basically uses *a template* for all systems it addresses: the many-body system, made of interacting parts and responsive to external actions. And there is no reasonable motive to think that, *at grass-root level*, that is – mind, heart, soul and their many troublesome consequences not included – a critical infrastructure or a human community is much more than 'a collection of interacting parts responsive to external influences'. At least, one may try thinking like that – and see what happens.

Measuring Vulnerability

The key *Credo* in Statistical Physics is that all measurable, macroscopic property and/or behavior of a material entity can be explained by interpreting the motions within its microscopic model, provided the interactions between parts (intensity, effectiveness radius, anisotropies etc.) and parts suceptibility to external influences are properly set and described. The challenge of the attempt to build a physical analogy relevant to the vulnerability/resilience of systems such as e.g. the Electric Power system, or the Energy System overall is to (i) devise a conceptual machine working on measurable indicators, that (ii) would eventually emerge as the observable, macroscopic expression of the inner dynamics of its microscopic, many-body, counterpart.

The Face of Things

It was shown (e.g. Gheorghe, Vamanu, 2004) that one possible macroscopic model for a system vulnerability metrics may be built on the following assumptions:

Assumption 1: One may see *vulnerability* as a system's virtual openness to lose its design functions, and/or structural integrity, and/or identity under the combined interplay of two sets of factors:

U – Internal, system-featuring factors; and

V – External, essentially management/governance-featuring factors.

All factors are supposed to be eventually quantifiable by appropriate indicators.

The U-factors feature the system part interactions, or exchanges – how intense, how far-reaching. The V-factors feature external influences acting upon the system and relate to system parts susceptibility to be affected by such influences.

Since both the U and V factors may comprise a wide variety of indicators having little or nothing in common as far as metrics a solution is in order, to make all these underscored by a common, non-dimensional metrics allowing comparison. That takes one to the next assumption.

Assumption 2: System's measurable/monitored indicators (parameters) may indeed be aggregated such that two monitoring variables U and V are obtained.

One submits that U and V are membership functions of the fuzzy-sets theory approach to impact indicators [4]. Accordingly, if X_i , i = *1, 2, ..., n* are the normalized system indicators, then one may see U and V as some generalized euclidian distances in the *n*-dimensional space of the indicators (Christen et al., 1995):

$$U = (\Sigma_k \, X(k;U)^{\,p_U})^{1/p_U} \qquad (1)$$
$$V = (\Sigma_k \, X(k;V)^{\,p_V})^{1/p_V} \qquad (2)$$

with p_U, p_V – two real exponents.

The normalized indicators X_i are obtained from the measurable, physical indicators Y_i as:

$$X_i = A_i \log_{10}(Y_i) + B_i, \qquad i = \textit{1, 2, ..., n} \qquad (3)$$

The constants A_i and B_i are derived from the assumed knowledge of two pair of values for the normalized and physical indicators: (Y_i^1, Xi^1) and (Y_i^2, X_i^2), where e.g. $X_i^1 = 0.2$ and $X_i^2 = 0.6$:

$$X_i^1 = A_i \log_{10}(Y_i^1) + B_i, \qquad (4)$$
$$X_i^2 = A_i \log_{10}(Y_i^2) + B_i, \qquad i = \textit{1, 2, ..., n}$$

wherefrom on gets:

$$A_i = (X_i^2 - X_i^1)/(\log_{10} Y_i^2 - \log_{10} Y_i 1) \qquad (5)$$
$$B_i = (X_i^2 \log_{10} Y_i^1 - X_i^1 \log_{10} Y_i^2)/(\log_{10} Y_i^1 - \log_{10} Y_i^2)$$

Assumption 3: Once U and V are determined, one theorize that these can make the aggregated monitoring variables of a two-state multi-component system. In Physics, the behavior of such a system is known as the *Ising Model*, covering macroscopic properties, stability issues and phase transitions in e.g. the ferromagnets, binary alloys, order-disorder phenomena and the like. Though no exact solution is available, a variety of approximations including the *Bethe-Peierls* solution, the

Bragg-Williams solution, and the *Onsager* solution are currently curricular in the trade (v.e.g. Huang, 1963).

According to one of these, the membership fractions in the two-state system can be obtained on certain assumptions on the probabilities of individual transitions between the two states – as shown in the next section. The interplay of the actual, 'physical', and potentially numerous system indicators will result in variations of the aggregated parameters, U and V, which in turn will drive the system 'state' in- and out of a region of instability.

In a conventional sense, an *operable* system may thereby appear as (Figure 1):
– *Stable*, and thereby featuring a *low vulnerability*;
– Critically unstable/vulnerable; or
– *Unstable*, and thereby featuring a *high vulnerability*.

Beyond these, the system may only be found *inoperable*.

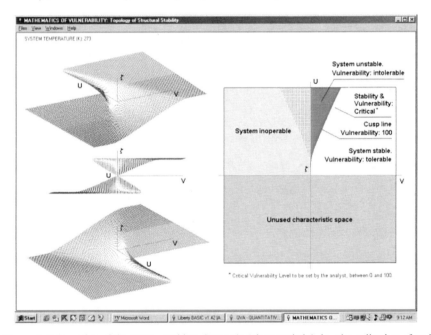

Figure 1. Schematics of the QVA machine. System's 'characteristic', i.e. the collection of real solutions of the 'equation of state'.

*Assumption 4***:** In consideration of the above, a 'Vulnerability Scale' may be defined, based on the assessment of the system state in the (U, V)-space. Obviously, such a definition is not univocal.

One possibility that was experimented on was to measure the *Vulnerability Index* by either the euclidian distance, or the relative abscissa, of the state (U, V) to the cusp line in the $U >= 0$, $V >= 0$ region of the (U, V)-plane; and then normalize the index such that, everywhere in the cusp region, the Vulnerability Index be equal to 100%, that is – reach its assumed maximum.

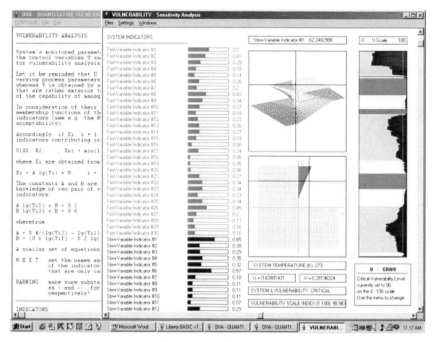

Figure 2. A computer-assisted QVA exercise. For the sake of the example, the system is described by 30 (generic) indicators of U-type, and 20 indicators of V-type. The variation of these, interactively performed by the analyst on acting the bars mid-screen (click-and-drag style) results in the aggregated variation of the U and V phase space parameters (cross-wires in the boxes near). The cross-wire follows the 'system state'.

In Figure 2, the 'V-gram' on the right is a histogram of the evolvement in time of the vulnerability, on the 0–100 scale as defined.

The borderline of the cuspidal region, known as the 'characteristic' (Thom) is described by the equation:

$$th((U \cdot \zeta + V)/\theta) = 2\zeta \tag{6}$$

th standing for the hyperbolic tangent.

The region of the characteristic's topological foil featuring a single solution to the equation of state is the *region of system stability*, whereas the region featuring three solutions (of which only two are taken to have a physical meaning and can be accessed) is the *region of system instability*.

The following rough equivalence is assumed:

System Instability   Highest/Intolerable Vulnerability
System Stability   Lower/Tolerable Vulnerability

In the sense of the definition, within the region of instability the *Vulnerability Index* is supposed to be uniformly 100, while it would gradually decrease away from the edge of the instability region.

How can one come to asserting that there *is* a cuspidal region in system 'phase space' U, V, allowing a vulnerability metrics definition; and how can one get in a natural manner to the cusp equation (6)? The next section answers these, indicating that, indeed, the model above is only the visible tip of an iceberg, molded by inner forces and external pressures down below, far less visible and perhaps less intuitive.

The Things in Depth, I

Assume a system consisting of a large number, M, of elemental constituents, or *members*. 'Elemental' should, in the context, be taken in the 'atomic' sense, i.e. a member, should be seen as a complex, and fully connected to its environment, yet indivisible 'black box'.

System members would interact with each other with, in principle, a varying intensity. To describe the interaction, a *coupling constant*, or *intrinsic parameter, U*, is assumed to be either known, or inferrable. Members state can also be influenced by factors exterior to the system, this being accountable via an *influence field*, or *extrinsic parameter, V*.

One assumes that a member of the system may take only two distinct *states*, say -1 and 2. The generic 'states' may be seen as opposite in respect with a given criterion of judgment (e.g. 'normal' vs. 'abnormal', 'up' vs. 'down', 'pro' vs. 'con', 'functional' vs. 'dysfunctional'), though this may not necessarily be the case. The only condition of essence is that states 1 and 2 be distinguishable from each other. At any given time, t, let M_1 members be in state 1 and M_2 members be in state 2. Since only two states are possible, one has

$$M_1 + M_2 = M \tag{7}$$

The overall state of the system may then be described via the pair of numbers (M_1, M_2), while the system dynamics, or 'motion' in its state space, will follow from variations in M_1 and M_2 that should be consistent with equation (7). The smallest transitions in the system's state would obviously involve alterations by one unit in the numbers of members:

$$\left(M_1 -1, M_2 +1\right)\frac{\leftarrow w_{12}}{w_{21} \rightarrow}\left(M_1, M_2\right)\frac{w_{21} \rightarrow}{\leftarrow w_{12}}\left(M_1 +1, M_2 -1\right) \tag{8}$$

Assume that the respective transitions are governed by the *probabilities* w_{12} and w_{21}, respectively, as indicated in the relationship (8) above. Admission of this process leads also to the recognition of a *function of distribution* of the system's states, $f(M_1, M_2)$, that would obey the master equation:

$$\partial f (M_1, M_2, t) / \partial t = w_{12}(M_1 - 1, M_2 + 1) \times f(M_1 - 1, M_2 + 1) +$$
$$w_{12}(M_1 + 1, M_2 - 1) \times f(M_1 + 1, M_2 - 1) - \tag{9}$$
$$(w_{12}(M_1, M_2) + w_{12}(M_1, M_2)) \times f(M_1, M_2)$$

The state (M_1, M_2) of the system can alternatively be described by the *membership fraction*

$$\zeta = (M_1 - M_2)/(2M), \tag{10}$$

defined such that, if all system members are in state 1, then $\zeta = 1/2$, whereas if all members are in state 2, then $\zeta = -1/2$.

Upon that, one notes that the master equation (9) involves the following states:

(M_1, M_2) ζ

$(M_1 - 1, M_2 + 1)$ $\zeta - 1/M$

$(M_1 + 1, M_2 - 1)$ $\zeta + 1/M$

so that the equation may be re-written as:

$$\partial f(\zeta)/\partial t = w_{21}(\zeta - 1/M)\, f(\zeta - 1/M) + w_{12}(\zeta + 1/M)\, f(\zeta + 1/M) \tag{11}$$
$$- (w_{21}(\zeta) + w_{12}(\zeta))\, f(\zeta)$$

The initial assumption that the number M of system members is large allows one a series expansion of all quantities in the second member of equation (11). Restricting the expansion to the second order in $(1/M)$ one obtains:

$$\partial f/\partial t + \partial J/\partial \zeta = 0. \tag{12}$$

Equation (12) is a continuity (conservation) equation for the state distribution function f, involving the 'current':

$$J = (1/M)\,(w_{21} - w_{12}) \cdot f - (1/(2\,M_2))\, \partial((w_{21} + w_{12}\ ) \cdot f)/\partial z \tag{13}$$

Looking for the stationary states of the system one now assumes:

$$\partial f/\partial t = 0, \tag{14}$$

which leaves one with the equation:

$$\partial J/\partial \zeta = 0. \tag{15}$$

having as solution:

$$J = \text{constant and, in particular,}$$
$$J = 0 \tag{16}$$

Using the expression (13) of the current J, equation (16) can immediately be integrated to give

$$f(\zeta) = const \cdot \frac{exp\left[2M_1 \int\limits_{-1/2}^{\zeta} \frac{w_{21}(\xi) - w_{12}(\xi)}{w_{21}(\xi) + w_{12}(\xi)} d\xi\right]}{w_{21}(\zeta) + w_{12}(\zeta)} \tag{17}$$

The constant in equation (17) can be determined setting $f(\zeta)$ to be normalized to 1:

$$\int\limits_{-1/2}^{1/2} f(\zeta)d\zeta = 1 \tag{18}$$

To normalize – that is, fully determine the distribution function $f(\zeta)$ – one need to make an assumption on the analytical form of the transition probabilities w_{12} and w_{21}.

The following expressions would correspond to the notion that the transitions are a co-operative phenomenon:

$$w_{12}(\zeta) = w\, M_1 \exp(-U \cdot \zeta + V)/\theta) \tag{19}$$
$$w_{21}(\zeta) = w\, M_2 \cdot \exp(U \cdot \zeta + V)/\theta)$$

where U is the coupling constant (internal parameter) and V is the influence field (external parameter) that were previously introduced, while θ is a generalized 'temperature' of the system.

One now makes the natural assumption that the values of the membership fraction ζ that make the distribution function $f(\zeta)$ reach its extremes would make the space of possible states (the 'characteristic') of the system. Taking the expressions (19) of the transition probabilities into equation (17), and requesting that the condition:

$$\partial f(\zeta)/\partial \zeta = 0 \tag{20}$$

be fulfilled, one has:

$$cth\, ((U \cdot \zeta + V)/\theta) = (1/2 - 1/(U/\theta - 2\, M))/\zeta, \tag{21}$$

where cth denotes the hyperbolic co-tangent function, $cth(x) = (exp(x) + exp(-x))/(exp(x) - exp(-x))$.

Again, using the fact that the number of members M in the system is large, the second term in the parenthesis in the right hand side of equation (21) is ignored, so that, finally, the space of system states (U, V, ζ) is given by the equation:

$$th((U \cdot \zeta + V)/\theta) = 2\zeta \tag{22}$$

where th denotes the hyperbolic tangent function, $th(x) = (exp(x) - exp(-x))/(exp(x) + exp(-x))$.

Depending on the degree of interaction between system constituents (members), reflected in the coupling constant U, and on the external influence on all system members – reflected in the field V, and also taking into consideration the temperature θ of the system, the equation (22) may display the following number of real solutions ζ that may be related to the overall system condition:

Number of Real Solutions	System Condition
1	Stable. Smooth transitions in population membership, between states 1 and 2. Low and/or acceptable vulnerability.
3, of which 2 identical	Critical. Sharp transitions in membership between states 1 and 2 are possible. Either state 1 or state 2 may suddenly become unprobable. System is critically vulnerable.
3, different from each other	Unstable. Sharp transitions in membership between states 1 and 2 are possible. Frequency of occurrence of states 1 and 2 are comparable. Though equation (22) has three real roots, the intermediate root is generally taken as having no physical meaning and is therefore discarded. System is dangerously/unacceptably vulnerable.

Figure 3 renders the situation. The boxes in the left-hand side present the cuspidal foil $\zeta_{sys} = \zeta(U, V)$, also known as system's 'characteristic', seen in perspecpective. The (U, V) plane in the right-hand side is color-coded to emphasize the different basins of the system's 'phase space'.

As it turns out, the aspect of the foil expressing the topology of the system's space of states would vary with the generalized 'temperature' θ. The concrete details would depend on the scaling adopted for the 'energy'-wise parameters involved. Indeed, drawing further upon the physical analogy behind the model one would have $\theta = k_B T$, with k_B a 'Boltzmann constant' relating to the energy per degree of freedom of a system member, and T the 'absolute temperature'. Likewise, with the Ising model of ferromagnets in mind, the coupling constant U would be reminiscent of the pair-exchange energy, while V would bring to mind an external magnetic field casting its influence on all the 'spins' that make up the system.

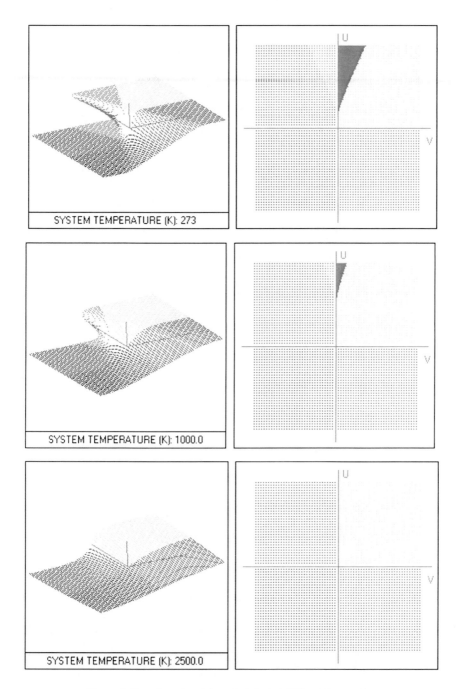

Figure 3. Generic system 'characteristics' at different temperatures.

For practical purposes, the exercise attempted has adopted a 'Boltzmann constant' equal to 1/273, while preserving the absolute 'temperature' scale, where the 0°C would correspond to T = 273.15. That would take parameters U and V in the convenient range of 1–15.

The notion of using the crease, or cuspidal foil catastrophe as an archtype for describing systems' structural vulnerability has an origin in an early work by Weidlich (1973). Several applications have meanwhile been suggested (Ursu et al. 1983, 1985). Beyond these, a fair recognition should be brought to the overall topical work by Thom (1975, 1983), Zeeman (1977), Gilmore (1981).

With equation (22) we are back to equation (6), on a reckoning based entirely on generic, rate equations applied to *coalitions* within the system (group 'M₁', group 'M₂'). While recognizing a microscopic (multi-part) structure to the system this approach does not, however, require a part-by-part analysis. Such an analysis would mean one step further down into the inner dynamics of the system; and the next section endeavors to show that the step is worth taking, because it inevitably reveals the cooperative behavior in template systems of the kind described and the, perhaps, cardinal aspect of this behavior when it comes to vulnerability and resilience: the *hysteresis* in system's response to externally-induced stresses.

The Things in Depth, II

Let us now close in on the representation of the system as a collection of interactive parts subject to external influences.

Let $S(i)$, $i = 1, 2, \ldots, M$ be the variable qualifying the functionality of part i, $S(i) = 1$ indicating a functional part and $S(i) = -1$ a dysfunctional part – which accommodates systems within the *Ising model* as observed in the preceding section (Gheorghe, Vamanu, 2008).

Let again be said that parts may switch from a functional to a dysfunctional state, and conversely, the process being assumed to be, in the final analysis, reversible, and *probabilistic* in nature.

Observant to the natural systems that are coherent enough – within their boundaries of definition – to feature a certain autonomy, or quasi-isolation of their own in respect with the remaining environment, the overall behavior of our model-system may be thought to be governed by *a variational principle*, applicable to system's total energy. According to such a principle, in a steady state of the system the individual states of the parts are such that the system 'energy',

$$E = -(1/2) \, \Sigma_{ij} \, \varepsilon_{ij} \, S(i) \, S(j) - H \, \Sigma_i \, \mu_i \, S(i) \qquad (23)$$

is a minimum for any given *temperature*.

The first term in equation (23) denotes the total *internal* 'energy' of the system of the parts exchanging an 'energy' ε_{ij}, $i = 1,2, \ldots, M$, $j = 1,2, \ldots, M$, whereas the second term features the total 'energy' imparted to the parts by their coupling to the external, compelling 'field' H.

Physicists will immediately note that, in a textbook rendering of an Ising or a Heisenberg model – that are at the origin of our analogy – the normal assumption is that both the coupling ('exchange') energy, ε_{ij}, and the field-coupling constant, μ_i do *not* depend on the individual parts i, j – a fact that has to do with the assumption that all parts are identical (and in effect indiscriminate) to each other. In this respect, equation (23) is a generalization to a many-body system of *non-identical* parts, the validity of which owes to the fact that, in effect, our ε_{ij} expresses only the strength of a *connectivity* between parts.

In applying the notion above, note that any part-i state-flip (from functional, 1, to dysfunctional, -1, or vice versa) entails a change in system's energy, of

$$\Delta E = -S(i)\, (\Sigma'_j\, \varepsilon_{ij}\, S(j) + \mu_i H) \tag{24}$$

where Σ'_j indicates a sum that, in practice, extends over a certain neighborhood of part i – although in principle it may extend over *all* the parts other than i.

Following the Ising model philosophy, a part's behavior is governed by the following set of rules, consistent with the assumptions above:

Rule 1:   If $\Delta E <= 0$ (i.e. taking system to a lower energy), then the part would always undergo a state-flip.

Rule 2:   If $\Delta E > 0$ (i.e. taking system to a lower energy), then the part flips only with a probability

$$P = \exp(-\Delta E/(k_B T)), \tag{25}$$

with T a 'system temperature', and k_B a 'Boltzmann constant', conveniently taken as 1.

In practice (see Sprott [10]), a *Metropolis algorithm* (Metropolis, Teller, Rosenbluth, see also [9]) is recommended for the implementation of *Rule 2*. It reads:

Let r be a random number between 0 and 1.
Then,
    if $r <= P$ (P given by (25)) then *do* flip;
    else, *do not* flip.

Under these terms, for any 'temperature' T there will, in principle, be M_1 system parts that would be functional and $M_2 = M - M_1$ parts that would be dysfunctional, so that one may again define a *system performance fraction*, ζ as:

$$\zeta = (M_1 - M_2)/(2M) \tag{26}$$

Definition (26) places performance fraction ζ between (-0.5) and $(+0.5)$, and favors the following assessment rule:

a system featuring $\zeta >= 0$ is *mostly functional*, whereas
a system featuring $\zeta < 0$ is *mostly dysfunctional.*

And the value-judgment placed on a system management policy/strategy relates to *an assessment of the extent the system is kept mostly functional.* Like in the interpretation given in the preceding section, the *macroscopic* behavior of a system, normally expressed via variations in a number of indicators of definition perceived as relevant, appears a result of system's *microscopic, co-operative* behavior, the macroscopic echo of which is the performance fraction, ζ.

To further explicit this relationship let it be noted that, assuming $S(i) = 1$ and performing the sum Σ over $M_1 - 1$ terms $S(j)$ equal to 1 and M_2 terms equal to -1, and in consideration of the definitions (10), or (26), of ζ, one has:

$$\Delta E = -\varepsilon(M_1 - M_2 - 1) - \mu H = -\varepsilon.2M\zeta - \mu H = U\zeta + V \qquad (27)$$

Similarly, for $S(i) = -1$,

$$\Delta E = \varepsilon(M_1 - M_2 - 1) + \mu H = \varepsilon.2M\zeta + \mu H = -(U\zeta + V) \qquad (27')$$

Here, since ε is the exchange (pairing) energy for any pair of interacting system constituents, $U = 2M\varepsilon$ relates to *the total interaction energy of the system's parts* – a quantity featuring the *internal* dynamics of the system. In turn, energy $V = \mu H$ features *the coupling of parts to the external 'field' H.* The parallel with the U and V defined in the macroscopic theory is immediate.

Equations (27) provide a consistent interpretation of the microscopic, state-flip probability (25) in terms of macroscopic, *overall system transitions.* Thus, the probability of a system transition from an overall state characterized as (M_1, M_2) – i.e. M_1 functional and M_2 dysfunctional, parts, to an overall state characterized as $(M_1 - 1, M_2 + 1)$ is

$$P_{12} = \exp(-(U\zeta + V)/(k_B T)), \qquad (28)$$

whereas the probability of a system transition from an overall state characterized as (M_1, M_2) to an overall state characterized as $(M_1 + 1, M_2 - 1)$ is

$$P_{21} = \exp((U\zeta + V)/(k_B T)), \qquad (28')$$

With the aggregated variables U, V, T and the probabilities thus defined, the gap between the microscopic vision of the system in this section and the macroscopic vision adopted in the preceding sections is now completely sealed, and the Statistical Physics protocol requesting a microscopic substantiation of the macroscopic, phenomenological models is observed.

The process described can be numerically simulated for any, generic, system of parts. The algorithm induces in the system the afore-described microscopic transitions, under the logics of an asynchronously-refreshed collection of cellular automata, cyclically stressing the system by external fields coupled to the parts.

Such stress tests give the system performance fraction ζ as a function of the applied field H (Figures 4–6).

Observing the response of the system one immediately notes its reactivity, or otherwise tendency to antagonize the drift imposed on it from the outside. In plain words,

- If a system is dominantly functional then it tends to maintain its level of functionality (performance) in spite of applied stresses threatening to make parts dysfunctional;
- If a system is dominantly dysfunctional then it tends to maintain low levels of functionality (performance) in spite of the applied stresses attempting to make parts functional again; and
- The transition from a dominantly functional to a dominantly dysfunctional system, and vice versa, tends to be abrupt (as opposed to gradual) and essentially depends on system's 'temperature'.

Systems showing such a behavior are known as featuring a *hysteresis*.

One deems that the reluctance to changes in the level of performance under applied stress, of systems with hysteresis, is an expression of their internal coherence through interactivity and can be construed as **resilience**. (Gheorghe, Vamanu, 2008). In this interpretation, a resilient system does not need to necessarily get back to its original parameters once the external constraint was removed – it has only to maintain a satisfactory level of functionality-as-per design while bearing with the changed, external conditions. This, it is believed, may be a more decent and realistic demand than asking for a rigorous, mathematical system stability – a quality that would, within limits, take the system back to its initial condition.

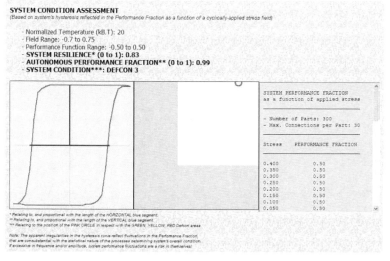

Figure 4. Hysteresis in a 300-part, 30-part links/part system, at a normalized temperature of 20 units: DEFCON 3.

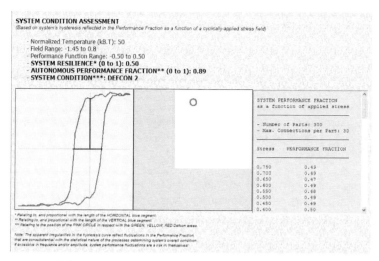

Figure 5. Hysteresis in a 300-part, 30-part links/part system, at a normalized temperature of 50 units: DEFCON 2.

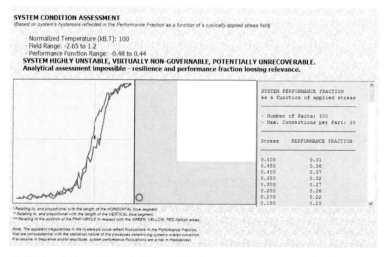

Figure 6. Hysteresis in a 300-part, 30-part links/part system, at a normalized temperature 100 units. System unstable, virtually non-governable and potentially unrecoverable.

One submits that, under the *adaptive* form of resilience described, a system can evolve and atune itself to, or at least tolerate, new operational terms cast upon him by the realities of its environment, while securing the delivery of the services it was meant for. The other face of resilience – the one relying on sheer 'stability' – is, by nature, *reactive*, revulsive to changes and, because of that – probably less sustainable in a reasonable long-term strategy.

These being said, a natural measure of the adaptive resilience turns out to be *the distance of the intersections of the hysteresis cycle with the abscissa* (see Figure 4).

Expressed in units of the applied stress (field), this quantity may be termed – by analogy with the Theory of Magnetism – a *'Coercive Force'*, or *'Coercivity'*. Further along the analogy, the maximum value of the performance function ζ, measured on the ordinate axis for a nil-stress may be termed *'Remanent Performance Level'* (v. *'remanent magnetization'*, or *'remanence'*), although an alternative and perhaps more appropriate term in the context may be *'Autonomous Performance Fraction'* (APF). AFP would thus qualify a desirable feature of systems: their capability to sustain operations even when most of the incentives and/or subsidies, (financial, logistic etc.) normally required to set the system in motion have been tuned down, or withdrawn.

In such terms, a system deemed *'in good order'*, or *'condition'* should display both

- a high *adaptive resilience* – indicating a good tolerance, or *absorptive capacity* (Cohen, Levinthal, 1990) for the effects of stress; and
- a high *autonomous performance fraction* – indicating an acceptable level of performance even in the absence of incentives/subsidies to maintain it.

This finding leaves one with the need to employ in the representation of the system condition *the cartesian product* of the said quantities in an X-Y plane, one choice being to place the resilience on the X-axis and the APF on the Y-axis. This manner of visualizing/monitoring a system condition would immediately call to mind the defense drills, that deals with readiness for appropriate response in threatening conditions in terms of 'DEFCONs'. In the context, one may, for instance, leave to the drill gamer the definition of boundaries between, say, three 'DEFCONs' of incremental degree of severity, the most severe featuring *the lowest system resilience*, OR *the lowest autonomous performance fraction* (APF). The conjunction – lowest resilience AND lowest AFP would then make the 'worst-case scenario'.

Glossing and Gaming on Energy System Strategies

Both the macroscopic and microscopic models described were, in actual fact, implemented in the guise of a web-based 'serious game', meant to provide graphic expressions to the proposed understanding of *resilience* and *autonomous performance capability* in systems (v. SDA Security and Defense Agenda (2010)).

Figure 7 and Table 1 illustrate an application of the *macroscopic model*. The table reproduces the segment of the interface listing the indicators selected for analysis – here grouped in 'tangibles' [T] (the U-type in model's terminology) and 'intangibles' [I] (the V-type), on a perception according to which the 'tangibles' are more structural – and thereby slow-varying, whereas the 'intangibles' are more volatile and pervasive and thereby featuring faster variations and playing the influence agents.

Relating to the data in the table and introduced as a 'vulnerability map', Figure 7 places in perspective the vulnerability condition of targeted system's parts – here the power plants in the generating system of a given country. A continual monitoring would make such synoptics a potentially helpful 'dashboard' for system managers and planners.

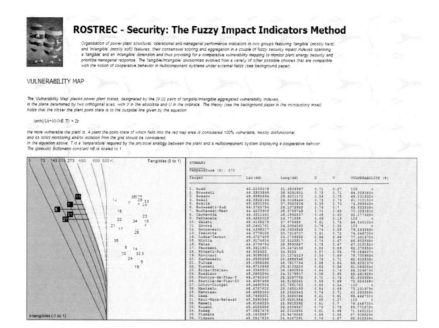

Figure 7. A vulnerability map of the power generation plants in a target-country.

TABLE 1. Indicators and scoring

ANALYTIC INDICATOR SCORING

TARGET: Arad

Latitude (deg): 46.2005078
Longitude (deg): 21.3508987

Assessment team is invited to consensually (i) set the low and high score bounds; and
(ii) score plant performance by indicator. Defaults are arbitrary and only indicative.
When ready, 'PROCESS INPUT' again.

– Maximum Assignable Score, for the Tangible Indicators: 10
– Minimum Assignable Score, for the Tangible Indicators: 0.01

– Maximum Assignable Score, for the Intangible Indicators: 10
– Minimum Assignable Score, for the Intangible Indicators: 0.01

- The normalized Xt1: 0.1
- The normalized Xt2: 0.9

- The normalized Xi1: 0.1
- The normalized Xi2: 0.9

- The Tangible Factor Exponent, pT: 50.0
- The Intangible Factor Exponent, pI: 50.0

VULNERABILITY IMPACT INDEXES

On the scores assigned to the analytical plant performance indicators, one may now compute a couple of aggregated Vulnerability Impact Indexes – one factoring the 'tangible', and the other factoring the 'intangible', analytic indicators in the vulnerability assessment.

To obtain these it was proposed (Gheorghe and Vamanu, 2004) that the concept of fuzzy impact indicators be borrowed from the Swiss (BUWAL) methodology of disaster consequence evaluation. Accordingly one has

The Tangible Factor Index, U, $U = \min\{1, \mathrm{Sum}(j = 1{-}50).$
$$X(Yt(j))^\wedge pT\}^\wedge (1/pT)\}, \qquad (1.1)$$
The Intangible Factor Index, V, $V = \min\{1, \mathrm{Sum}(k = 1{-}53).$
$$X(Yi(k))^\wedge pI\}^\wedge (1/pI)\}, \qquad (1.2)$$

where $X(Yt(j))$, $j = 1,...,nT$ and $X(Yi(k))$, $k = 1,...,nI$ are normalized scores of the analytic indicators of types 'T' and 'I', obtained from the assigned absolute scores $Yt(j)$ and $Yi(k)$, respectively, as

$$X(Yt(j)) = At.lg(Yt(j)) + Bt, \text{ with } 0.01 <= Yt() <= 10, \qquad (2.1)$$
$$X(Yi(j)) = Ai.lg(Yi(j)) + Bi), \text{ with } 0.01 <= Yi() <= 10 \qquad (2.2)$$

The constants A and B are, in turn, derived from the assumed knowledge of two pair of values
for the normalised and physical indicators, as follows:

If Yt1 = 0.01 then Xt1 = user-assumed value between 0 and 1   (3.1)
If Yt2 = 10 then Xt2 = user-assumed value between 0 and 1 and Yt2 > Yt1   (3.2)

and, similarly,

If Yi1 = 0.01 then Xi1 = user-assumed value between 0 and 1   (3.3)
If Yi2 = 10 then Xi2 = user-assumed value between 0 and 1 and $|Yt2| > |Yt1|$ (3.4)
With equations (3.1–3.4), equation (2) gives:

$A = (Xi2 - Xi1)/(\log10(Yi2) - \log10(Yi1))$

$B = (Xi2.\log10(Yi1) - Xi1.\log10(Yi2))/(\log10(Yi1) - \log10(Yi2)).$

Quantities pT, pI are exponents of two pseudo-euclidian 'distances' in the spaces spanned by the indicators.

The application of the algorithm above gives the following:

THE NORMALIZED INDICATORS

Condition of fueling equipment [T]   0.33
Condition of process water, including makeup [T]:   0.47
Condition of steam generators [T]:   0.41
Condition of steam management equipment [T]:   0.53
Condition of turbines and ancillary equipment [T]:   0.62
Condition of power generators and ancillaries [T]:   0.63
Condition of cooling equipment [T]:   0.52
Security and threat prevention technology [T]:   0.60
Protection to weather-related hazards – thunderstorms, icestorms, iceslurry, coal pile freezing [T]:   0.65
Mitigation of design and-or wear-related hazards – earthquake [T]:0.40
Mitigation of design and-or wear-related hazards – fireproofing [T]:0.48
Mitigation of design and-or wear-related hazards – drainage and spill control [T]:   0.66
Mitigation of design and-or wear-related hazards – flood protection [T]:
                                                                      0.63
Mitigation of design and-or wear-related hazards – unit spacing [T]:0.63

Condition of control room [T]:   0.65
Condition of gas detection systems [T]:   0.49
Condition of fire detection system [T]:   0.45
Condition of isolation and sealing – compartmental, valves, piping [T]:
                                                                      0.52
Condition of alarm management system [T]:   0.41
Condition of blow-out system [T]:   0.55
Condition of firebrigade [T]:   0.61
Condition of fixed firefighting system [T]:   0.63
Condition of mobile firefighting system condition [T]:   0.66
Plant status – current power delivery capability [T]:   0.63
Manpower – manning level [T]:   0.49
Manpower – training level [T]:   0.68
Manpower – education level [T]:   0.66
Manpower – experience [T]:   0.67
Manpower – adequacy of hiring practice and recruting standards [T]:0.65

Manpower – level of know-how on site [T]:   0.43
Safe Work Practice – shift management [T]:   0.68
Safe Work Practice – safety manuals, meetings, task forces [T]:   0.62
Safe Work Practice – bypass procedures [T]:   0.43
Safe Work Practice – mitigation of safety system impairments[T]:0.65
Incident management – tracking and trending [T]:   0.65
Incident management – investigation and reporting culture and follow up on recommendations [T]:0.56
Incident management – near miss reporting culture [T]:   0.57
Adequacy of budget for safety improvements [T]:   0.49
Emergency plan coverage, updating level, and quality [T]:   0.68
Emergency drills, simulating activities, adequacy of [T]:   0.63
Organization – manning level of key managerial positions [T]:   0.60
Organization – manning level of key technical positions [T]:   0.68
Organization – conflict level [T]:   0.58
Organization – authority shutdown incident level [T]   0.61
Software Process Improvement and Capability Evaluation (SPICE) procedures (ISO 15504)t [T]:   0.31
Standard specifications and code of practices for IT Service Management (ISO 20000 0 [T]:   0.64
Quality Assessment and Control technology (QA-QC) [T]:   0.62
Sound Cost-Benefit Analysis techniques [T]:   0.42
Strength, Weaknesses, Opportunities, and Threats (SWOT) technology [T]:
                                                                 0.66
Prioritization techniques (e.g. ABC Analysis, Preference Matrix, Criteria Catalogue, [T]:   0.56
Response to situations of reduced primary energy resources [I]:   0.26
Response to situations of reduced capital resources [I]:   0.18

Emergency preparedness to mitigate natural disasters [I]:   0.18
Safe-proof operations against disruption by terrorism [I]:   0.26
Security of operation continuity in abnormal circumstances – pandemics; civil unrest; war [I]:   0.18
Mng. of systemic risks rel.to connectivity, abnormal behavior/performance, regulatory breaches [I]: 0.23
Preventive/response capability to face economic, administrative, legal and societal challenges [I]:   0.23
Physical protection of plant infrastructure – assets, hardware, software, transportation, ICT, [I]:   0.23
Ability to identify structural and operational vulnerabilities [I]:   0.23
Means to correctly assess risks and potential mitigating costs, ALARA-based [I]:   0.23
Control of operational risk regarding process error, failed execution, and system failure [I]:   0.18
Control of operational risk relating to fraud, and failure of security and physical protection [I]:   0.23

Legal claims mitigtn. entailing monetary damage; defense costs, injunctive relief, penalties [I]:   0.10

Positive attitude towards system's shift towards unbundling and deregulation [I]:   0.10

Maintenance of market buoyancy through securing adequate shares in demand, customer adherence [I]:   0.23

Ability to identify market trends, developments, and opportunities [I]:
                                                                                                              0.23

Prepare for unexpected markets moves in terms of speed, dircction, severity or correlation [I]:   0.18

Maintenance of a sound liquidity and funding management [I]:   0.10

Maintenance of Company's market rating and public image [I]:   0.23

Ability to retain qualified personnel [I]:   0.23

Personnel Competence, Fitness, Loyalty, and cultural compatibility [I]:
                                                                                                              0.18

Power operator culture of an efficient, responsible and alert professional and societal servant [I]:   0.18

Monitoring of power system partner performance and behaviour [I]: 0.18

Plant status – history of availability [I]:   0.10

Plant status – history of unplanned shutdowns [I]:   0.23

Plant status – technical age [I]:   0.18

Plant status – upgrading effort over current lifetime [I]:   0.10

Level of operational costs – non-personnel-non-safety-relating [I]: 0.18

Level of operational costs – personnel-relating [I]:   0.26

Level of operational costs – safety-relating [I]:   0.26

Level of promotional costs [I]:   0.26

Level of breakdown-relating maintenance [I]   0.23

Level of essential process-relating maintenance [I]   0.10

Maintenance – adequacy of budgeting level [I]   0.18

Maintenance – inspection program quality [I]:   0.23

Housekeeping [I]:   0.18

Ergonomics [I]   0.18

Risk Management – process hazard analysis policy, expertise and performance [I]:   0.23

Risk Management – process hazard analysis team and training [I]: 0.18

Risk Management – tracking, follow up recommendation and budgeting [I]:
                                                                                                              0.23

Incident management – loss history, of site [I]:   0.10

Safety Management Audits – frequency and adequacy of internal and external audits [I]:   0.18

Tracking and follow up of audit recommendations [I]:   0.26

Level and quality of communication between senior technical staff and the operation personnel [I]:   0.10

ROSTREC - The Resilience Issue

Testing large systems behavior under stress.
Glossing on cooperative effects, system stability and hysteresis.

System Condition Tags

System Condition is to be assessed, next, in consideration of system's response to a cyclically applied stress.
Two quantities featuring the response hysteresis are monitored: the RESILIENCE; and the AUTONOMOUS PERFORMANCE FRACTION
(see background notes), both normalized to cover a range of 0 to 1. The plane defined by these is divided in three adjustable regions:
DEFCON 3, 2, and 1, indicating a normal (green), tolerable (yellow), and intolerable (red) system condition, respectively.
A condition beyond DEFCON 1 is also defined, in cases when system's response to stress shows frequent and/or high-amplitude oscillations -
an occurence possible with small (low number of parts), weakly-coherent (low number of part links and/or low-strength links and part susceptibilities)
systems, and with virtually any system at high temperatures. In such cases the system is declared 'highly unstable, virtually non-governable,
and potentially unrecoverable'

Please set accordingly the green-to-yellow, and the yellow-to-red DEFCON limits.

'DEFCON 3 (Normal, Green)', down to : 0.75
'DEFCON 2 (Tolerable, Yellow)', down to : 0.35

Accept

SYSTEM CONDITION ASSESSMENT
(Based on system's hysteresis reflected in the Performance Fraction as a function of a cyclically-applied stress field)

- Normalized Temperature (kB.T): 20
- Field Range: -0.95 to 0.85
- Performance Function Range: -0.50 to 0.50
- **SYSTEM RESILIENCE* (0 to 1): 0.88**
- **AUTONOMOUS PERFORMANCE FRACTION** (0 to 1): 1.00**

 SYSTEM CONDITION*: DEFCON 3**

* Relating to, and proportional with the length of the HORIZONTAL blue segment.
** Relating to, and proportional with the length of the VERTICAL blue segment.
*** Relating to the position of the PINK CIRCLE in respect with the GREEN, YELLOW, RED Defcon areas.

Note: The apparent irregularities in the hysteresis curve reflect fluctuations in the Performance Fraction,
that are consubstantial with the statistical nature of the processes determining system's overall condition.
If excessive in frequence and/or amplitude, system performance fluctuations are a risk in themselves!

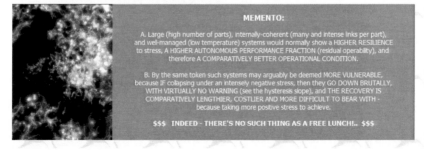

MEMENTO:

A. Large (high number of parts), internally-coherent (many and intense links per part),
and well-managed (low temperature) systems would normally show a HIGHER RESILIENCE
to stress, A HIGHER AUTONOMOUS PERFORMANCE FRACTION (residual operability), and
therefore A COMPARATIVELY BETTER OPERATIONAL CONDITION.

B. By the same token such systems may arguably be deemed MORE VULNERABLE,
because IF collapsing under an intensely negative stress, then they GO DOWN BRUTALLY,
WITH VIRTUALLY NO WARNING (see the hysteresis slope), and THE RECOVERY IS
COMPARATIVELY LENGTHIER, COSTLIER AND MORE DIFFICULT TO BEAR WITH -
because taking more positive stress to achieve.

$$$ INDEED - THERE'S NO SUCH THING AS A FREE LUNCH!.. $$$

Figure 8. Interface web page, of an interactive game implementing the microscopic model.

Quality mngm. via preventive actions, evidence of compliance with procedures (ISO-9000-1994) [I]:   0.18

Process effectiveness metrics, process improvmnt., customer satisfactn. tracking (ISO-9000-2000)[I]:   0.10

Best practice recommendations on Information Security Management (ISO-IEC 17799) [I]:   0.18

Total Quality Management (Deming's TQM) principles and practices [I]:
0.18

Creativity techniques (Brainstorming, Brainwriting, Method 635, Delphi, CNB, Morphology etc.) [I]:   0.18

Goal-Finding methods [I]:   0.18

Safety awareness of management – perceptiveness and mentality [I]: 0.18

Safety awareness of workforce – compliance, performance, participation, rigor acceptance [I]:   0.23

Safety culture, adequacy of the level of investment in Safety [I]: 0.10

As a consequence, cf. equations (1.1 and 1.2) one has:

– THE TANGIBLE VULNERABILITY INDEX, U: 0.705667462
– THE INTANGIBLE VULNERABILITY INDEX, V: 0.270070852

A distinct module was dedicated to implement the microscopic vision, with a more explicit emphasis on the hysteresis effects and the interpretation given, in this context, to the notion of adaptive resilience.

After some enduring experiments with changing coupling constants, susceptibilities, and temperature one may end up with a 'feeling' on how systems with hysteresis behave. Here are the chief findings:

THE OBVIOUS

The facts that every intuitive observer will expect are:

- Large and internally coherent systems tend to show a higher adaptive resilience and APF and thereby a higher-grade condition in comparison with small, poorly-coherent systems, and also feature more stable (fluctuation-free) operation regimes. That would validate the integrative features provided for by the EU's 20-20-20 initiative: higher grid connectivity; redundance; power highways, shared structural (e.g. ITC) and operational standards etc.;
- Systems subject to poor/negligent/lax management/governance in terms of maintenance, monitoring, updating, corporate spirit, truthful self-assessment, ethics, as well as unfair business climate etc. – which translates as 'disorder', or '*higher temperatures*' show *degraded resilience and/or performance fractions*, down to complete collapse. Without prejudice, this aspect would mostly concern the EU's newbies of later waves of accession, now still bogged down in what seems to be a painful, wobbling, asymptotic transition towards the Western reference standards.

THE INTRIGUING

Less intuitive yet not incomprehensible is the following finding:

- Changes occurring, or forced upon the *strength of internal couplings* between system constituents are considerably more consequential in terms of resilience and capability for autonomous performance, in comparison with the temperature-induced effects.

This aspect is definitely consequential in regard with the substantive action core of the 20-20-20 EU initiative, which outstandingly requires an important volume of *technology substitution* as a vehicle to implement the sought policies.

Indeed, running the microscopic model under the assumption that a certain fraction of the traditional parts were replaced by new parts – hardware, software, management and mindware included – the exotism in comparison with the old being reflected in considerably weaker couplings with the 'old guard', (grid connectivity, load regimes, power traffic, other exchanges, compatibilities, interfaces etc.) would immediately result in changes in the hysteresis pattern and the consequent cartesian product of *adaptive resilience* by the *autonomous performance fraction* (AFP) – the more dramatic the changes the more 'exotic' the newcomers are.

Thus, if one can bear with a simplistic and, probably, overconservative reckoning according to which a 20% GHG reduction, plus a 20% of renewables in covering demand, plus another 20% in conservation would amount to, roughly, a 3-times-20%, i.e. 60% substitution in the technology pool serving EU's energy system, one would be faced with the change in system's response pattern and consequent resilience and AFP depicted in Figure 9.

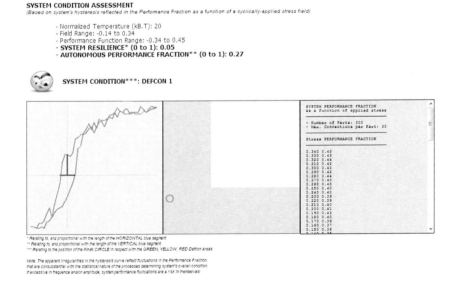

Figure 9. System response to cyclic stress (hysteresis pattern) for a 60% technology substitution.

SYSTEM CONDITION ASSESSMENT
(Based on system's hysteresis reflected in the Performance Fraction as a function of a cyclically-applied stress field)

- Normalized Temperature (kB.T): 20
- Field Range: -0.27 to 0.22
- Performance Function Range: -0.44 to 0.43
- **SYSTEM RESILIENCE* (0 to 1): 0.18**
- **AUTONOMOUS PERFORMANCE FRACTION** (0 to 1): 0.64**

 SYSTEM CONDITION*: DEFCON 1

* Relating to, and proportional with the length of the HORIZONTAL blue segment
** Relating to, and proportional with the length of the VERTICAL blue segment
*** Relating to the position of the PINK CIRCLE in respect with the GREEN, YELLOW, RED Defcon areas

Note: The apparent irregularities in the hysteresis curve reflect fluctuations in the Performance Fraction that are consubstantial with the statistical nature of the processes determining system's overall condition. If excessive in frequence and/or amplitude, system performance fluctuations are a risk in themselves!

Figure 10. System response to cyclic stress (hysteresis pattern) for a 40% technology substitution.

SYSTEM CONDITION ASSESSMENT
(Based on system's hysteresis reflected in the Performance Fraction as a function of a cyclically-applied stress field)

- Normalized Temperature (kB.T): 20
- Field Range: -0.48 to 0.37
- Performance Function Range: -0.50 to 0.49
- **SYSTEM RESILIENCE* (0 to 1): 0.45**
- **AUTONOMOUS PERFORMANCE FRACTION** (0 to 1): 0.85**

 SYSTEM CONDITION*: DEFCON 2

* Relating to, and proportional with the length of the HORIZONTAL blue segment
** Relating to, and proportional with the length of the VERTICAL blue segment
*** Relating to the position of the PINK CIRCLE in respect with the GREEN, YELLOW, RED Defcon areas

Note: The apparent irregularities in the hysteresis curve reflect fluctuations in the Performance Fraction that are consubstantial with the statistical nature of the processes determining system's overall condition. If excessive in frequence and/or amplitude, system performance fluctuations are a risk in themselves!

Figure 11. System response to cyclic stress (hysteresis pattern) for a 20% technology substitution.

Assuming that only two of the policy directives will be commissioned simultaneously, entailing a 40% technology substitution, the effects will look like in Figure 10.

Finally, if a prioritized timing for the implementation of the three directives would be considered, implying only a 20% technology substitution at any single step the situation will improve as shown in Figure 11. To be noted that numerical trials indicate that a mere rise in 'temperature', i.e. in the business and political climate may only marginally improve this outlook.

THE OUTRAGEOUS

- Finally, there may be also a speculative finding that is almost counterintuitive and, some may say, intellectually insulting, according to which *highly resilient and performant systems, that is – systems commonly appreciated as high-grade tend to be... unexpectedly vulnerable in* a specific, if odd, sense. This kind of vulnerability relate to the near-ideal shape of their hysteresis cycle (Figures 4 and 8): quasi-rectangular and covering a large expanse in the performance vs. stress X-Y plan. For indeed, such a shape may encourage the foul feeling that 'things are all right' even if negligence, or external circumstances – like e.g. a prolonged recession – would take the normal, residual positive stress normally consisting of developmental and maintenance costs plus 'facilities', or 'stimuli' (financial, logistic, intelligence, etc.) into the red, in the negative-stress realm, because, on the account of system's internal, quality-interactivity (smooth operations) *even there the performance fraction is still on the higher plateau* – isn't it? Well, it is – *yet perhaps only for a while,* for the unsuspecting system finds itself dangerously-close to the precipice that, if reached by a mere further, apparently insignificant decrement or fluctuation in the stress level, *will take the entire structure down into a full-fledged collapse*

Aggravating factors of such a perverse form of vulnerability are:

- *a virtually-complete lack of early-warnings on the imminence of a collapse* (system stands steady at a high(est) level of performance although its environment is clearly deteriorating);
- *the brutality of the collapse, if/when it happens* (v. the steep slope of the hysteresis) that would dramatize the entire scenario; and – perhaps equally important,
- *the remarkably long and costly way to a full system recovery* (see the length of the lower hysteresis cycle plateau).

And, in such a case, the only good thing about the bad things is that, if/when the recovery point – the lower-right corner of the cycle – is finally reached, the ensuing process is expected to be swift and effective.

Should the obsessive disruptions of the century's first decade – 9/11 and the Recession – be not with us, these authors would have been much more reluctant in giving such a speculation to print. Yet, can one *really* dismiss the thought that both developments were signs of a brutal descent on the left-hand slope of the hysteresis loop featuring *a highly resilient and performant, yet overconfident and oblivious of hidden faults, System*? And can one wave aside the suspicion that the apparent ineffectiveness of all that money pumped into the System by the Administration to stop Recession and assume the long-expected climb-up to better times is due to the fact that the turning point suggested, in a symbolic sense, by the Physical analogies – *'the lower-right corner of the hysteresis cycle'* – has not yet, in spite of all efforts, been reached?

Conclusion

As often observed, large and complex inanimate systems may have much in common with the living bodies. Among other things, doctoring both takes wisdom, moderation, and compassion. From this metaphorical perspective and on a more relaxed chord – if we may – one may look at resilience like at cholesterol: like with the latter, *one needs a 'good' (adaptive) resilience, in a certain amount, as opposed to a 'bad' (reactive) resilience and, in fact, any resilience in an excessive amount*. Physical analogies popular not only in the trade itself, but also in such disparate fields like Biology, Physiology, Neuroscience, Sociology, Econometrics, Electrical Engineering and Electronics, the ITC, the Psychology (see, for a starter Wikipedia 2010) may help one understand that the implementation of such ambitious a plan as 20-20-20, endeavoring to take EU to the next level as far as energy security via, virtually, a crash campaign in technology substitution – hardware, software, management, mindware and all – requires *a consolidated capability to withstand and absorb changes while maintaining performance by continual adaptation, one component of which is permanently securing (i) system's absorptive capabilities; and (ii) the acceptability of policies and their net effect on people.* The border between adaptive resilience and reactive resilience is not thin, as the saying usually goes – it is fuzzy. While modeling it is an ambitious task yet to be assumed, one may say with a sufficient degree of confidence based on casuistry (v.e.g. Iran's 'Westernization' in the 1970s) that 'too much, too soon' is a certain way to failure.

From an even broader perspective, contemporary remedies to the shortcomings of a World *en route* to globalization tend to assume one out of three grades: *doctoring; reforming;* and *revolutionizing.* To these authors, incentives and stimuli fall under *doctoring* – as a prerequisite of reforming; liberalization, unbundling and deregulation in the energy field aim at *reforming* – as a prerequisite of revolutionizing; and massive technology substitution goes to *revolutionizing* – one does not need nuclear fusion going commercial to revolutionize an energy system: 20% renewables plus 20% low-GHCs plus 10% biofuels is just about enough. With the necessary wisdom, moderation and compassion in mind, it is for the managers of this World to decide what is *appropriate*, and what is *timely* among these – and *choose, for us all.*

To be sure, any progress in modeling complexities comes at the cost of a fresh cohort of new and perplexing questions. One of these, that is long with us by now is – *how much resilience is enough resilience*? That is – beyond what threshold a highly resilient and performant society gets oblivious of 'intangibles' such as the principles of good governance, management, ethics, work discipline and culture, moderation, and compassion for the lesser-adapted – on the account of the absence of early warnings that its transgressions are just about to be sanctioned by the blind, dispassionate justice of System Dynamics?

It is believed that reconciling the quest for sound, resilience-conscious changes *and* for an acceptable degree of stability; and implementing policies endowed with sufficient intrinsic warning features to prevent disruptive developments and setbacks may be *the* outstanding challenge that the governance bodies, in Europe and elsewhere, are facing, in the decade to come.

References

Christen P., Bohnenblust H., Seitz S.(1995). *How to Compare Harm to the Population with Damage of the Environment? A Quantitative Multi-Attribute Approach for Risk Analysis based on Fuzzy Set Theory*. In *Loss Prevention and Safety Promotion in the Process Industries*, Vol.1. Mewis J.J. Pasman H.J., De Rademaeker E.E., Editors. Elsevier Science B.V.

Cohen W.M, Levinthal D.A. (1990). *Absorptive Capacity: A New Perspective on Learning and Innovation*. Administrative Science Quarterly, Volume 35, Issue 1 pg. 128–152.

EC Environment (2010). The EU Climate and Energy package. http://ec.europa.eu/environment/climat/climate_action.htm.

Gheorghe A.V., Vamanu D.V. (2004). *Towards QVA – Quantitative Vulnerability Assessment: a generic practical model*, Journal of Risk Research, vol. 7, Issue 6, September 2004

Gheorghe A.V., Vamanu D.V. (2008). *Mining Intelligence Data in the Benefit of Critical Infrastructures Security: Vulnerability Modeling, Simulation and Assessment*. International Journal of System of Systems Engineering, Vol. 1, Nos. 1/2, 2008, pp. 189–221.

Gilmore R. (1981). *Catastrophe Theory for Scientists and Engineers*. Wiley, 1981.

Huang K. (1963). *Statistical Mechanics*. John Wiley & Sons, Inc., New York, London.

SDA Security and Defense Agenda (2010). *Is Europe's Energy Security Policy a Reality or an Ambition?* May 26, 2010, Bibliotheque Solvay, Brussels.

Sprott J. (1993). *Strange Attractors: Creating Patterns in Chaos*, M&T Books, New York.

Thom, R. (1975). *Structural Stability and Morphogenesis*. The Benjamin/Cummings Publishing Company, Inc.

Thom, R. (1983). *Mathematical Models of Morphogenesis*. Ellis Horwood Limited Publishers, Chichester. Halsted Press: a division of John Wiley and Sons, New York, Brisbane, Chichester, Ontario.

Ursu I, Purica I.I., Vamanu D. (1983). *Towards More Safety: Observing Synergisms in Reactor Behaviour*. In *Nuclear Power Experience*, Vol.IV, paper IAEA-CN-42/139, p.255. International Atomic Energy Agency, Vienna.

Ursu I., Vamanu D., Gheorghe A., Purica I.I. (1985). *Socioeconomic Risk in Development of Energy Systems*. Risk Analysis 5, 315.

Weidlich, W. (1973), in *Synergetics—Cooperative Phenomena in Multicomponent Systems*, edited by H. Haken (Teubner, Stuttgart).

Wikipedia (2010). *Hysteresis*. http://en.wikipedia.org/wiki/Hysteresis.

Zeeman E.C. (1977). *Catastrophe Theory*, Addison-Wesley.

STOCHASTIC SIMULATION OF CRITICAL INFRASTRUCTURES FOR ELECTRIC POWER TRANSMISSION

ENRICO ZIO*

Ecole Centrale Paris-Supelec, Paris, France
Politecnico di Milano, Milano, Italy

Abstract. Critical infrastructures of electric power transmission are considered. Their distributed dynamic characteristics, common to other critical infrastructures e.g. of transportation and communication, demand innovative approaches of analysis in order to identify vulnerabilities, for effective protection against extensive failures. The uncertainties associated to the system behavior which emerges from initially local failures call for appropriate models and methods of stochastic simulation.

Introduction

Generally, a Critical Infrastructure (CI) is 'a network of independent, large-scale, man-made systems (set of hard and soft structures)…that function collaboratively and synergistically to produce a continuous flow of essential goods and services' [Ellis et al., 1997]. They are dynamic, complex systems which are also highly interdependent, both physically and through a pervasive use of information and communication technologies. They can be subject to multiple threats (technical, intentional or unintentional human, physical, natural, cyber, contextual) and may pose risks themselves. According to [Dueñas-Osorio et al., 2007] these systems are made of "a large number of interacting components (real or virtual), show emergent properties difficult to anticipate from the knowledge of single components, are characterized by a large degree of adaptability to absorb random disruptions and are highly vulnerable to widespread failure under adverse conditions." Indeed, small perturbations can trigger large scale consequences in CIs; furthermore, disruptions may also be caused by targeted attacks.

For the above reasons, these lifeline systems are termed 'critical' in the sense that their incapacity or destruction would have a debilitating impact on the health, safety, security, economics, and social well-being of a country. An extended failure within one of these infrastructures, with loss of its continuous service, can significantly damage many aspects of a society, including its economy; furthermore,

---

* To whom correspondence should be addressed: Enrico Zio, Ecole Centrale Paris-Supelec, Paris, France Politecnico di Milano, Milano, Ital, e-mail: enrico.zio@polimi.it

A. Gheorghe and L. Muresan (eds.), *Energy Security: International and Local Issues,*
Theoretical Perspectives, and Critical Energy Infrastructures,
DOI 10.1007/978-94-007-0719-1_9, © Springer Science+Business Media B.V. 2011

the cascading across CI boundaries has the potential for multi-infrastructural collapse with unprecedented, transnational consequences.

The electric power transmission network is the CI object of the present contribution, and indeed serves as a good illustrating example of a CI facing greater and tighter integration (also of new intermittent power sources) following the liberalization of most national markets and being closely interconnected with other infrastructures, particularly the information and communication network.

The role of dependencies among infrastructures (so called interdependencies) and the intrinsic difficulties arising in their modeling have been highlighted in empirical studies. For example, the database in [Luiijf et al., 2009] has been built from public reports of disruptions of CIs from open sources like newspapers and internet news. Events have been classified as "cascade initiating" (i.e., an event that causes an event in another CI), "cascade resulting" (i.e., an event that results from an event in another CI) and "independent" (i.e., an event that is neither a cascade initiating nor a cascade resulting event). The information in the database shows in particular that: (i) "cascade resulting" events are more frequent than believed, and that "cascade initiators" are about half as frequent; (ii) the dependencies are more focused and directional than often thought; (iii) energy and telecommunication are the main "cascading initiating" sectors.

As an example, interdependencies among telecommunication systems, trans-portation systems, and power distribution grids played a negative role in the small telecommunication blackout which took place in the suburbs of Rome in January 2004 [IRRIIS, 2007]. A flooding due to a break of a metallic pipe carrying cooling water of an air-conditioning plant in a major telecommunication service node led to a communication blackout which, besides affecting the communication/ transmission operations in the area, hit the country's biggest printed news agency transmissions, stopped the check-in, ticketing, and luggage acceptance and switching services at the International Fiumicino Airport, disturbed post offices and banks operations, and caused problems to the communication network connecting the main Italian research institutions. Furthermore, the telecommunication blackout had effects on the operation of the power distribution grid causing the disconnection of two control centers and resulting in the loss of control on a number of remote substations in the area of Rome.

Similar occurrences [Zimmermann, 2001] point to the need for identifying infrastructure interdependencies [Rinaldi et al., 2001] and determining the design and operational conditions which can prevent the onset and propagation of cascading failures.

Given the above observations and considerations, investigating the risks and vulnerabilities of the electric power transmission CI has become a national and international priority.

Here, the concept of risk is intended to quantitatively combine the probability of occurrence of undesired/adverse events leading to losses, damages and/or injuries, and the magnitude of such consequences. The uncertainties associated to these quantities need to be quantified, also. Although not explicitly treated here, it is recognized that there is a non-technical dimension of risk related to aspects of societal and psychological experience and perception, which are contextual in nature and which may greatly influence the decision making which follows the technical analysis.

The concept of vulnerability is here intended to complement that of hazard, accounting of the fact that a hazard of low intensity could have severe consequences, while a hazard of high intensity could have negligible consequences: the level of vulnerability makes the difference.

Then, the complete focus is on:

- magnitude of losses and damages due to the impact of a hazard (technical dimension);
- degree of exposure to the hazard, i.e., likelihood of being exposed to the hazard and susceptibility of suffering losses and damages;
- degree of resilience, i.e., ability of the system to anticipate, cope with/ absorb, resist and recover from the impact of the hazard [Bouchon, 2006].

In this view, the vulnerability of the electric power transmission system is analyzed with respect to the changes in the network characteristics which may occur following faults or attacks on its nodes and links, and the scale and duration of the associated losses: 'translated' in practical terms, the frequency of major blackouts [events per year] and the associated severity [MW or MWh].

Taking into account the need for considering the properties of the individual infrastructure, and its interdependencies with others, the steps of vulnerability analysis are:

1. list the set of events and event sequences that can cause damages and losses;
2. identify the relevant set of "initiating events" and evaluate their cascading impact on the system;
3. identify the interdependencies and evaluate the impact of cross-cascading of failures through them.

The performance of these steps relies on the analysis of the system, its parts and their interactions within and outside the system; the analysis must account for the environment which the system lives and operates in, for the functions the system is expected to deliver and for the physical and functional interdependencies with other systems. The aim is that of assessing the influences and limitations which interacting infrastructures impose on the individual system operating conditions, for avoiding fault propagation by detecting and recognizing threats, designing redundancies, and implementing alternative modes of operations [Zimmermann, 2001].

The two main outputs of a CI vulnerability assessment are the quantification of system vulnerability indicators and the identification of its critical elements. The information they provide is complementary: while vulnerability indicators are parameters encompassing the static and/or dynamic characteristics of the whole system, the identification of critical elements comes from their ranking with respect to their individual connectivity efficiency and/or their contributions to the propagation of failures, with their effects, through the system. A number of approaches can be undertaken for the vulnerability assessment of CIs depending on the type of system, the objective of the analysis, and the available information. Given the distributed topology of such systems and the emergent properties which characterize their dynamics, these approaches need to go beyond the usual cause-consequence analysis to be able to focus on spill-over clusters of failures in case of strong interdependencies [IRGC, 2006]. Indeed, the behavior of a complex system cannot be described as the sum of the behavior of its individual elements. This renders questionable the suitability of classical risk analysis methods, e.g. fault tree/event tree analyses, which are typically founded on a decomposition of the system into subsystems and basic elements, and their subsequent re-composition for quantification. Pre-defined causal chains, e.g. identified by event tree analysis, seem inappropriate to identify the hidden risks and vulnerabilities emerging in a complex infrastructure.

On the other hand, stochastic simulation techniques are apt to functioning as 'emergent scenario generators', but their computational cost needs to be controlled for applications on real-size systems. In other words, one is confronted with the usual inevitable compromise between model adherence to reality and the budget of resources/costs available for the analysis. The availability of data for estimating the model parameters also plays a decisive role, and this brings in the picture the issue of uncertainty in the outcomes of the vulnerability quantifications. In this regard, a recent Nuclear Science Foundation (NSF) workshop report points at the fact that uncertainty is pervasive in the modeling of complex systems [Guckenheimer and Ottino, 2008]; building reliable predictions of the system behavior in the face of the large numbers of uncertain parameters entering its modeling is a major challenge; quantifying this uncertainty and determining how it propagates throughout the system is a key aspect of reliable prediction and control of cascading failures in critical infrastructures.

In the last 10 years, a number of researchers have focused on the study of interdependent infrastructures and developed a variety of methodologies for analyzing interdependent infrastructures. According to [Rinaldi, 2004], these modeling and simulation techniques can be grouped into six broad categories: (i) aggregate supply and demand tools, which evaluate the total demand for infrastructure services in a region and the ability to supply those services [Adachi and Ellingwood, 2008; Apostolakis and Lemon, 2005; Dekker, 2005; Helseth and Holen, 2009; Lee et al., 2007; Min et al., 2007; Piwowar et al., 2009]; (ii) dynamic simulations, which examine infrastructures operations, the effects of disruptions, and the associated downstream consequences [Carreras et al., 2007; Duenas-Osorio et al, 2007 and Duenas-Osorio and Vemuru 2009; Johansson and Jönsson, 2009; Newman et al., 2005; Ouyang et al., 2009; Svendsen and Wolthusen, 2007]; (iii) agent-based models, which allow the analysis of the operational characteristics and physical states of infrastructures [Barton and Stamber, 2000; Casalicchio et al., 2007; Panzieri et al., 2004; Schläpfer et al., 2008]; (iv) physics-based models, which analyze physical aspects of infrastructures with standard engineering techniques (e.g., power flow and stability analyses for electric power grids or hydraulic analyses on pipeline systems) [Chen and McCalley, 2005; Kodsi and Canizares, 2007]; (v) population mobility models, which examine the movement of entities through geographical regions [Casalicchio et al., 2009; Germann et al., 2006; Hong et al., 1999]; (vi) Leontief input-output models, which in the basic case provide a linear, aggregated, time-independent analysis of the generation, flow, and consumption of various commodities among infrastructure sectors [Cagno et al., 2009, Haimes and Jiang, 2001; Jiang and Haimes, 2004; Haimes et al., 2005a and b; Reed et al., 2009].

In the present contribution, stochastic simulation of the type of category (ii) is discussed in the framework of the vulnerability analysis of electric power transmission systems. Due to the broad nature of the topic, it is not possible to be complete and exhaustive in its treatment and in the detailed discussion of the methods, nor in the reference to the specialized literature. The intent is to provide the reader with a general view on the concepts underlying this type of simulation models. For exemplary purposes, two modeling approaches are illustrated; the choice of the approaches presented is not motivated by any declaration of alleged superiority in comparison to the many other methods proposed in the literature, but by the need to rely on the experience of the author in their development and application.

A Random Walk Model for the Analysis of Electric Power Transmission Networks

Deterministic load flow (LF) analysis is typically used to study the behavior of electric power transmission systems under specific operating scenarios. The analysis allows finding nodal voltages and line flows around the network. On the other hand, uncertainties exist with respect to the actual conditions that the network will be operating in, at different instances during its life (e.g. because of variation in bus power values and network configurations). To account for this, the LF problem has also been addressed by probabilistic methods, where the input quantities for the calculations are treated as random variables [Borkowska, 1974; Dopazo et al., 1975; Sobjerajski, 1978]. The outcomes of such analyses are given in terms of adequacy indices like the probability of a line flow being greater than its design rating, the probability of a bus voltage being outside its operational constraints, etc. These indices are typically obtained under steady state operating conditions.

In general, the probabilistic LF analyses only account for uncertainties in the load and power generation data, whereas the network configuration is considered fixed. In practice, however, any change in the network of links of the electric power transmission system modifies the configuration and, consequently, the functional relations among inputs and outputs. In this respect, a relevant source of variation which needs to be given due account relates to the availability of transmission lines, transformers, etc, which at any point in time may be subject to outages due to faults and maintenance.

In this section, a random walk model previously proposed by the authors [Zio and Piccinelli, 2010] is briefly illustrated as an example of a probabilistic model for the analysis of electric power transmission systems.

The electric power transmission system is modeled as a network consisting of N nodes (also called vertexes) and K edges (also called arcs or links): the buses of the electric grid are represented as nodes interconnected by undirected edges representing the transmission lines; NS nodes are power sources (generators), NT nodes are targets (loads) and the rest are transmission nodes. The N×N adjacency matrix $\{a_{ij}\}$ defines the topological structure of the network, i.e., the pattern of connectivity among its nodes, with the matrix entry a_{ij} being equal to 1 if there is an edge linking nodes i and j, and 0 otherwise; the entries on the diagonal elements, a_{ii}, are undefined and for convenience they are set equal to 0. The matrix $\{q_{ij}\}$ defines the probabilities of failure of the links over the period of interest.

The capacities of the links are assumed to vary stochastically, to account for the uncertainties inherent in their behavior and operation; then, to each capacity value w_{ij} is associated a probability distribution $\pi(w_{ij})$ of the possible values.

The model based on random walks gives proper consideration to the following variabilities:

- each link connecting two nodes is characterized by a transmission capacity which cannot be exceeded;
- the capacities of the network lines are assumed to stochastically vary, to account for the inherent uncertainties;
- not only the links on the direct and shortest paths are considered in the analysis of the transmission of flow; this is achieved by a randomization of the direction of the flow in output from a node; the randomization is driven by the capacities of the outgoing links, with the highest capacity links most probably channeling the flow;
- the network interconnecting links are assumed fallible, with given probabilities;
- source generation and load demands are assumed to vary stochastically, to account for the fluctuations inherent in the network behavior and operation.

These variabilities are accounted for in three nested cycles of randomization, as in the following procedural steps:

- Sample the fault configuration of the transmission network on the basis of the failure probabilities of each element (node or arc) of the system.
- Sample the production from the sources, the demand at the targets and the capacity of the arcs.
- Build the discrete cumulative distribution function of the capacities of the arcs leaving each one of the source nodes.
- Sample the arcs along which the flow will leave each of the source nodes.
- Develop the flow propagation cycle from each source node.

1. the random walk of flow follows the arc sampled on the basis of the actual capacities of the arcs departing from the successive nodes traversed by the flow;
2. if the flow goes into an isolated node with no departing connections, the cycle ends;
3. the flow between a pair of nodes is accounted once (repeated flows between the same pair of nodes are neglected);

4. once the flow arrives at a target node, the capacities of the incoming arcs are checked: if their sum is larger than the maximum capacity of the node, an overload is recorded;
5. if the flow does not reach the target, a new random walk is started. If no flow arrives at any of the targets, then a blackout is recorded.

The IEEE 14 BUS network system is used to show the model in application; the network represents a portion of the American Electric Power System and consists of 14 bus locations connected by 20 lines and transformers. The transmission lines operate at two different voltage levels, 132 kV and 230 kV, with three 230/132 kV tie stations and voltage corrective devices, and two generating units.

To carry out the analysis, each network component is transposed into a node or edge of the representative topological network. Three different physical types of nodes are considered: source nodes (where the electricity is fed into the network), load nodes (where customers are connected) and transfer or transmission nodes (without customers or source). The failure probability of each edge in the network has been defined from literature data of failure rate and lines length (two line lengths of 48 and 50 km have been considered); lines containing transformers are considered to be zero-length lines. Source generation and target demand are sampled from a normal distribution with a mean value of 30 and a variance of 100, in arbitrary units (a.u.). The values of the capacities of the network links are assumed all distributed according to a normal distribution of mean value 100 a.u. and a standard deviation of 10 a.u. The flow is sampled on the actual capacities of the arcs. once the flow arrives at a target node, the capacities of the incoming arcs are checked: if their sum is larger than the maximum capacity value of the node, an overload is recorded. The targets are absorbing nodes: the flow stops and the received flow is recorded for evaluating the network lost load and the network service efficiency. If no flow reaches any target, a service blackout is recorded.

The following network performance characteristics are computed at the end of the simulation (Table 1):

blackouts and overloads (the average value of flow that does not reach the targets or that exceeds the capacities of the transmission lines, respectively);
network demanded load (the sum of the power generated from all the sources);
network received load (the sum of the flow reaching the targets);
the network lost load (the difference between demanded and received loads);
the network service efficiency (the ratio between received and demanded loads);

TABLE 1. Network indicators

Blackout (%)	0.44
Overload (%)	$3.33.10^{-4}$
Network service efficiency	0.60
Network demanded load (a.u.)	59.93
Network received load (a.u)	36.24
Network lost load (a.u.)	23.70

As an indicator of the criticality of the network elements, the centrality betweenness has been calculated based on the proposed random walk model. Roughly speaking, the random walk betweenness of a node i is equal to the number of times that a random walk starting at s and ending at t passes through i along the way, averaged over all s and t. This measure reasonably describes the fact that current will flow along all paths from source to target, and nodes that lie on no path from source to target get a betweenness of zero [Zio and Piccinelli, 2010]. The results reasonably point out that the lower half of the network, which contains the generating units, has higher values of betweenness than the upper half.

A Simulation Model for the Analysis of Cascading Failures in Electric Power Transmission Networks

In electric power transmission networks, cascades of failure events leading to blackout usually occur on time scales of minutes to hours and are completed in less than 1 day [Dobson, Carreras et al. 2007]. Abstract modeling paradigms for analyzing the system response to such cascading failures are being developed to investigate the cascade dynamics and its dependence on network topology and element characteristics, in order to protect and mitigate the evolution of the cascade. Despite their apparent simplicity, these models provide indications on the elements critical for the propagation process [Zio and Sansavini, 2010] and on the actions that can be performed in order to prevent or mitigate the undesired effects.

MODELING FAILURE CASCADES IN SINGLE CIS, INITIATED BY RANDOM FAILURES

Various abstract models of cascading failures have been applied to simulate the propagation process in single CIs, differing for both the logic of redistribution of the failure load and the nature of the cascade triggering event, i.e. either a random failure or a targeted intentional attack [Motter and Lai 2002; Dobson et al. 2005; Zio and Sansavini 2008]. The choice of the most suitable algorithm for modeling the spreading process taking place in a given CI must be performed carefully, considering the type of service provided by the CI, as further explained below.

In this section, an exemplary approach to abstract failure cascade modeling in single CIs initiated by random failures, is presented [Zio and Sansavini, 2008].

As in the previous section, the CI is still modeled as a graph G(N,K), like in the static analysis. The N components are assigned random initial loads sampled uniformly between a minimum value Lmin and a maximum value Lmax. The system is operated so that the initial component loadings vary from Lmin to Lmax = Lfail = 1 (after normalization). Then, the average initial component loading L = (Lmin + 1)/2 can be increased by increasing Lmin.

For simplicity, all components are considered identical, with the same limit of operation Lfail, beyond which they are failed. To model the cascading effects of random disturbances on a network of components, the cascade is started by imposing on each component an additional load D. If the sum of the initial load Lj of component j and the disturbance D is larger than the component load threshold Lfail, component j fails. This failure occurrence leads to the redistribution of an additional load P on the neighboring nodes which may, in turn, get overloaded and thus fail in a cascade which follows the connection pattern of the network system. If there is no working node in the neighborhood of a failed component, the cascade spreading in that "direction" is stopped.

In the end, the damage caused by the cascade is quantified in terms of the number of network components which have failed, i.e. the cascade size S, when the propagation comes to an end.

The cascade propagation algorithm is embedded in a Monte Carlo simulation, in which a large number of cascades is triggered for the same range of initial load, [Lmin, Lmax], in order to obtain statistically significant results. The damage caused by the cascades for any initial load level, [Lmin, Lmax], is quantified in terms of the number of network components which have failed on the average, i.e. the average cascade size, $\bar{S}$. The simulation can be repeated for different ranges of initial load, [Lmin, Lmax], with Lmax = 1 and Lmin varying from 0 to 1, and a point $(L, \bar{S})$ is drawn in a load-size diagram, so that the average critical load, Lcr, at which the phase transition between the absence of cascades and the emergence of cascades with significant size ($\bar{S} \geq S_{cr}$, e.g. involving a relevant fraction of network components) in the system can be identified.

Other hypotheses of failure propagation can be easily accommodated in this general modeling framework. For example, the actual load carried by a component can be transferred onto the other (neighboring) components in the network, upon its failure. Under this formulation, upon failure of a component its load is uniformly shared among its neighbors.

MODELING FAILURE CASCADES IN INTERDEPENDENT CIS, INITIATED BY RANDOM FAILURES

The modeling of cascading failures in interdependent CIs, initiated by random failures, can be carried out in a conceptual framework similar to the one for single CIs previously presented. In such a framework, interdependencies are modeled as links connecting nodes of the interdependent systems; these links are conceptually similar to those of the individual systems and can be bidirectional with respect to the "flow" between the interdependent networks. Cascading failures are then assessed considering the local propagation of the overload originated from a failure to first-neighbors and to the interdependent set of components linked to the failed one [Newman et al. 2005; Zio and Sansavini, 2010]. The number of interdependency links and the load of flow transferred over the interdependency links are essential features characterizing the "coupling energy" between the interdependent systems.

To account for the dynamics of changing connections between the two network systems under developing failure cascade processes, a Monte Carlo simulation can be performed in which the number of interdependency links, M, and the load transferred over the interdependency links, I, are kept constant but the interdependency connections among components are randomly rewired at each Monte Carlo trial.

The effects of the interdependencies between the two systems can be analyzed in terms of the average cascade size, $\overline{S}_i$, i.e., the number of failed components in the i-th system at the end of the cascade spread, versus the average initial load in the system, $\overline{L}_i$, which represents the system operating level in the system. For each value of $\overline{L}_i$, several Monte Carlo realizations are repeated, corresponding to different sampled patterns of the M interdependency links.

The interdependencies cause a shift to lower values of the loading threshold for which the cascading phenomenon starts appearing with significance. To quantitatively assess the effects of the interdependency, a threshold can be set representing the maximum allowable cascade size, Sicr, in the i-th system, which identifies the critical load, Licr, beyond which the threshold is exceeded in such system. The maximum allowable cascade size, Sicr, is interpreted as the maximum number of components which can be lost in system i without affecting the global service provided by the infrastructure. This threshold can vary from system to system and is a distinguishing feature of the provided service.

The critical load, Licr, is a relevant feature of a network system since it identifies, together with the continuous change in gradient, a type-two transition between the cascade-safe region and the onset of disrupting cascades in terms of the loading conditions. Along with the average cascade size, it gives essential information on the system vulnerability towards cascading failures and it can help identifying safety margins of system operation.

To understand the effects on the cascade process of the parameters characterizing the interdependency between the two systems, a further sensitivity analysis can be performed under several operating conditions which reflect real system operations. First, it is important to assess the extent to which an interdependent system working at different, fixed load levels influences the coupled network system with respect to its vulnerability towards cascading failures. Then, the effects of the number of interdependency links on the vulnerability to cascading failures can be assessed under different system operating conditions. The characterization of this relationship is relevant in the definition of cascade-safe operating regimes for the interdependent systems: for a fixed number of interdependency links in the system, a critical loading level can be identified below which the systems are safely operated.

Modeling Failure Cascades In Single Cis, Initiated By Targeted Intentional Attacks

Simulation frameworks can be developed to investigate also the effects of targeted intentional attacks at network components. To abstractly model the cascade dynamics which follows, one can assume that at each time step one unit of the quantity processed by the network is exchanged along the shortest paths connecting every pair of components. The load at a component is then the total number of shortest paths passing through that component to be compared with its capacity which is the maximum load that the component can handle. The capacity is usually limited by technological limitations and economic considerations. In this view, one can assume that the capacity is dimensioned proportionally to its nominal load at which it is designed to operate initially [Motter and Lai, 2002], the proportionality constant representing the capacity tolerance parameter. When all the components are working, the network operates without problems; on the contrary, the occurrence of components failures leads to a redistribution of the shortest paths in the network and, consequently, to a change in the loads of the surviving components: if the load on a component increases beyond capacity, the component fails and a new redistribution of the shortest paths and loads follows, which, as a result, can lead to a cascading effect of subsequent failures.

When examining the potential for cascading processes triggered by the removal of a single component, two situations are expected: if prior to its removal the component is operating at a relatively small load (i.e. if a small number of shortest paths go through it), its removal will not cause major changes in the balance of loads and subsequent overload failures are unlikely; however, when the load of the component is relatively large, its removal is likely to affect significantly the loads of other components and possibly start a sequence of overload failures. Intuitively, then, the following behavior is expected [Motter and Lai, 2002]: global cascades occur if the network exhibits a highly heterogeneous distribution of loads and the removed component is among those with highest loads; otherwise, cascades are not expected.

In the modeling scheme illustrated, the distribution of loads is highly correlated with the distribution of links: networks with heterogeneous distribution of links are expected to be heterogeneous with respect to the load, so that on average components with large number of links will have high loads. This behavior confirms the robust-yet-fragile property of heterogeneous networks, first observed in [Albert et al., 2000] with respect to the attack on several components.

Conclusions

Two main outputs of a CI vulnerability assessment are the quantification of system vulnerability indicators and the identification of critical elements. These are obtained by a thorough system analysis comprising the representation and adequate modeling of its properties (e.g. physical and logical structure, and operation modes), the identification and modeling of the dependencies within the system and the interdependencies with other systems, the identification of the hazards and threats, the modeling of the dynamic effects (e.g. cascading failures).

In the quantification task of a CI vulnerability assessment, stochastic simulation plays a fundamental role as a 'what-if scenario generator' addressing the challenges posed by the many uncertainties on system characteristics, properties, and responses to changes (e.g. due to failures of or attacks on its components).

The present contribution has touched upon the use of stochastic simulation for the study of electrical power transmission systems. Two examples have been provided of simulation models for analyzing the structural characteristics and criticalities of such type of CI, and for following the dynamics of cascading failures so as to derive conclusions on its critical loadings and operation settings.

The simulation framework allows considering the interplay between system structural characteristics and process dynamical aspects, which however makes the modeling and analysis very complicated since the load and capacity of each component, and the flow through the network are often highly variable quantities both in space and time.

In simulation of complex systems, the usual inevitable need arises for a compromise between model adherence to reality and the computational costs of the analysis. Indeed, the functional analysis of CIs is a complex task, because of the complexity of the system itself and the lack of accurate and complete information on the infrastructures. Functional models of CIs require, in fact, the knowledge of a very large amount of data; the network topological characterization must be complemented by a number of information consisting of the technical characteristics of lines and nodes, load requirements, etc. These data are often unavailable as they are treated as confidential information from the stakeholders.

In the particular case of CIs such as the electric power transmission networks, the large numbers of uncertain parameters in the models poses an additional computational challenge of uncertainty propagation.

The functional models of the type presented as examples in this contribution can help shed light on the way transmission networks react to faults and attacks, evaluating their consequences when the dynamics of flow of the physical quantities in the network is taken into account. The response behavior often results in a dramatic cascade phenomenon due to avalanches of node breakings; the cascade can spread across boundaries of interconnected CIs, with significant impacts on the individual system operating conditions.

Abstract modeling paradigms for simulating the response of a CI to cascading failures, such as the exemplary one here presented, can be used to guide a successive detailed simulation focused on the most relevant physical processes and network components. The need for such an analysis tool is even stronger for systems like those for electric power transmission here considered, in which the cascade dynamics is rapid and modifications are actuated onto the network in order to mitigate the evolution of the cascade.

Despite their apparent simplicity, these types of simulation models can provide indications on the elements critical for the propagation process and on the actions that can be performed in order to prevent or mitigate the undesired effects.

As improvements in network components alone cannot ensure system robustness or protection against disproportionate cascading failures, topological changes are typically needed to increase cascading robustness at the required tolerance levels. In the end, managing the risk of network service unavailability requires a combination of redundant topology, increased flow-carrying capacity, controlled-loading, and other non-conventional consequence reduction strategies, such as layout homogenization and the deliberate inclusion of weak links for network islanding.

Finally, dependencies and interdependencies among different CIs have to be modeled for assessing the influences and limitations which interacting infrastructures impose on the individual system operating conditions, for avoiding fault propagation by designing redundancies and alternative modes of operations, and for detecting and recognizing threats. Infrastructure interdependency stems from the functional and logical relations among individual components in different distributed systems. In developing modeling and simulation frameworks that allow the coupling of multiple interdependent infrastructures, it is important to know that simply linking existing infrastructure models together fails to capture the emergent behavior arising in interdependent infrastructures, a key element of interdependency analysis.

In order to characterize the extent to which a contingency affecting an infrastructure is going to weaken, and possibly disrupt, the safe operation of an interconnected system, it is necessary to model the relations established through the connections linking the multiple components of the involved infrastructures. The modeling of interdependencies among network systems and of their effects on failure propagation can be carried out within the simulation framework of failure cascade processes; the sensitivity of the coupling parameters defining the interdependency strength is of particular interest for the definition and prescription of cascade-safe operating margins in interdependent CIs.

Acknowledgments

This work has been partially funded by the "Foundation pour une Culture de Securite' Industrielle" of Toulouse, France, under the research contract AO2006-01.

References

[Adachi and Ellingwood, 2008] T. Adachi and B. R. Ellingwood, "Serviceability of earthquake-damaged water systems: effects of electrical power availability and power backup systems on system vulnerability", Reliability Engineering and System Safety, 93(1), pp. 78–88, 2008.
[Albert et al., 2000] Albert R., Jeong H. and Barabasi A.-L., Error and Attack Tolerance of Complex Networks, Nature, Vol. 406, 2000, pp. 378–382.
[Apostolakis and Lemon, 2005] Apostolakis, G.E., Lemon, M.D., "A Screening Methodology for the Identification and Ranking of Infrastructure Vulnerabilities Due to Terrorism", Risk Analysis, Vol. 25, No. 2, 2005.
[Barton and Stamber, 2000] D. C. Barton and K. L. Stamber, "An agent-based microsimulation of critical infrastructure systems", Conference: International Energy Foundation's ENERGEX

2000 - 8th International Energy Forum, Las Vegas, NV (US), 07/23/2000-07/28/2000, SANDIA REPORT SAND2000-0808C, 2000.

[Borkowska, 1974] Borkowska, B. "Probabilistic Load flow", IEEE Trans.,1974., PAS-93, pp 752–759.

[Bouchon, 2006] S. Bouchon, The Vulnerability of Interdependent Critical Infrastructures Systems: Epistemological and Conceptual State-of-the Art, EUR-report, 2006

[Cagno et al., 2009] E. Cagno, M. De Ambroggi, O. Grande and P. Trucco, "Risk analysis of underground infrastructures in urban area: time-dependent interoperability analysis", Reliability, Risk and Safety: Theory and Applications – Briš, Guedes Soares & Martorell (eds), Proceedings of ESREL 2009 Europe Annual Conference, 6–11 September 2009, Prague, Czech Republic, Taylor & Francis Group, London, pp. 1899-1906, 2009.

[Carreras et al., 2007] B. A. Carreras, D. E. Newman, P. Gradney, V. E. Lynch and I. Dobson, Interdependent risk in interacting infrastructure systems, Fortieth Hawaii International Conference on System Sciences, Hawaii, January 2007.

[Casalicchio et al., 2007] E. Casalicchio, E. Galli and S. Tucci, Federated Agent-based Modeling and Simulation Approach to Study Interdependencies in IT Critical Infrastructures, 11th IEEE Symposium on Distributed Simulation and Real-Time Applications, 2007.

[Casalicchio et al., 2009] E. Casalicchio, E. Galli and V. Ottaviani, "MobileOnRealEnvironment-GIS: A Federated Mobile Network Simulator of Mobile Nodes on Real Geographic Data", Proceedings of the 2009 13th IEEE/ACM International Symposium on Distributed Simulation and Real Time Applications, pp. 255–258, 2009.

[Chen and McCalley, 2005] Q. Chen and J. D. McCalley, "Identifying high risk N-k contingencies for online security assessment", IEEE Transactions on Power Systems, vol. 20, no. 2, 2005.

[Dekker, 2005] Dekker A. H., "Simulating Network Robustness for Critical Infrastructure Networks", Conferences in Research and Practice in Information Technology, Proceedings of the 28th Australasian Computer Science Conference, The University of Newcastle, Newcastle, Australia, vol. 38, V. Estivill-Castro, Ed., 2005.

[Dobson et al., 2005] I. Dobson, B. A. Carreras and D. E. Newman, "A loading-dependent model of probabilistic cascading failure", Probability in the Engineering and Informational Sciences, vol.19, no.1, pp. 15–32, 2005.

[Dopazo et al., 1975] Dopazo, J.F., Klitin, O.A. and Sasson, A.M., "Stochastic load flow method", IEEE Trans., 1975, PAS-94, pp. 1551–1556.

[Duenas-Osorio et al, 2007] L. Duenas-Osorio, J. I. Craig, and B. J. Goodno, "Seismic response of critical interdependent networks", Earthquake Engineering and Structural Dynamics, 2007. 36(2): p. 285–306.

[Duenas-Osorio and Vemuru, 2009] L. Dueñas-Osorio and S. M. Vemuru, "Cascading failures in complex infrastructure systems", Structural Safety, vol. 31, pp. 157–167, March 2009.

[Ellis et al., 1997] Ellis, J, Fisher D, et al. Report to the President's Commission on Critical Infrastructure Protection, S.E. Institute. Editor Carnegie Mellon University, 1997.

[Germann et al., 2006] T. C. Germann, K. Kadau, I. M. Longini, Jr. and C. A. Macken, "Mitigation strategies for pandemic influenza in the United States", Proc. The National Academy of Sciences of the USA, vol. 103, no. 15, pp. 5935–5940, April 2006.

[Guckenheimer and Ottino, 2008] Guckenheimer J. and Ottino, J. M. Foundations for Complex Systems Research in the Physical Sciences and Engineering, Report from a NSF Workshop, September 2008 http://www.math.cornell.edu/~gucken/PDF/nsf_complex_systems.pdf.

[Haimes and Jiang, 2001] Y. Haimes and P. Jiang, "Leontief-based model of risk in complex interconnected infrastructures", Journal of Infrastructure Systems, vol. 7, no. 1, pp. 1–12, March 2001.

[Haimes et al., 2005a] Y. Y. Haimes, B. M. Horowitz, J. H. Lambert, J. R. Santos, C. Lian and K. G. Crowther, "Inoperability Input-Output model for interdependent infrastructure sectors. I: theory and methodology", Journal of Infrastructures Systems, vol. 11, no. 2, pp. 67–79, 2005.

[Haimes et al., 2005b] Y. Y. Haimes, B. M. Horowitz, J. H. Lambert, J. R. Santos, C. Lian and K. G. Crowther, "Inoperability Input-Output model for interdependent infrastructure sectors. I: case studies", Journal of Infrastructures Systems, vol. 11, no. 2, pp. 80–92, 2005.

[Helseth and Holen, 2009] A. Helseth and A. T. Holen, "Structural vulnerability of energy distribution systems: Incorporating infrastructural dependencies", Electrical Power and Energy Systems, vol. 31, pp. 531–537, 2009.

[Hong et al, 1999] X. Hong, M. Gerla, G. Pei and C.-C. Chiang, "A group mobility model for ad hoc wireless networks", Proceedings of the 2nd ACM international workshop on Modeling, analysis and simulation of wireless and mobile systems, Seattle, Washington, United States, pp. 53–60, 1999.

[IRGC, 2006] White Paper on Managing and Reducing Social Vulnerabilities from Coupled Critical Infrastructures, IRGC, 2006.

[IRRIIS, 2007] EU project IRRIIS, Deliverable D222, "Tools and techniques for interdependency analysis", pp. 45–51, June 2007. Available: http://www.irriis.org/File.aspx?lang=2&oiid= 9138&pid=572.

[Jiang and Haimes, 2004] P. Jiang and Y. Y. Haimes, "Risk management for Leontief-based interdependency systems", Risk Analysis, vol. 24, no. 5, pp. 1215–1229, 2004.

[Johansson and Jönsson, 2009] J. Johansson and H. Jönsson, "A model for vulnerability analysis of interdependent infrastructure networks", in Safety, Reliability and Risk Analysis: Theory, Methods and Applications, Proc. ESREL 2008 and 17th SRA-Europe Conf., Valencia, September 2008, Taylor & Francis Group, London, pp. 2491–2499, 2009.

[Kodsi and Canizares, 2007] S. K. M. Kodsi and C. A. Canizares, "Application of a Stability-constrained Optimal Power Flow to Tuning of Oscillation Controls in Competitive Electricity Markets", IEEE TRANSACTIONS ON POWER SYSTEMS, 22(4), pp. 1944, 2007.

[Lee et al., 2007] E. E. Lee, J. E. Mitchell and W. A. Wallace, "Restoration of services in interdependent infrastructure systems: a network flows approach", IEEE Transactions on Systems, Man, and Cybernetics-Part C (Applications and Reviews), 37(6): p. 1303–17, 2007.

[Luijf et al., 2009] E. Luiijf, A. Nieuwenhuijs, M. Klaver, M. van Eeten and E. Cruz, "Empirical Findings on Critical Infrastructure Dependencies in Europe", in Critical Information Infrastructure Security, Lecture Notes in Computer Science, Springer Berlin/Heidelberg, 2009.

[Min et al., 2007] H.-S. J. Min, W. Beyeler, T. Brown, Y. J. Son and A. T. Jones, "Toward modeling and simulation of critical national infrastructure interdependencies", IIE Transactions, vol. 39, no. 1, pp. 57–71, January 2007.

[Motter and Lai, 2002] A. E. Motter and Y.-C. Lai, "Cascade-based attacks on complex networks", Physical Review E, vol. 66, no. 6, p. 065102, December 2002.

[Newman et al., 2005] D. E. Newman, B. Nkei, B. A. Carreras, I. Dobson, V. E. Lynch and P. Gradney, "Risk Assessment in Complex Interacting Infrastructure Systems", Proc. Thirty-Eight Annu. Hawaii International Conf. on System Sciences, January 3–6, 2005, Computer Society Press, 2005.

[Oyuang et al., 2009] M. Ouyang, L. Hong, Z.-J. Mao, M.-H. Yu and F. Qi, "A methodological approach to analyze vulnerability of interdependent infrastructures", Simulation Modeling Practice and Theory, 17(5), p. 817–828, 2009.

[Panzieri et al., 2004] S. Panzieri, R. Setola and G. Ulivi, "An agent based simulator for critical interdependent infrastructures", Securing Critical Infrastructures, CRIS2004 : Conference on Critical Infrastructures, October 25 - 27, 2004, Grenoble, FRANCE, 2004.

[Piwowar et al., 2009] Piwowar, J., Chatelet, E. and Laclemence, P., "An efficient process to reduce infrastructure vulnerabilities facing malevolence", Reliability Engineering and System Safety, vol. 94, pp. 1869–1877, 2009.

[Reed et al., 2009] D. A. Reed, K. C. Kapur and R. D. Christie, "Methodology for assessing the resilience of networked infrastructure", IEEE Systems Journal, vol. 3, no. 2, pp. 174–180, 2009.

[Rinaldi et al., 2001] S. M. Rinaldi, J. P. Peerenboom and T. K. Kelly, "Identifying, understanding and analyzing critical infrastructures interdependencies", IEEE Control System Magazine, vol. 21, no. 6, pp. 11–25, December 2001.

[Rinaldi, 2004] S. M. Rinaldi, "Modeling and simulating critical infrastructures and their interdependencies", Proc. Thirty-Seventh Annual Hawaii International Conf. on System Sciences, January 5-8, 2004, Computer Society Press, 2004.

[Schläpfer et al., 2008] M. Schläpfer, T. Kessler and W. Kröger, "Reliability Analysis of Electric Power Systems Using an Object-oriented Hybrid Modeling Approach"in Proc. 16th Power Systems Computation Conf., 14-18 July, Glasgow, 2008.

[Sobjerajski, 1978] Sobjerajski, M., "A method of stochastic load flow calculation",Archiv für Elektrotecnik (1978), pp.37–40.

[Svendsen and Wolthusen, 2007] N. K. Svendsen and S. D. Wolthusen, "Connectivity models of interdependency in mixed-type critical infrastructure networks", Information Security Technical Report, 12(1): p. 44-55, 2007.

[Zimmermann, 2001] Zimmerman R., "Social Implications of Infrastructure Network Interactions", Journal of Urban Technology, vol. 8, no. 3, pp 97–119, December 2001.

[Zio and Piccinelli, 2010] Zio, E., Piccinelli, R., "Randomized flow model and centrality measure for electrical power transmission network analysis", Reliability Engineering and System Safety, 95 (2010) 379 – 385.

[Zio and Sansavini, 2008] E. Zio and G. Sansavini, "Modeling Failure Cascades in Network systems due to Distributed Random Disturbances", ESREL 2008, European Safety and Reliability Conference, September 22–25, 2008, Valencia, Spain, pp. 1861–1866.

[Zio and Sansavini, 2010] E. Zio and G. Sansavini, "Modeling Interdependent Network Systems for Identifying Cascade-Safe Operating Margins" submitted to IEEE Transactions on Reliability Special Issue on Complex Systems, 2010.

CONVERSION OF RENEWABLE KINETIC ENERGY

OF WATER: SYNTHESIS, THEORETICAL MODELING,

AND EXPERIMENTAL EVALUATION

ION BOSTAN
Technical University of Moldova, Chisinau, Moldova

ADRIAN V. GHEORGHE[*]
Old Dominion University, Norfolk, VA, USA

VALERIU DULGHERU, VIOREL BOSTAN,
ANATOL SOCHIREANU, ION DICUSARĂ
Technical University of Moldova, Chisinau, Moldova

Abstract. The current paper deals with concepts for conversion of the kinetic energy of fluent water into mechanical and finally electrical energy. A distinct objective is the study of the energetic potential of the Republic of Moldova's rivers – the Nistru, Prut, and Răut, in view of assuring energy security for limited kind of applications. The study of the energetic potential of the Prut River – an affluent of the Danube, that flows through the Republic of Moldova's territory has been carried out. On the basis of the performed study, the future installation place of the pilot plant of microhydrostation was chosen, that satisfies the following conditions: adequate harvest of low level speed of the water flows; existence of the energy consumers nearby; geological characteristics of the water front, which would permit the connective foundation installation of the microhydrostation with the river bank.

Introduction

Energy, a complex resource, is the key in searching of a continuous sustainable development for human society. The effects of the air pollution and of the climatic changes over the surrounding environment impose, in a striking way, the increasing necessity of exploring renewable energy resources. For the Republic of Moldova the usage of the renewable energy resources is important both from economical and political points of view, because the country does not possess its

---

[*] To whom correspondence should be addressed: Adrian V. Gheorghe, Old Dominion University, Norfolk, VA, USA, e-mail: adriangheorghe9145@gmail.com

A. Gheorghe and L. Muresan (eds.), *Energy Security: International and Local Issues,*
Theoretical Perspectives, and Critical Energy Infrastructures,
DOI 10.1007/978-94-007-0719-1_10, © Springer Science+Business Media B.V. 2011

own fossil fuels. The energy system of the Republic of Moldova is connected to the energy system of Ukraine. The internal system lines of the high voltage installations have a total length of 7,245 km and are distributed as follows: 214 km coupled to 400 kV, 530 km to 330 kV, 4,104 km to 110 kV and 2,397 km to 35 kV. The low voltage network of internal distribution includes 57,698 km of wire and 2,174 km of electrical cable. Between 1991 and 2001 the consumption of the primary energy resources for the Republic of Moldova was reduced by 70%. The intensity of national economical energy was reduced from 2.3 t.c.f./1,000 USD in 1991 to 1.9 t.c.f./l,000 USD in 2002. At present the Republic of Moldova depends, to a great extent, on the delivery of the natural gas from Russia.

The economical development depends on the capacity of assuring the demand of the mechanical and thermal electrical energy, to a great extent. A major importance, regarding the energy offerings in the Republic of Moldova, will be given to the capacity of distinguishing between various energy resources. The renewable energy resources, as well as technologies with less negative impacts on the environment, constitute a preference for the future.

The current paper deals with concepts for conversion of the kinetic energy of fluent water in mechanical and electrical energy. A distinct objective is the study of the energetic potential of the Republic of Moldova's rivers – the Nistru, Prut, and Răut. The study of the energetic potential of the Prut River – an affluent of the Danube, that flows through the Republic of Moldova's territory has been carried out. On the basis of the performed study the future installation place of the pilot plant of microhydrostation was chosen, that satisfies the following conditions:

- explorable speed of the water flows;
- existence of the energy consumers nearby;
- geological characteristics of the water front, which would permit the connective foundation installation of the microhydrostation with the river bank.

On the basis of the carried out studies a conceptual scheme of the microhydrostation, based on the usage of the rotor with aerodynamic profile blades, was created.

The Energetic Potential of the Prut River

In the Republic of Moldova there are three rivers with the explorable hydro energetic potential: the Nistru, Prut, and Răut. The usage of the potential energy of waters through dam building is not opportune for the Republic of Moldova, because it seriously endangers the ecological equilibrium, which is actually instable (deforestation of woods, in the past, created the situation that only about 20% of the lands are forested), and will lead to the occupation of new lands by water storage dams.

For the project realization, at the beginning, the Prut River was identified (Figure 1). The choice was dictated by the following reasons:

- The Prut River is a border river between the Republic of Moldova and Romania, which, since 2007, is a component part of the European Union and NATO.
- On both banks of the Prut River the density of settlements are quite high and a is home to historically unique economical system, temporarily separated as a result of a territorial rape. This fact will permit the extension of the obtained results within the project over the bordering zones with Romania. There are already demands on behalf of the mayoralty of Iaşi city (Romania).

Prut, an affluent of the Danube River, begins on the north-eastern coasts of the Carpathian Mountains, at a height of 1,580 m and flows through the Moldo-Basarabean Plateau. The total length of the river is 950 km with a reception basin of 28,400 km². On the distance of 900 km from the inflow, the river represents a natural border between the Republic of Moldova, Romania, and Ukraine. The section of the Prut River, from the outflow to the inflow in the mountains, has a

Figure 1. Map of the hydroenergetic potential of the Prut River.
● – Stoieneşti village, where the pilot plant of the microhydrostation will be installed
● – zones, where the water speed measuring on the Prut River was done ($v > 1$ m/s)

high flow. This portion of the river flows in a mountain valley. The length of this portion is of 200 km. Near the city of Cernăuți (Ukraine) the portion with the average flow of the river begins, flowing into a valley. The valley width in this region is of 5–6 km. The banks are low and easily flooded. The river high tide, in its middle part, is powerful, in the flood period it is dense and creates a new riverbed.

The region with the average flow extends to the city of Ungheni and has a length of 380 km. The descendent portion of the Prut River, from the city of Ungheni to the river inflows, has a length of 396 km. This region has less important valleys, with a width of 10–12 km. In a large portion with a reduced flow, the river floods its beds. Preventing floods, the river creates channels on a certain portion; in the low tide period can be observed drain channels can be observed as well as partial water erosion. The affluent system of the river is unimportant. It was created from the outbreak in radiation isthmuses. In this region there are often landslides, sometimes very serious, for example, in 1981, when a landslide around the Taxobeni locality covered nearly the entire river. The river sinuosity is marked with an average coefficient of the bends equal to two (2). The river ground is irregular, covered here and there with gravel, where separated boulders or even stones accumulations can be often found. A variety of trees and bushes grow along the river banks.

The temperature instability characterises the river frost period, during the winter months. On the biggest part of the river, temporary ice plates can be observed. More stable ice plates were observed 2–3 times in 10 years. The ice breaks mainly at the end of February, and its appearance can be observed as early as the end of November. The ice thickness is on average of 0.26–0.35 m, and in very cold winters it can reach 60 cm. The river is navigable for an average of 266 days, 50–60 days of these the river overflows, and 190–210 days the river is calm.

The wind speed is relatively low, and its annual movement is weakly expressed. Approximate 35% of the year the days are windless. The average wind speed varies between 3 and 4.9 m/sec. The average annual temperature of the air in the mountainous zones of the river basin is approximately +7°C. In the hilly region of the basin the temperature is +10°C. The absolute maximum of the basin is +40°C, the absolute minimum is −31°C.

In the mountainous zone of the river, annual precipitations reach 800 mm. In the region of the other portion of the river the precipitations level varies from 600 to 300 mm. The biggest part of the precipitations devolves to the spring – summer season. The atmospheric precipitations are the main water source from the Prut River region. The water level of the river increases especially thanks to floods in the summer period. Its level is increased in the summer – autumn period with 50% from annual precipitations as the result of the rains. According to the hydro-meteorological data, the following information was accumulated:

- The biggest height is of 55 m (Criva village region).
- The hydroenergetic potential can be explored on the Prut River portion from Criva village to village of Giurgiulești from the south of the country, where the water level is of 20–27 m compared to sea level.

- On this portion of the Prut River there are about 80 rural settlements with a population of about 200,000 people.

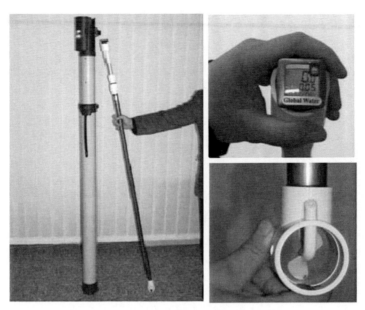

Figure 2. FP 201 Global water flow probe.

To undertake the water speed measurements on different portions of the Prut River a water speed measuring device *FP 201 Global Water Flow Probe Digital Velocity Meter*, IRIS Instruments, USA (Figure 2) was used. The telescopic construction of the device permits the speed measurement of water flow at different depths.

The water speed measuring device *Flow Probe FP201*, adding to the capabilities of the *FP201*, is designed to measure the maximum and average speed of water flow; the *FP201* is calibrated and certificated by the producer.

The Prut Basin characteristic is (S = 28,396 km^2; L = 952.9 km; Q_o = 86 m^3/s). Although the Prut River's outflow originates in the Carpathian mountains, about 80% of its basin area unfolds in the forest steppe and steppe zone of the Podolo-Moldovan Plateau. In this area, the typical phases of the rivers from the hilly regions of European climatic province are thus instilled. This leaves the mountainous region of Deleatin, where the valley expands in Neocene molasses formations, and downstream the Prut passes through the plateau zone, feeding itself with the water from affluent steams from both sides. In this large valley, to the Moldovan border, is a powerful alleviation, where the river waterside has a phreatic water abundance of a good quality.

Having analyzed the acquired information under many facets, it was finally decided to locate the pilot-station of the micro-hydro-power plant mounting on

Prut River in the sector near the village of Stoieneşti because of the following reasons:

- The water flow rate is 1.0–1.3 m/s, which makes it exploitable from an economical point of view
- In the close vicinity of the pilot-station of the micro-hydro-power station mounting zone are three big ponds for water storage that can be used for irrigation.
- The former farmland in this area is about 1000 ha of irrigated agricultural lands which sometimes need draining. This is the reason why it is not in agricultural use for the last 5 years.
- Free access up to the river bank is available as well as there are good conditions to build the foundation for anchoring.
- Not far from the mounting place of the pilot-station of the micro-hydro-power plant is a frontier guard post and a building, whose spaces will be used for the location of the Centre for Renewable Energy Conversion Systems Testing. The Centre was set up by the Department of Theory of Mechanisms and Machine Parts as a testing ground and storage of the high quality equipment.
- The place for the pilot-station of the micro-hydro-power plant mounting is very close to the village of Stoeneşti, a relevant electricity consumer.

Conceptual Design of a Floating Mini-Hydro-Electric Power Plant with Horizontal Axis and Helical Turbine

ANALYTIC MODEL FOR HELICAL AND MULTI/BLADE TURBINE

The design of the helical turbine (Figure 3) includes axis *1* where two blades are stiffly fixed in a constantly paced helical line. The wing profile (Figure 4) is characterized by a blunted foreside and sharp backside. Its central line is the geometric place of circuit centres inscribed in the profile. The main geometric parameters of the profile are:

- relative thickness of the profile $\vec{c}$, which was determined as the relation of the peak thickness of the profile c towards the chord length b, $\vec{c} = c/b$;
- relative hollow $\vec{f}$ which was determined as the relation of peak bending-deflection of the axial curve f towards the chord length b, $\vec{f} = f/b$;
- the camber, which was determined through the bending angle of the central line ε; that is the angle between the tangent lines at the central line of the profile in its foreside and backside.
- Positions $\vec{c}$ and $\vec{f}$ were determined through the relative abscissas: $\vec{x}_c = x_c/b$ and $\vec{x}_f = x_f/b$. The converse position of the profile in

the reticle is characterized by pace t, position angle Θ (the angle between the chord of the profile and the flank of the reticle), and angles φ_1 și φ_2 between the tangents at central line of the profile in its points, and the flank of the reticle. The relative pace of the reticle was determined by the relation of pace t towards the chord length b, $t=t/b$. The reticular density, which is the inverse value of the relative pace was determined from the relation $\tau = 1/\overrightarrow{t} = b/t$.

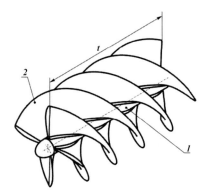

Figure 3. Helical Turbine.

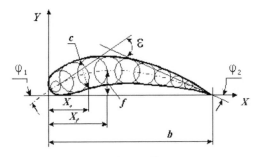

Figure 4. Wing Profile.

The elaborated mathematical design allows the determination of the basic kinetic and energy parameters of the helical turbine. In order to do this, a motion equation system of the perfect incompressible and isoenthropic fluid was taken, that describes water movement around the helicoidal rotor with a rather high accuracy:

- $\rho \operatorname{div}\overline{V} = 0$ – continuity equation,

- $\rho\dfrac{\mathrm{d}V}{\mathrm{d}t} = -\operatorname{grad}p + \rho f$ -pulse equation,

- $\rho\dfrac{\mathrm{d}e}{\mathrm{d}t} = -p\operatorname{div}V$ energy equation $\qquad (1)$

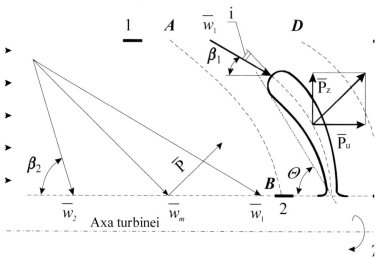

Figure 5. Diagram of forces generated by fluid currents.

The solutions to these equations comply with the limit conditions on the turbine rotor propeller, and at big distances in undisturbed limits of the fluid. Generally, the setting of these conditions presents certain difficulties related to the constructional form and the operating conditions of the water helicoidal turbine. This is why it was the determination for certain optimal working conditions was made, where the speed value in the outflow of the rotor and of the induced speed in the propeller blade were known. Thus, through the given integral equation system (meant for the examined helicoidal turbine) the immediate calculus of the aerodynamic characteristics was possible.

The following works were carried out in order to determine the water speed and the forces generated by them. The effect force of the current over a unitary profile in cross direction to the figure plan was determined at careening of an infinite reticle profile by a continuous parallel air flow. Sections *1* and *2* (Figure 5), parallel to the figure flank, were pointed out in the current and spaced from the reticle at a distance that permitted the acceptance of constant speed and pressure in each section, that is, where the current will not be disturbed. The current lines *AB* and *CD* were laid out at the distance of reticle pace *t*.

Motion quantity equation was applied to ABCD space:

$$F\Delta T = m\vec{w_2} - m\vec{w_1}.\qquad(2)$$

The resultant projections of all forces which act in this space on Z axis and U reticle flank are:

$$P'_U = M\left(-w_2\cos\beta_2 + w_1\cos\beta_2\right) = M\left[-w_{2U} + w_{1U}\right],\qquad(3)$$

where P'_U is the resultant force projection on U axis;

$M = m / \Delta T$ – air-mass which passes per second through the reticle with pace t and unitary length (in cross direction to the figure plan).

From the continuity equation:

$$M = \rho_1 w_{1Z} \cdot t \cdot 1 = \rho_2 w_{2Z} \cdot t \cdot 1. \tag{4}$$

For an incompressible gas:

$$\rho_1 = \rho_2 = \rho$$
$$w_{1Z} = w_{2Z} = w_Z, \tag{5}$$

and the resultant force projection on Z reticle axis is determined by the relation:

$$P'_Z + (\rho_1 - \rho_2) \cdot t \cdot 1 = M(w_{2Z} - w_{1Z}) = 0. \tag{6}$$

$\overrightarrow{P}$ force projection, which acts over the profile – wing of unitary length:

$$P_U = -P'_U = -M(w_{1U} - w_{2U}) = -\rho w_Z t(w_{1U} - w_{2U}); P_Z = -P'_Z = (\rho_1 - \rho_2)t. \tag{7}$$

Thus, $\overrightarrow{P}$ is the resultant force that acts over the profile, but $\overrightarrow{P'}$ – the force applied to the computational load.

According to the Bernoulli equation:

$$p_1 + \rho w_1^2 / 2 = p_2 + \rho w_2^2 / 2, \tag{8}$$

where p_1 and p_2 are static pressures in sections 1 and 2;

$\rho w_1^2 / 2$ și $\rho w_2^2 / 2$ – dynamic pressures in sections 1 and 2 accordingly.

Thus:

$$p_1 - p_2 = \rho / 2 \left(w_2^2 - w_1^2 \right) = \rho / 2 \left(w_{2U}^2 + w_{2Z}^2 \right) - \rho / 2 \left(w_{1U}^2 + w_{1Z}^2 \right) = \rho / 2 \left(w_{2U}^2 - w_{1U}^2 \right) \tag{9}$$

Let us determine the speed circulation on $ABCD$ contour accepting the counter-clockwise direction as positive:

$$G_{ABCD} = G_{AB} + G_{BC} + G_{CD} + G_{DA} \tag{10}$$

As AB and CD current lines are congruent, and the speed distribution on them is the same, then:

$$G_{AB} = -G_{CD}$$

$$G = G_{ABCD} = \oint_{ABCD} c \cdot \cos\left(\overrightarrow{c}, \overrightarrow{s}\right) ds = -w_{2U} \cdot t + w_{1U} \cdot t = t(w_{1U} - w_{2U}). \tag{11}$$

The medium geometric vector of $\overrightarrow{w}_m$ speed is determined by the relation:

$$\overrightarrow{w}_m = \left(\overrightarrow{w}_1 + \overrightarrow{w}_2 \right) / 2. \tag{12}$$

The projection of this vector on U axis is equal to $\left(w_{1U} + w_{2U} \right)/2$, and on Z axis respectively:

$$\left(w_{1Z} + w_{2Z} \right) = 2 w_Z / 2 = w_Z$$

The direction of the medium geometric speed is determined:

$$ctg\,\beta_m = \frac{w_{mU}}{w_{mZ}} = \frac{w_{1U} + w_{2U}}{2w_Z} = \frac{1}{2}\left(\frac{w_{1U}}{w_Z} + \frac{w_{2U}}{w_Z} \right) = \frac{1}{2}\left(ctg\,\beta_1 + ctg\,\beta_2 \right). \quad (13)$$

Thus the result of all velocities, which reacts on the reticulum from the part of the incompressible gas current, is equal to the product between the density, medium geometric speed, and speed circulation around the profile. Its agency direction is perpendicular to the medium geometric speed vector. In order to determine the direction of P force vector W_m is rotated with an angle of 90° counter-clockwise.

ELABORATION OF A 3D DYNAMIC MODEL OF THE HELICAL TURBINE

Development of 3D dynamic simulation models, by means of computer design makes the designing process easier at certain phases, in contrast with the rather costly process of product launching.

There is a large range of commercial CAD software products. Among the most known on the market are: CATIA, SolidWorks, ProINGINIRING, ANSYS, Mechanical Desktop, etc. All these CAD design software systems have the same working concept and differ only in possibilities and specialisations. After long periods of practicing with some of these design software, the researchers chose SolidWorks. This software, along with large design possibilities, disposes simplicity in operation and a friendly interface. SolidWorks allows integration of various modules for analysis, such as the calculation of joints at load action, dynamic analysis, library of standard elements, etc.

On the basis of the obtained results, a computer model of the helicoid rotor with four (4) inceptions and an aerodynamic profile of the blades in normal section was elaborated. Later, the model was simulated in the CFD environment by varying the geometrical and kinetostatic parameters of the fluid. As the working element of the micro-hydro power station represents a complex body it was necessary to carry out spatial interaction simulation (3D) of the working element with the fluid. According to theoretical research and to the justification of the functional parameters of the helicoid turbine, dynamic models have been designed by varying functional parameters. In Figures 6a,b, the dynamic models designed in the SolidWorks environment are shown by varying the geometrical parameters.

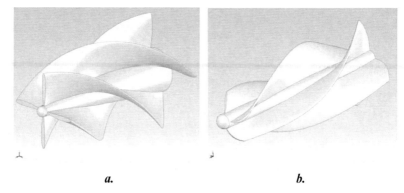

a.   *b.*

Figure 6. (**a**) – Helical rotor with constant profile and conical core; (**b**) – helical rotor with variable profile and constant core.

MODELLING OF BLADES AND FLUID INTERACTION

To investigate the influence of the fluid characteristics (kinetostatic parameters, density, etc.) a series of simulations on the helicoid rotor (Figures 7a, b; 8a, b; 9a, b) were carried out utilising CFX software. The modelling, carried out by using the CFX5.7 software, allowed the determination of some optimum geometrical parameters which secure both maximum energetic effect and reduction of flow turbulence. On the basis of previous modelling, more conceptual diagrams have been elaborated concerning floatable electrical hydraulic stations with horizontal axels and one or more helical turbines (Figure 10a, b). The solutions have been patented (Bostan et al., 2005 – Patents).

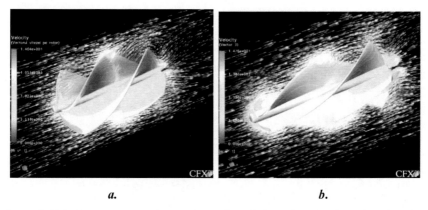

a.   *b.*

Figure 7. Velocity vector on a plane surface oriented along the axis of rotation, CFX Post: (**a**) Constant profile on the direction of rotation; (**b**) Variable profile on the direction of rotation.

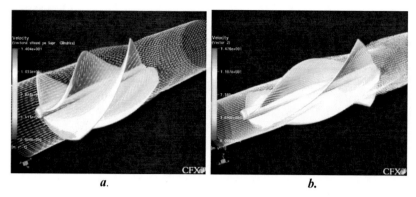

Figure 8. Velocity vectors on a circular surface oriented along the axis of rotation, CFX Post: (**a**) Constant profile on the direction of rotation; (**b**) Variable profile on the direction of rotation.

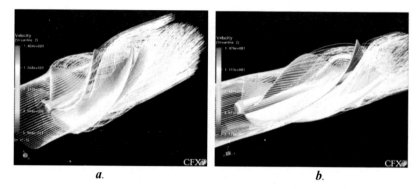

Figure 9. Liquid flow form around the rotor (projection on the cylindrical surface coaxial with the rotor axis), CFX Post: (**a**) Constant profile on the direction of rotation; (**b**) Variable profile on the direction of rotation.

In order to optimise the working parameters of the helicoid multi-blade turbine, a number of computer simulations have been carried out by utilising CFX5.7 software. The solutions of these equations satisfy the boundary conditions on the propellers of the turbine rotor and of the fluid unperturbed domains at big distances.

The determination of these conditions, in a general case, presents certain difficulties related to the constructive form, and to the functioning regimes of the helicoid turbine. Therefore their definition was referred to for certain optimum functioning conditions, for which the flow speed values of the rotor and the speed induced on the propeller blade are already known. Thus, by integrating the given system of equations (transcribed for the examined helicoid turbine), direct calculation of aerodynamic specifications has been done. On the basis of the obtained results, a computer model of the helicoid rotor with four inceptions and aerodynamic profile of the blades in normal section has been elaborated (Figure 11). Later, the model was simulated in the CFD environment by varying the geometrical and kinetostatic parameters of the fluid. As the working element of the micro-hydro power station represents a complex body, it was necessary to carry out spatial interaction simulation (3D) of the working element with the fluid.

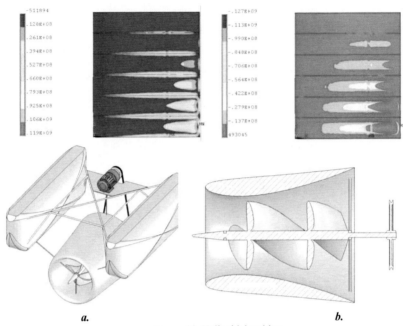

a.   b.

Figure 10. Helicoidal turbine.

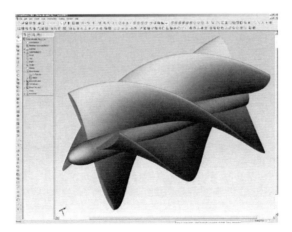

Figure 11. Helicoidal profile.

According to theoretical research, and to the justification of the functional parameters of the helicoid turbine, dynamic models have been designed by varying functional parameters. Below, the dynamic models designed in the SolidWorks environment are shown by varying the geometrical parameters (number of inceptions – 4, (Figure 11); propeller angle; constant propeller pitch (Figure 11) and variable propeller pitch (Figure 12a); with anterior and posterior straight abuts of the propeller (Figure 12a) and with countersunk abuts (Figure 12), etc.). In order to carry out research on the interaction of the helicoid rotor with the fluid environment by using CRF5.7 software the blade profiles have been discretised into a net of finite elements (Figure 12b).

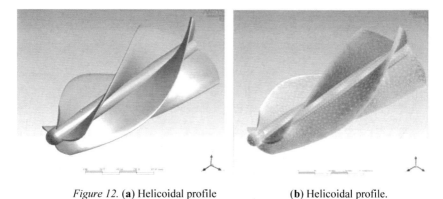

Figure 12. (**a**) Helicoidal profile   (**b**) Helicoidal profile.

Conceptual Design of the Electrical Micro-Hydro-Power Station for the Conversion of Flowing Water Kinetic Energy into Mechanical and Electrical Energy

To avoid the construction of dams, the kinetic energy of rivers can be utilised by means of exploiting water stream turbines. This type of turbine is easily mounted, is simple in operation, and its maintenance cost is suitable. The 1 m/s current velocity represents an energetic density of 500 W/m^2 of the crossing section, but only a part of this energy can be drawn off and converted into useful electrical or mechanical energy. This depends on the type of rotor and blades. Velocity is especially important as a double increase in the water velocity can result in an increase (8×) of energetic density. The Prut River has a section equivalent to 60 m^2 and an average velocity in explorable zones of (1–1.3) m/s, which is equivalent to an approximate theoretical energy of (30–65) kW. Taking into account the fact that a turbine can occupy only a portion of the river bed the generated energy might be much smaller. There are various conceptual solutions, but the issue of increasing the conversion efficiency of water kinetic energy is in the view of further work. The analysis of constructive versions of floatable micro-hydro power stations previously examined was not satisfied from the point of view of conversion efficiency of water kinetic energy. In a classical hydraulic wheel horizontal axle (Figure 13a) the maximum depth at which one of blades is sunk makes approximately 2/3 of the blade height *h*. Namely, only this area participates in the transformation of water kinetic energy into mechanical one. As well, the prior blade covers approximately 2/3 of the blade surface sunk utmost in the water *(h″≈ 2/3h′)*. This fact significantly the water stream pressure significantly on the blade. The blade that comes next to the blade that is sunk maximally into water is covered completely by it and practically does not participate in the conversion of water kinetic energy. Therefore, the efficiency of such hydraulic wheels is small.

The work leading to the elaboration and patenting of some advanced technical solutions for floatable micro-hydro power stations, based on the hydrodynamic effect, generated by the hydrodynamic profile of blades, and their orientation at optimum positions (concerning the water streams) with the account of energy conversion in each phase of the turbine rotor rotation is shown in Figure 13a [1]. It was necessary to perform a large volume of multi-criteria theoretical research concerning the selection of optimum hydrodynamic profile of the blades, and the design of the orientation mechanism towards the water streams.

The basic advantages of these types of micro-hydro power stations are as follows:

- Small impact on the environment
- Civil constructions do not have to be carried out
- The river does not change its natural course
- Includes the possibility of utilizing local knowledge in order to produce floatable turbines.

Another important advantage is the fact that along the river's course it is possible to mount a series of micro-hydro stations at small distances (approximately 30–50 m) because the influence of turbulence provoked by the adjacent installations can be excluded.

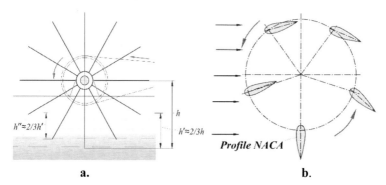

Figure 13. (a) Conceptual scheme of clasic multiblade rotor; (b) Elaborated conceptual multiblade rotor with hydrodynamic profile.

The results of the research carried out by the current authors, concerns the water flow rate in the location selected for the micro-hydro power stations mounting, the geological prospecting of the river banks in the place of anchoring foundation mounting, the energetic needs of the consuming potential, represent initial data for the conceptual design of the micro-hydro power stations and its working element.

Conceptual design of the micro-hydro power station constructions with a hydrodynamic profile of blades was carried out on the basis of three conceptual schemes (Figure 13b):

- micro-hydro power station with pintle and blades mounted on vertical axles anchored by the metallic structure
- floatable micro-hydro power stations with pintle and blades mounted on vertical axles
- floatable micro-hydro power stations with horizontal spindle and blades mounted on horizontal axles.

Aiming at an increase of the conversion coefficient of the water kinetic energy (known as the Betz coefficient), a number of structural diagrams of floatable micro-hydro power plants have been designed and patented [1–4]. They comprise a rotor with pintle and vertical blades and a hydrodynamic profile in normal section. The blades are interconnected by an orientation mechanism towards the direction of the water streams. The motion of rotation of the rotor with pintle is multiplied by a mechanical transmission system and is transmitted to an electrical generator or to a hydraulic pump. The mentioned knots are fixed on a platform, mounted on floatable bodies. The platform is linked to the bank by a hinged metallic truss and by straining cables.

An important aspect in the functional optimization of micro-hydro power plants is the selection of optimum hydrodynamic profile of the blades which allows increasing the conversion coefficient (Betz coefficient). Due to the hydrodynamic upward forces the increase in the conversion level is reached by means of ensuring the optimum position of the blade towards the water streams in various phases of rotor rotation by utilizing blades orientation mechanism. Thus, practically all blades (even those which move opposite the water streams) participate simultaneously in the generation of summary torque moment. The blades which move along the water streams utilize both hydrodynamic forces and water pressure exercised on blade surfaces for the generation of the torque moment. The blades which move opposite the water streams utilize only hydrodynamic upward forces for the generation of the torque moment. Due to the fact that the relative velocity of the blades toward water streams at their motion opposite water streams is practically twice as large, the hydrodynamic upward force is relatively big and the generated torque moment is measurable to the one generated by the water pressure. This effect forms the basis of all patented technical solutions.

- In the process of designing industrial prototypes of micro-hydro power plants for the conversion of river water kinetic energy, the following criteria and requirements have been taken into consideration:

- the elimination of dam constructions and, implicitly, of the negative impact on the environment;
- minimal costs;
- construction simplicity and operation;
- high reliability at dynamic overstressing caused by operating conditions;
- utilization of resistant composite materials including increased humidity conditions; and
- automatic control of micro-hydro power plant platform position at water level variation.

The results of the research undertaken on the water flow rate in the place chosen for the micro-hydro-power station mounting, geological prospecting of the river banks in the place of the anchoring foundation and potential consumers needs represent initial data for the conceptual elaboration of the micro-hydro-power station and the working element.

Conceptual elaboration of the micro-hydro-power station construction has been done on the basis of three conceptual designs:

- micro-hydro-power station with pintle and blades fixed on horizontal axles;
- micro-hydro-power station with pintle and blades fixed on vertical axles; and
- floatable micro-hydro-power station with horizontal axle and helical turbines.

THEORETICAL RESEARCH AND ELABORATION OF ROTOR WITH BLADES WITH NACA AERODYNAMIC PROFILE

On the basis of preliminary research the construction of two rotors, with 5 and 3 blades (Figure 14) with NACA profiles, was elaborated. The blades are oriented at a setting angle α, which is variable concerning the action line of the flowing water speed vector. Theoretical research reduced the optimization of construction parameters of blades with various symmetrical NACA profiles (0012, 0014, 0016, 0018, 63012, 63015, 63018, 66015, 66018, 67015 – 32 profiles were researched in total), with account of the maximal moment of torsion of the rotor shaft.

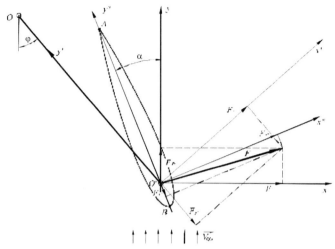

Figure 14. NACA orientation and angles.

Numerical Modelling of the Hydrodynamic Profile Blades and Derivation of the Optimal Geometric Characteristics

Consider a symmetric profile of the blade in a fluid flow with uniform velocity $\vec{V}_\infty$ (Figure 14). Points A and B correspond to the trailing and the leading edges, respectively. In the fixing point O' of the symmetric blade with the boom OO' three coordinate systems are considered, namely: the $O'xy$ system with axis $O'y$ oriented in the direction of velocity vector $\vec{V}_\infty$, and axis $O'x$ normal to this direction; the $O'x'y'$ system with axis $O'y'$ oriented along the boom direction OO', and axis $O'x'$ normal to this direction, and finally the $O'x''y''$ system with axis $O'x''$ oriented along the profile's chord toward the trailing edge and axis $O'y''$ normal to this direction. The angle of attack α is the angle between the profile chord AB and $\vec{V}_\infty$, and the positioning angle φ is the angle between the boom $O'O$ and $\vec{V}_\infty$. The hydrodynamic force $\vec{F}$ has its components in directions $O'x$ and $O'y$, named lift and drag forces, respectively, given by:

$$F_L = \frac{1}{2}C_L\rho_\infty V_\infty^2 S_p, \quad F_D = \frac{1}{2}C_D\rho_\infty V_\infty^2 S_p, \tag{14}$$

while the pitching moment is given by:

$$M = \frac{1}{2}C_M\rho V_\infty^2 c S_p,$$

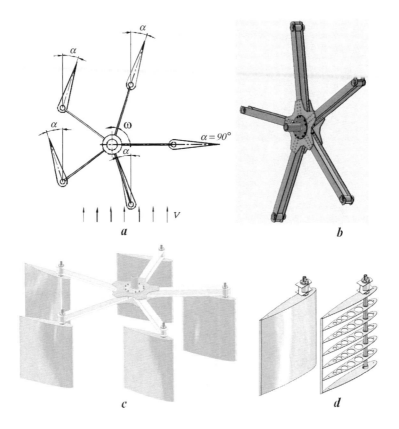

Figure 15. Various aerodynamic shape of the blade's profile.

where ρ_∞ is the fluid density, V_∞ is the flow velocity, $S_p = ch$ (c is the chord length, h is the blade height) representing the lateral surface area of the blade, and C_L and C_D are the dimensionless hydrodynamic coefficients, named lift coefficient and drag coefficient. The hydrodynamic coefficients, C_L, C_D and C_M are dependent on the angle of attack α, the Reynolds Re number and the aerodynamic shape of the blade's profile (Figures 15a, b).

An inviscid – boundary layer method is used to perform an analysis of C_L, C_D and C_M. A high order panel method (linear distribution of sources and vortexes) was used to compute the velocity distribution along the surface of the blade's profile. Lift and moment coefficients were computed from it. In order to compute the drag coefficient for a given angle of attack the viscous boundary layer analysis is performed. Using velocity distribution provided by the panel method, an integral boundary layer method was implemented. For the laminar part a two equation formulation was used, for the turbulent part Head's model was used. The drag coefficient was then computed with the Squire-Young formula. The computations of the hydrodynamic coefficients were performed in Matlab using 120 panels for velocity distribution and 250 grid points for the boundary layer analysis. After computing the hydrodynamic coefficients for a rack profile standard, and, in particular, a NACA profile with chord $c = 1.3$ m. By applying the computation methods described above in order to calculate the coefficients corresponding to NACA 0016 profile with length chord $c_{ref} = 1$ m: $C_{L,ref}, C_{M,ref}$ and $C_{D,ref}$. The coefficients that correspond to profile with length chord 1.3 m were computed from the relations;

$$C_L = C_{L,ref} \cdot 1.3, \; C_M = C_{M,ref} \cdot (1.3)^2, \; C_D = C_{D,ref} \cdot 1.3. \qquad (15)$$

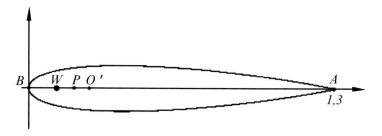

Figure 16. Blade profile.

Due to the fact that the hydrodynamic force does not have the application point in the origin of the blade axis system O', the pitching moment is produced. This moment is determined with respect to a certain reference point. As a reference point, point P is situated at a distance ¼ of the chord from the leading edge B (Figure 16) coordinate system $O'x''y''$ are given by:

$$F_{x''} = F_L \cos \alpha + F_D \sin \alpha,$$
$$F_{y''} = -F_L \sin \alpha + F_D \cos \alpha. \qquad (16)$$

By using the values of F_L and F_D obtained from previous computations, $F_{x''} = 1601.2$ N and $F_{y''} = -413.8$ N.

Therefore $|O'P| = |M|/|F_{x''}| = 0.0249$ m ≈ 25 mm.

In order to assure the stability of the blade's movement, the fixing point W should be chosen such that 25 mm $\leq |O'W| \leq H$, where $H_{min} \leq H \leq H_{max}$. The values of H_{min} and H_{max} should be derived in the future work such that the friction force in the cinematic coupling of the orientation mechanism is minimal.

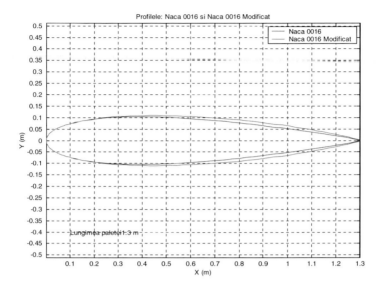

Figure 17. Blade profile – A computational representation.

In order to maximize the torsion moment developed by the micro-hydro power plant, the shape optimization of the blade's profile was carried. The blade's profiles were chosen from the NACA 4- and 5-digits library with a shape expressed as a function of three parameters: maximum thickness, maximum camber, and maximum camber location. As a shape parameter only maximum thickness is considered. Due to the use of symmetric profiles maximum camber shape parameter is taken to be zero, while the maximum camber location is arbitrary. The angle of attack is considered to be the second parameter. The goal of the shape optimization is to maximize the lift force, while ensuring the pitching moment and drag coefficient is not too large.

The following design optimization problem is considered:

Maximize $C_L = C_L(\theta, \alpha)$ subject to bounds on C_D and C_M,

where θ is the maximum thickness and α is the angle of attack. The values of the bounds are derived as follows: the maximum negative value for the pitching coefficient is chosen to correspond to the solution at zero angle of attack. The maximum value for the drag coefficient is chosen to correspond to the solution at angle of attack $\alpha = 18°$. Bounds on the parameters themselves were also added, so that the optimization was performed in the space of reasonable profiles: $10\% \leq \theta \leq 20\%$ and $0° \leq \alpha \leq 20°$. In order to find the optimal values of a given function $f = f(x_1, \ldots, x_n)$ the variable metric iterative methods can be used.

While given precision is not attained:

Solve $B_i s_i = -\nabla f(x_i)$

$$x_{i+1} = x_i + \alpha_i s_i$$

End do

where α_i are step multipliers and B_i are positive definite approximations to the Hessian of f. The derivative of f with respect to the ith component can be approximated by the central difference formula:

$$\frac{\partial f}{\partial x_i}(x) = \frac{f(x + he_i) - f(x - he_i)}{2h}, \text{ where } e_i \text{ is the } i\text{th basis vector.}$$

The shape optimization is performed within the Matlab optimization toolbox: a Sequential Quadratic Programming algorithm with a line search and a BFGS Hessian update. The quadratic subproblems were solved with a modified projection method. The gradients of $C_L = C_L(\theta, \alpha)$ were computed with central difference formulas with a constant step size h = 1e−04. As initial point for the optimization the symmetric NACA 0016 profile was used and considered at angle of attack $\alpha = 18^o$. The initial and optimal profile shapes are shown in Figure 17. About 30 iterations were needed in optimization subprogram to achieve the suitable convergence.

Numerical Modelling of the Fluid Flow Action on the Rotor Blades and the Establishment of the Optimal Position of the Blades in Order to Minimize the Energy Losses

In order to establish the optimal position of the blades we compared the torsion moment developed by one blade, and the total torsion moment developed by all blades for different angle of attacks. The results are presented in Figure 18.

It can be seen that the optimal angle of attack is $17^\circ \leq \alpha \leq 18^\circ$, therefore the torsion moment is stable with respect to angle of attack. All computations were performed for the optimized profile NACA 0016M for the flow velocity 1 m/s.

The performance of the rotor with 3, 4 and 5 blades was analyzed, in order to choose the most suitable rotor configuration. The total torsion moments developed by all blades for these configurations are presented in Figure 19. As future work , the turbulence in the rotor area should be studied by carrying out a number of computer simulations in CFX5.7 software. In other words, the spatial interaction between the rotor and fluid will be investigated for various functional parameters in order to increase the efficiency of the turbine and decrease the energy loss.

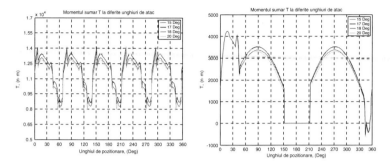

Figure 18. Computational results for torsion momentum.

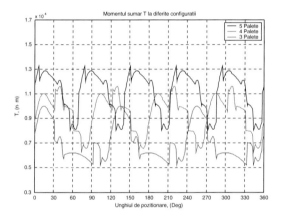

Figure 19. Torsion momentum for all blades configurations.

Moment of Torsion and Power Applied to the Rotor with Hydro-Dynamic Profile Blades

Next the hydro-dynamic coefficients for a rack profile standard are calculated, and, in particular, a NACA profile with chord $c = 1.3\,\text{m}$. By applying the calculation methods described previously, the coefficients corresponding to NACA 0016 profile with length chord $c_{ref} = 1\,\text{m}$: $C_{L,ref}, C_{M,ref}$ and $C_{D,ref}$ provided by formulas (18), (19) and (33), are computed, respectively. The coefficients that correspond to profile with length chord $1.3\,\text{m}$ are calculated from the relations:

$$C_L = C_{L,ref} \cdot 1.3, \ C_M = C_{M,ref} \cdot (1.3)^2, \ C_D = C_{D,ref} \cdot 1.3. \tag{17}$$

Figure 20a shows the lifting power and drag coefficients dependent on the entering angle. Figure 20b shows the hydro-dynamic power modulus $\vec{F}$, which actuates over the blade, and its tangential and normal components $F_{x'}, F_{y'}$ dependent on the angle of positioning. Figure 21a shows the moment T_{ri} developed by a blade dependent on the angle of positioning calculated by formula (4), and Figure 21b represents the summary moment $T_{r\Sigma}$ dependent on the positioning angle calculated by formula (14).

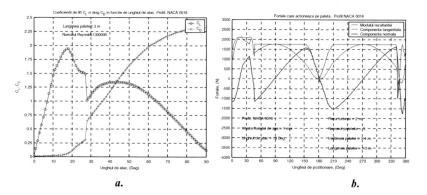

a.   b.

Figure 20. Lifting power and drag coefficients.

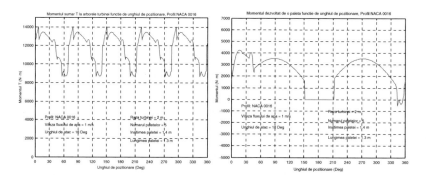

Figure 21. Summary moment $T_{r\Sigma}$

Minimization of the Turbulence Effects on the Technical Characteristics of the Multiblade Rotor by Using the Constructive Optimization of the Blades

An inviscid–boundary layer method was used to perform an analysis of hydro-dynamic coefficients. A high order panel method (linear distribution of sources and vortexes) was used to compute the velocity distribution along the surface of the blade's profile. Lift and moment coefficients were computed from it. In order to compute the drag coefficient for a given angle of attack the viscous boundary layer analysis was performed. Using velocity distribution provided by the panel method, an integral boundary layer method was implemented. For the laminar part a two equation formulation was used, for the turbulent part Head's model was used. The transition from the laminar flow to the turbulent flow was localized using Michael's criterion. The drag coefficient was then computed with the Squire-Young formula.

Since the software does not handle the turbulent separation well, Java Foil software developed by Martin Hepperle was used in order to study the turbulence regions. Figure 22 contains the velocity field around the profile NACA0016 at an angle of attack 18°.

Figure 22. The velocity field at the angle of attack 18°.

The Reynolds number is:

$$\mathrm{Re} = \frac{\rho c \vec{V}}{\eta} = \frac{c \vec{V}}{v}$$

where the density of the fluid $\rho = 998.4 \ \mathrm{kg/m^3}$ at $20^{\circ} \mathrm{C}$, the kinematic viscosity $v = 1.012 \cdot 10^{-6} \ \mathrm{m^2/s}$ and the chord's length $c = 1.3 \ \mathrm{m}$.

For flow velocities $\vec{V} = 1$ and 2 m/s we have the following values for Reynolds numbers Re = 1284600, 1798400, 2312300. Figure 23 contains the turbulence flow separation on the lower/upper surfaces of the profile. T.U. and T.L. stand for transition point from laminar to turbulent on the upper and lower

surface, respectively. S.U. and S.L. stand for separation point on the upper and
lower surface, respectively.

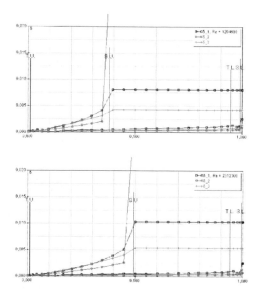

Figure 23. Transition points and separation points at flow velocities 1 and 2 m/s.

It can be seen that in all cases the transition from the laminar flow to
turbulence, the transition will occur on the upper surface near the stagnation point
and the flow separation is predicted to happen at 40–50% of the chord's length.
On the lower surface, the transition from laminar to turbulent as well as the
separation will happen near the trailing edge.

Numerical Modelling of the Interaction Between Fluid Flow and Working Elements

Using ANSYS 8.0 software the deformations and tensions that will appear in the
blade immersed in the water stream were studied. The hydrostatic pressure and the
resultant hydrodynamic force (corresponding to the flow velocity of 2 m/s) acting
on the blade are taken into account. Various thicknesses of the blade are considered. According to the analysis of the obtained results the following resistance
structure of the blade shown in the Figure 24.

In order to perform the deformation and stress analysis, the ANSYS 8.0
software was used. The hydrodynamic blade was considered under the action of
the hydrostatic pressure and the hydrodynamic force (lift and drag components)
corresponding to the flow velocity of 2 m/sec. Thus the maximal value of the
force acting on the blade is approximately 11 kN. Discretization was done with
shell-type elements and the mesh was locally refined. The material used was the

aluminium alloy H37 with $E = 1.97 \cdot 10^{11}$, $\nu = 0.27$. Several design parameters and configurations of resistance structure have been considered. First the scheme with three slides of resistance structure was considered and is presented in Figure 24.

The side walls thickness was taken to be (a) 1 mm and (b) 1.5 mm. After performing the finite element analysis the maximal displacement of approximately 8 mm (for thickness 1.5 mm) was obtained, which has a negative effect on the functional performance of the blade, therefore the proposed scheme is not admissible. Then, the scheme from Figure 25 with 3 slides was considered. Similar to the previous case, two thicknesses were analyzed, 1 mm and 1.5 mm. Again the maximal displacement was not acceptable, even for thickness 1.5 mm. Finally, the constructive scheme presented in Figure 26 was considered with 5 slides and side walls thickness of 1 mm and 1.5 mm. The displacement values of the deformed shape are also presented in Figure 26. The maximal value of the displacement in this case was of approximately 3 mm. Still, this value was not considered acceptable, since the deformation of the blade's profile will lead to negative effects on the functional performance of the turbine. Therefore, a new constructive scheme is proposed, which will be a modification of the scheme presented in Figure 25, based on the utilization of composite materials.

Two variants of the resistance structure of the blade were examined: with 3 and 5 cross stiffenings. The adaptive digitization in finite elements was made with *shell63* elements. Figures 25 and 26 present the deformation of the blade. By analysing the numerical state of the deformation (Figures 25 and 26) of the cover of the blade with a width of 1 mm with 3 and 5 cross stiffenings were established so that the deformation of the cover in the maximum downwarped areas is of 7.8 and 3.5 mm. These local profile shiftings can negatively influence the flow conditions of the fluid in the zone adjacent to the hydrodynamic profile, but implicate the efficiency of the conversion of the kinetic energy of the water flow in profitable energy. That is the reason why the profiles with a width of the cover of 1 mm were abandoned. At the same time it was ascertained that the shiftings of the cover with a width of 1.5 mm at the blades with 3 and 5 transversal plates was diminished by 2.1 and 2.6 times, now were representing 3.7 and 1.3 mm. The blades that have the width of the cover of 1.5 mm and 5 cross stiffenings correspond to a maximum deformation of 1.3 mm, which is acceptable from the **perspective** of minimizing the negative impact on the efficiency of the conversion of the kinetic energy of the water flow in profitable energy. This is the reason why only the resistance structure with 5 cross stiffenings and cover with the width of 1.5 mm made out aluminium alloy H37 will be examined.

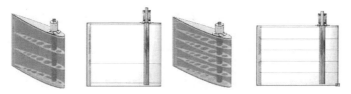

Figure 24. Resistance structure of the blade.

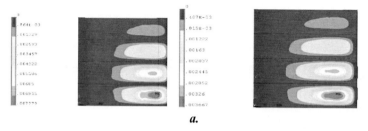

a.

Figure 25. The shiftings of the cover of the blade with hydrodynamic profile NACA 0016 (mm) with 3 cross stiffenings with the width of the cover S = 1 mm (**a**) and S = 1.5 mm (**b**).

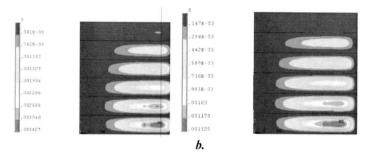

b.

Figure 26. The shiftings of the cover of the blade with hydrodynamic profile NACA 0016 (mm) with 5 cross stiffenings with the width of the cover S = 1 mm (**a**) and S = 1.5 mm (**b**).

Increasing the number of the cross stiffenings or the width of the cover will reflect a higher cost and a greater weight of the blades. The efficiency of the conversion resulting from the hydrodynamic forces applied to the blade depends on respecting the geometrical form established by the project not only by the length of the cord but also by the height of the blade. That is why it is necessary to evaluate the values of the shiftings on the length of the cord and the height of the blade in the areas that have maximum stress from hydrostatic pressure and hydrodynamic forces. Figure 27a presents the cover shiftings (mm) on the length of the cord in section A, situated in the middle from 2 cross stiffenings. According to Figure 27a, the maximum shiftings are situated at the mark of 235 mm and represent 1.3 mm. The placement of the shifting field corresponds to the portion of the cover with the maximum radius of curvature of the profile in section A. The area with shiftings greater than 1 mm is extended between the marks of 130 and 530 mm. This leads to modification of the angle of adjustment with ±0.29° which is considered admissible from the hydrodynamic perspective. Figure 27b are presents the shiftings in section B (situated at the mark of 235 mm from the rear edge of the blade), depending on the marks on the blade height (0–250 mm) situated between the last transversal plates. Shiftings greater than 1 mm are situated between the marks of height of 78÷172 mm.

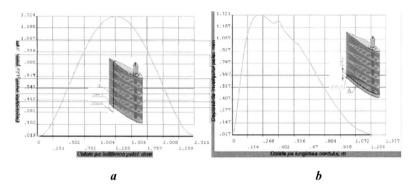

Figure 27. The shiftings in the cover (mm) along the length of the cord in section A (**a**) and in section B (**b**).

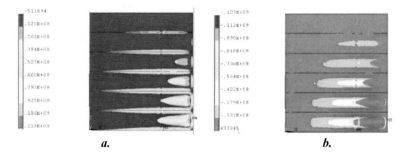

Figure 28. The main tensions (N/m²) σ_1 (**a**) and σ_3 (**b**) in the cover of the blade with the hydrodynamic profile NACA 0016 and 5 cross stiffenings with the width of the cover S = 1.5 mm.

For appreciating the state of tension of the cover of the blade with hydrodynamic profile NACA 0016 and width of 1.5 mm (Figure 28a, b) it is considered that the main tensions σ_1, σ_2 and σ_3, which are the proper values of the tension tensor arranged in decreasing order. Figure 28 presents the main tensions σ_1 (a) and σ_3 (b). Also the intensity of the tension is taken in consideration, calculated according to the formula:

$$\sigma_i = \frac{1}{\sqrt{2}}\left[\left(\sigma_1 - \sigma_2\right)^2 + \left(\sigma_2 - \sigma_3\right)^2 + \left(\sigma_1 - \sigma_3\right)^2\right]^{1/2}$$

and the Von Mises deformation, according to the formula:

$$\varepsilon_e = \frac{1}{1+\nu}\left(\frac{1}{2}\left(\left(\varepsilon_1 - \varepsilon_2\right)^2 + \left(\varepsilon_2 - \varepsilon_3\right)^2 + \left(\varepsilon_1 - \varepsilon_3\right)^2\right)\right)^{1/2},$$

where v is the Poisson coefficient, and $\varepsilon_1, \varepsilon_2$ and ε_3 are the main values of the tensor of deformation. Figure 29a presents the intensity of tension σ_I (N/m^2), and Figure 29b represents the Von Mises deformation of the cover of the blade in section A, with a maximum stress through the forces of hydrostatic and hydro-dynamic pressure. Due to the fact that the stress caused by the hydrodynamic forces is much smaller than the one caused by the hydrostatic pressure, a calculation was also taken for the restriction of the pressure of the hydrodynamic effect that is uniformly spread on the surface of the blade taken with its maximum value. From the analysis of the variation of the intensity of tension and implicit by the Von Mises deformations in the area situated at the mark of 402 mm on the length of the cord. This mark is situated at the border of passing of the profile NACA 0016 from the zone with a small radius of curvature (marks 1337–402 mm) to the zone with a greater radius of curvature (marks 42–0 mm). This behaviour of the tension intensity and Von Mises deformation must be taken into consideration for the layers from a composite material, giving a variable width to the cover of the blade in the zone adjacent to the mark 402 mm on the length of the cord.

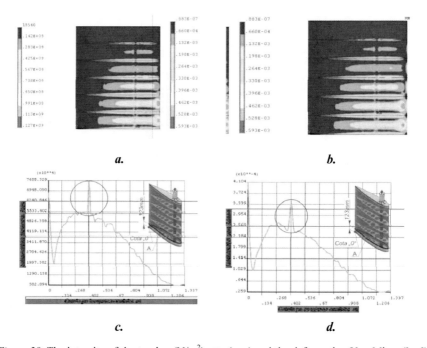

a.   *b.*

c.   *d.*

Figure 29. The intensity of the tension (N/m^2) σ_I (**a, c**) and the deformation Von Mises (**b, d**) in the cover of the blade in section A.

Assessment of the Energy Potential of the Multiblade Rotor with Vertical Axis and Hydrodynamic Profile Blades

Energetic potential of the water flow and the generated power at the rotor shaft with vertical axis with hydrodynamic profile blades is estimated for the following design parameters (Figure 30):

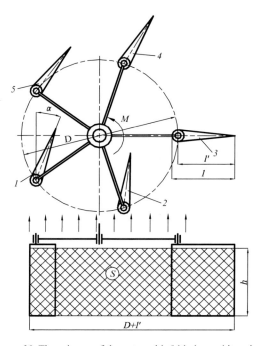

Figure 30. The scheme of the rotor with 5 blades and its orientation.

– rotor with 5 blades with hydrodynamic profile NACA 0016;
– effective embedment of the blade in water $h = 1.4$ m;
– length of the blade cord $l = 1.3$ m;
– effective length l' of the blade cord $l' = 1$ m;
– rotor diameter (diameter of the circle for blades vertical axis placement) $D = 4$ m;
– setting angle $\alpha = 18°$;
– setting angle of the 3rd blade positioned in the neutral zone $\alpha = 90°$.

The energetic potential of the water flow estimated in the rotor dimensions section:

$$P = \frac{1}{2}\rho \cdot V^3 \cdot S,$$

and the generated power by the water flow at the rotor shaft:

$$P_a = P \cdot K,$$

where: ρ – water density, kg/m^3;

V – water flow speed, *m/s*;

S – transversal surface of the water flow comprised in the rotor dimensions limits;

$S = H(D + l')$;

K – energy conversion efficiency.

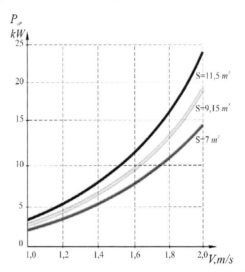

Figure 31. The generated power of the rotor.

In Figure 31 the dependence of the generated power P_a at the rotor axis depending on the water flow speed having the energy conversion efficiency 50% (K = 0.5) is presented.

The generated power P_a by the rotor driven by the water flow:

$$P_a = M_\Sigma \cdot \omega,$$

where: M_Σ – summary torque moment applied to the rotor shaft, Nm.

ω – angular speed of the rotor, sec^{-1}.

To induce the generated torque moment by each blade in particular and by rotor in general, the software centred and elaborated on the determination of the de fluid-blade interaction forces and of the contribution share of each blade to formation of the summary torque moment applied to the rotor shaft In Figure 32, the dependence of the summary torque moment M_Σ applied to the rotor shaft depending on its apex angle at various water flow speeds is presented.

$$P_a = M\Sigma\cdot\omega,$$

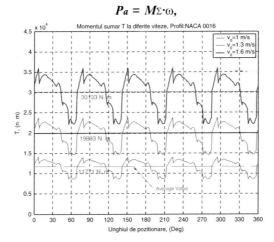

Figur 32. Summary torque moment versus apex angle at flow velocities 1, 1.3 and 1.6 m/s

For example, for the water flow speed, $V = 1.3$ m/s, $M_\Sigma = 19,893$ Nm.

In Figure 32, the power at the rotor shaft estimated from the energetic potential of the running water at the speed of 1.3 m/s, $P = 3.84$ kW.

The angular speed of the rotor will be $\omega = \dfrac{P}{M} = \dfrac{3,840}{19,893} = 0.1913$ sec^{-1},

and the frequency of rotation of the rotor is $n = \dfrac{30\omega}{\pi} = 1.843$ min^{-1}.

Elaboration of Micro-Hydro-Power Plant with Pintle and Hydro-Dynamic Profile Blades

ELABORATION OF PILOT-STATION OF THE MICRO-HYDRO-POWER PLANT WITH PINTLE AND HYDRO-DYNAMIC PROFILE BLADES

On the basis of the previous theoretical research and previous computer simulations the conceptual pilot-station of the bi-functional micro-hydro-power plant (Figure 33) was elaborated. The pilot-station will serve as site for experimental research on more technical solutions and working elements in natural conditions, but, in particular, research on efficient blades with hydro-dynamic profile. The pilot-station will be anchored to the bank with adjustable cables on its length via communicating bridge.

The rotor with hydro-dynamic profile blades, precessional multiplier, centrifugal pump, and electrical generator are fixed on the housing made from a metal framing. The framing, in their turn, are mounted on four landing pontoons.

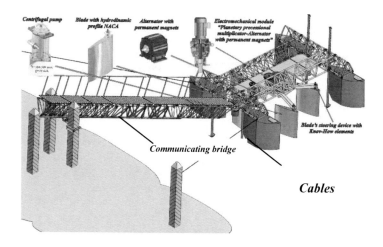

Figure 33. Hydroenergetic aggregate plant.

The communicating bridge and the mode of fixing to the bank by cables allow the pilot-station to float depending on the changing level of flowing water. The device for blades position control is not shown.

The concept of the micro-hydro-power station mounted on two pontoons is shown in Figure 33. This concept will be proposed for implementation and marketing.

The pilot plant of the micro-hydro-power station (Figure 33) includes the following fundamental aggregates:

- multiblade rotor;
- centrifugal pump of low speed;
- constant magnet generator of low speed; and
- a multiplier.

After the manufacturing of the centrifugal pump, constant magnet generator, and the multiplier, the experimental research was carried on the stands from the testing laboratories of the Technical University of Moldova and of the JSC "Hidrotehnica" industrial unit of Chişinău.

MANUFACTURING OF THE MICRO-HYDRO-POWER STATION AGGREGATES

Elaboration of Rotor's Blades Technology Fabrication

The basic working part that influences the efficiency of the conversion of the kinetic energy is the blade. In order to assure the efficient conversion of the rotor it is important to preserve the geometrical and shape parameters, as well as the structural resistance parameters of the blade during work in real conditions. Also, it is necessary to assure the low cost of the produced energy. Initially, a fabrication technology of the blades was proposed, consisting in a resistance structure coupled with the side envelope made of aluminium alloy H37 with wall thickness of 1 mm and 1.5 mm (Figure 34). The deformation analysis carried out has shown an unacceptable level of deformation. Thus, the functional performance of the blade will be reduced.

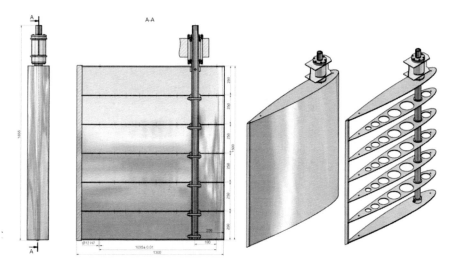

Figure 34. System of blades.

The goal of this research is to elaborate on a modern fabrication technology for the blades based on the composite materials. The study of modern technologies, allowed the researchers to elaborate a new fabrication technology of the rotor blades of the mini-hydro-power station with the usage of plastic materials armed with glass fibers.

The elaborated technology consists of the following stages:

- elaboration, design and execution of the mould for blades profile;
- elaboration, design and execution of the resistance structure of the blade;
- fabrication of the multi-layer side envelope made from plastic reinforced with fiberglass; and
- placement of the mould of the plastic envelope, resistance structure, and filling the inside with an expandable material (polyurethane).

Based on the theoretical research performed, the hydrodinamic profile of the blades was established. The resistance structure should be adapted to the fabrication technology. In order to do this, several versions of the structure were proposed and studied. Finally, the optimal (from the point of view of structural resistance and cost efficiency) version have been chosen: the division of hydrodynamic profile of the blade into two equal parts (symmetry), separate execution of the blades edge (Figure 35a). This configuration offers economy of the material and execution force while assuring the quality of fabrication and preservation of the geometrical parameters obtained earlier.

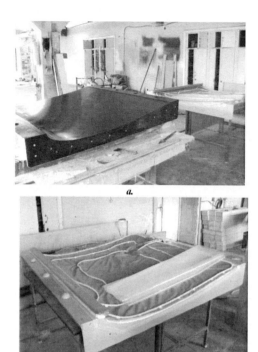

Figure 35. Hydrodynamic profile of the blades.

Materials used for the fabrication of the form mould for the blades envelope (Figure 36):

- MDF plates, thickness 30 mm, 16 mm, 4 mm;
- Glue Cleiberit D3;
- NOVAL APP;
- Polyurethane primer, polyurethane paint;

Polisher OSKAR'S M100, M50;

- Separation agent OSKAR'S W50;
- Meriguards MirorGlassm nr.88; and
- NorSlip 9600.

a   *b.*

Figure 36. Fabrication of form mould.

Materials used for the fabrication of blades envelope:

- Solidifier MEKP (methyl-exil-keton-pirexit);
- Luperox K1 standard, ATOFINA, France;
- Polyurethane resin CRYSTIC
- Gelcoot (white and black colours);
- Fiber glass Chapped Strand Mat M440 300/1250E, Chapped Strand Mat M440 450/1250E;
- Scrint Gobain Vetrotex;
- Woven Rowind;
- Spray UpRowind;
- Alluminium Hydroxide ATH; and
- Polyurethane glue IMFI, France.

The blades have been executed using the infusion through vacuum technology. In order to achieve this, the blade have been divided in three parts (Figure 35b).

1. Technological stages of the fabrication of the blades envelope:

1. Preparation of the mould;
2. Technological maintenance (4 layers, 6 h each);
3. Application of the polyvinyl material (Norslipp 9600);
4. Technological maintenance (15 min);
5. Application of the gelcout substance;
6. Technological maintenance (drying);
7. Fabrication of the material reinforced with fiberglass;
8. Installation in the mould:
   - 1st layer Woven Rowind;
   - 2nd layer Multimatelite 300/600/300;
   - 3rd layer Shaped Strand Mat M440/450;
   - 4th layer Phil ply.

9. Installation of the equipment for vacuum and infusion of the resin;
10. Pasting up of the mould, sealing;
11. Cutting out of the membrane (polyamide multilayer film);
12. Pasting up of the membrane;
13. Infusion of the resin;
14. Technological maintenance (until the polymerization of the resin, 2–3 h);
15. Technological maintenance for the elimination of the remainder stresses;
16. Extracting from the mould; and
17. Elimination of the surpluses.

The same technological process is used for the fabrication of the frontal cups of the blade.

Figure 37. Frontal cups of the blades.

2. Technological stages of the fabrication of the blades frontal cups of the blades:

1. Preparation of the mould;
2. Technological maintenance (4 layers, 6 h each);
3. Application of the polyvinyl material (Norslipp 9600);
4. Technological maintenance (15 min);
5. Application of the gelcout substance;
6. Technological maintenance (drying);
7. Fabrication of the material reinforced with fiberglass;
8. Installation in the mould:
9. Infusion of the resin;
10. Technological maintenance (until the polymerization of the resin, 2–3 h);
11. Technological maintenance for the elimination of the remainder stresses;
12. Extracting from the mould; and
13. Elimination of the surpluses.

3. Assembling of the blade

All components – two symmetric parts of the envelope of the blade, two front cups, the blades edge, resistance structure with the axis are assembled:

1. technological joints are cleaned;
2. technological holes are bored;
3. the inner space of the blade is filled with polyurethane;
4. technological holes are filled up; and
5. the surface of the blade is polished.

Figure 37 depicts the blade fabricated according to the proposed technology.

Carrying Out of Small-Scale Tests of the Micro-Hydro-Power Station Pumps

Investigation and optimisation of the manufactured pumps geometrical parameters (Figures 38, 39, and 40) were based on the assurance of the following functional parameters of the pump:

1st version:

- speed of rotation -500 min^{-1};
- productivity -40 m^3/h;
- pressure head of water -10 m.

2nd version:

- speed of rotation -300 min^{-1};
- productivity -30 m3/h;
- pressure head of water -10 m.

Experimental research on micro-hydro-power station pump was carried on a specialised stand at the JSC "Hidrotehnica" industrial unit, that allowed for: variation of speed of rotation in limits of $n = 300...600$ min^{-1}; determination of pumping productivity; and modelling of the pressure head of water.

The results of the experimental researches are presented in Figure 41 and represent the spectrums of the pressure head of water H, mechanical efficiency η, power P and of the current I by the productivity Q for the speed of rotation $n = 300$ min^{-1}; $n = 500$ min^{-1} and $n = 600$ min^{-1}. The analysis of the obtained results points out the fact that mechanical efficiency is higher at the pump with a speed of rotation of $n = 500$ min^{-1}. The pressure head of water of $H = 10$ m, necessary for pumping of water from Prut River in those three storage reservoirs from the village of Stoeneşti is assured at this speed of rotation. The pumping productivity $Q = 40$ m^3/h is satisfactory; this fact t permits the pumping of about 1000 m^3 of water in 24 h.

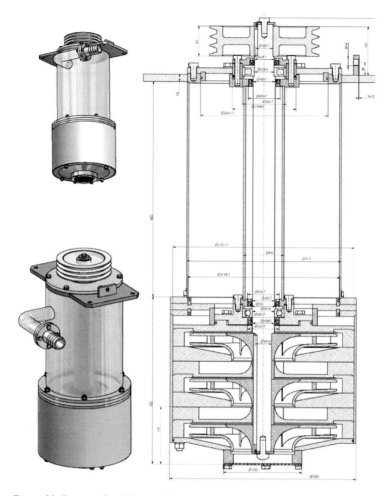

Figure 38. Computerized 3D model and Assembly design of the centrifugal pump.

A comparative analysis in the form of diagrams of those three essential parameters – H, η and P for a constant value of productivity $Q = 40 \text{ m}^3/\text{h}$ is presented in Figures 39 and 40, in order to explain better the technical performances of the tested pumps. The maximal pressure head of water, as was expected, is obtained at the highest speed of rotation – of $n = 600 \text{ min}^{-1}$. The maximal power is developed by the pump with the maximal speed of rotation. The results of the measurements of the pump basic qualitative values Q, H, P, I at the speed of rotation of $n = 300$, 500, 600 min^{-1} are presented in the graphs in Figure 41a–c.

Figure 39. Submersible centrifugal pump
Technical charasteristics:
Speed of rotation n = 300 min^{-1};
Mechanical efficiency of the pumps η = 0.7;
Productivity Q = 30 m^3/h.

Figure 40. Submersible centrifugal pump
Technical charasteristics:
Speed of rotation n = 500 min^{-1};
Mechanical efficiency of the pumps η = 0.7;
Productivity Q = 40 m^3/h.

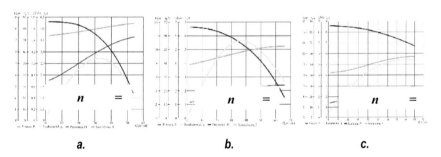

a.   b.   c.

Figure 41. Computational results.

The overall efficiency of the pump is calculated in the following relation:

$$\eta_p = {0.54}/{0.81 \times 0.95} = 0.7 \cdot$$

where 0.81· 0.95 represents the stand efficiency.

Also similar investigations have been carried out for the pump with the speed of rotation of 300 min^{-1}. As a result, the following parameters were obtained: pumping productivity – 30 m^3; pressure head of water – 10 m. Both pumps will be installed on the pilot plant of the micro-hydro-power station and will be tested in real conditions.

On the basis of the experimental research outcomes, the following conclusions can be formulated: at the rotational frequency of the micro-hydro-power station rotor of $n \approx 1.75$ min^{-1}, multiplied to $n \approx 500$ min^{-1} at the hydraulic pump shaft, the functional parameters, obtained within the experimental researches, are satisfactory.

Fabrication of the Pilot Micro-Hydro-Power Station Binding Mechanism to Link with the Bank

As a basis for the elaboration of the pilot micro-hydro-power station binding mechanism to link with the river bank, the condition that the platform should be self-adjustable concerning the water level, which is usually variable during the year, was stipulated. For the purpose of simulating the functioning of the binding mechanism according to water level fluctuations, a 3D model of the binding mechanism was built (Figure 42). It is composed of a metallic structure that is joined to the platform and to the foundation on the bank (the joint has two grades of liberty). The majority of components are made of metal (Figures 43 and 44). The model is shown in the final stage. Moreover, the platform is also connected to the bank by means of two cables with adjustable lengths. These cables have already been purchased.

On the Prut River near the village of Stoieneşti, the foundation has been prepared for the mounting of micro-hydro-power station (Figure 45).

Figure 42. 3D model of the binding mechanism.

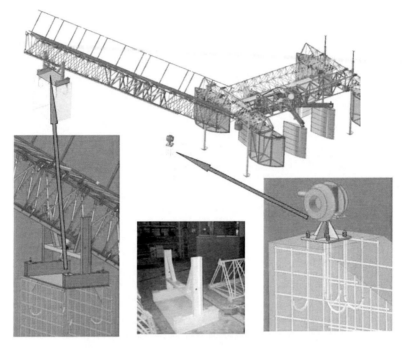

Figure 43. Micro-hydro power station schematic.

Figure 44. Micro-hydro-power-station.

Figure 45. People at work.

Elaboration of Industrial Prototype of the Micro-Hydro-Power Plant with Pintle and Hydro-Dynamic Profile Blades

The concept of industrial prototype of the micro-hydro-power station mounted on two pontoons is shown in Figure 46. This is concept that will be proposed for implementation and marketing. On the basis of the kinematic diagram (Figure 46), two conceptual industrial prototypes of the micro-hydro-power plant have been elaborated. Its aggregated utilization is described above. The rotor with hydro-dynamic profile blades, precessional multiplier, centrifugal pump, and electrical generator with permanent magnets are fixed on the housing made from a metal frame. The frame, in their turn, is mounted on four landing pontoons. The communicating bridge and the mode of fixing to the bank by cables allows the micro-hydrostation to float depending on the fluctuating levels of flowing water. The device for the blade position control is not shown.

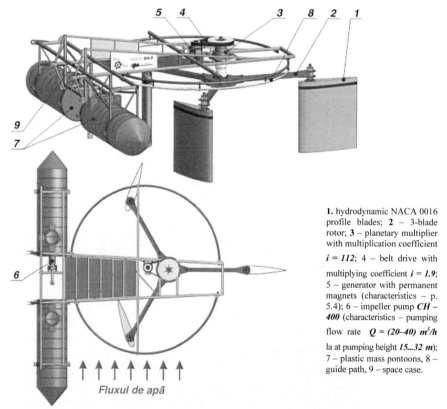

1. hydrodynamic NACA 0016 profile blades; **2** – 3-blade rotor; **3** – planetary multiplier with multiplication coefficient $i = 112$; 4 – belt drive with multiplying coefficient $i = 1.9$; 5 – generator with permanent magnets (characteristics – p. 5.4); 6 – impeller pump $CH – 400$ (characteristics – pumping flow rate $Q = (20\text{–}40)\ m^3/h$ la at pumping height $15...32\ m$); 7 – plastic mass pontoons, 8 – guide path, 9 – space case.

Fluxul de apă

Figure 46. Micro-hydro power plant with hydrodynamic rotor river water kinetic energy conversion into electrical and mechanical energy (rotor diameter $D = 4\ m$, water-submersed blade height $h = 1.4\ m$, length of the blade cord $l = 1.3\ m$) (MHCF D4×1.5 ME).

The adopted technical solutions have resulted in an ample theoretical and experimental research carried out at the Centre for Renewable Energy Conversion Systems Design, Department of the Theory of Mechanisms and Machine Parts. To justify the constructive and functional parameters, supplementary digital modelling and simulation have been carried out using ANSYS CFX5.7 software. Sub-programs developed by authors for the MathCAD, AutoDesk MotionInventor, etc. software, were used, namely simulation of the interaction "flow-blade" of the floatable steadiness and also the optimization of blades hydrodynamic profile, with the purpose to increase the river water's kinetic energy conversion efficiency for different velocities by using 3 and 5 blade rotors. In the process of micro-hydro power plants design, the experience gained at research-design-manufacturing of the pilot plant was utilized.

The efficiency of micro-hydro power plant operation by private consumers for special purposes depends on the right selection of the micro-hydro power plant constructive configuration and of the functional characteristics of the component aggregates participating in the process of flowing water kinetic energy conversion into useful energy.

In order to satisfy the objectives and consumers demand for micro-hydro power plants, and also for the increase in the flowing water kinetic potential conversion efficiency in certain zones of the river, the authors have designed the following constructive and functional concepts based on modular assembling:

- micro-hydro power plant with hydrodynamic rotor for river water kinetic energy conversion into mechanical energy – for water pumping (MHCF D4×1.5 M);
- micro-hydro power plant with hydrodynamic rotor for river water kinetic energy
- conversion into electrical and mechanical energy (MHCF D4×1.5 ME);
- micro-hydro power plant with hydrodynamic rotor for river water kinetic energy conversion into mechanical energy at small rotations (MHCF D4×1.5 ME);
- micro-hydro power plant with hydrodynamic rotor for river water kinetic energy conversion into electrical energy (MHCF D4×1.5 E).

The micro-hydro power plants, conceived as modular ones, allow the modification of destination and functional characteristics by replacing certain aggregates with others (generator, pump, blades with different hydrodynamic profile, 3–5 blades rotor).

Micro-hydro power plants have similar resistance structure as constructions calculated from the point of view of resistance and rigidity for dynamic demands. Floatability and maintenance of the perpendicularity of micro-hydro power plant rotor spindles for a variable river water level are ensured by technical solutions protected by patents. The instant orientation mechanism of blades for a constant entering angle concerning the direction of the water flow represents Know-How and is not described. The main working element on which the quantity of kinetic energy converted into useful energy depends on is the blade with a hydro-dynamic profile of NACA 0016, developed on the basis of the digital modelling that was performed. Two types of rotors with 3 and 5 blades have been designed for the mentioned micro-hydro power plants. The installed capacity of micro-hydro power plants with diameter $D = 4$ m, water-submersed blade height $h = 1.4$ m and the length of the blade cord $l = 1.3$ m for water flowing velocity $V = 1...2$ m/s can be within $P = 2...19$ kW.

The micro-hydro power plant with a constructive configuration of MHCF D4×1.5 ME for river water kinetic energy conversion into electrical and mechanical energy is poly-functional and can be used for street illumination, heating, water pumping for irrigation by weeping, and for drainage of agricultural areas adjacent to rivers.

The assembling of blades 1 with a NACA 0016 profile in hydrodynamic rotor 2 and its mounting on the inlet shaft of the multiplier 3 are done in the same manner as for micro-hydro power plant MHCF D4×1.5 M. The kinematics and constructive peculiarities of MHCF D4×1.5 ME plant are the following: rotation motion of hydrodynamic rotor 2 (Figure 47a) with angular speed ω_1, by means of multiplier 3 and of belt drive 4 having an effective multiplying coefficient $i = 212.8$, is being multiplied up to angular working speed of the generator with permanent magnets with small rotations 5:

$$\omega_3 = \omega_1 \cdot i_1 \ (s^{-1}).$$

Torque moment T_3, applied to rotor 5, is:

$$T_3 = \frac{T_1 \cdot \eta_1 \cdot \eta_2 \eta_r}{i}, (Nm),$$

where: η_1 is the mechanical efficiency of the multiplier ($\eta_1 = 0.9$);

η_2 – mechanical efficiency of the belt drive ($\eta_1 = 0.95$);

η_r – mechanical efficiency of the hydrodynamic rotor bearings ($\eta_1 = 0.99$).

i – effective multiplication coefficient equal to the composition of multiplying ratios of the planetary multiplier and of the belt drive. Diagrams of the summary torque moment T_1 at the hydrodynamic rotor shaft with blades for different water flow velocities is shown in Figure 47.

The electric energy produced by the generator with permanent magnets 5 (Figure 47b) can be used both for private consumer needs of power and for supplying electricity to impeller pump 6 (CH 400), for water pumping into irrigation systems by means of weeping or drainage of agricultural areas adjacent to the rivers (by relocation of the impeller pump 6). In the case of electric energy production, the energy utilization efficiency with account of mechanical losses in the kinematics chain of the micro-hydro power plant and in the generator with permanent magnets makes up (at the generator terminal): $\eta_\Sigma = \eta_1 \eta_2 \eta_r \eta_g = 0.9 \times 0.95 \times 0.99 \times 0.87 = 0.736$ and in the case of water pumping (at the shaft of the impeller pump):

$$\eta_\Sigma = \eta_1\eta_2\eta_r\eta_g\eta_{me} = 0.9 \times 0.95 \times 0.99 \times 0.87 \times 0.91 = 0.67 \text{ where: } \eta_g \text{ is generator}$$
efficiency and η_{me} – efficiency of the hydraulic pump of the electric motor.

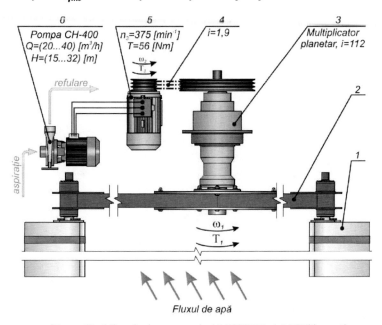

Figure 47a. Micro-hydro power plant MHCF D4×1.5 ME kinematics.

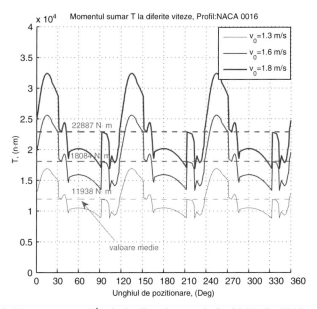

Figure 47b. Torque moment T^1 at hydrodinamic rotor shaft with NACA 0016 profile blades.

On the basis of the conceptual diagram designed above, technical document-
ation was developed and two industrial prototype of micro-hydro power plant sfor
river water kinetic energy conversion into electrical and mechanical energy was
manufactured (Figure 48a, b). Thus, micro-hydro power plant MHCF D4×1.5 ME

a.

b.

Figure 48. Industrial prototype of micro-hydro power plant.

provides conversion of up to 73.6% and 67% of useful energy for electricity production and for water pumping from the energy potential of flowing water entrapped by the hydrodynamic rotor.

Dissemination of Research Results

Monographs

Bostan I., Dulgheru V., Sobor I., Bostan V., Sochirean A. Renovable Energies Conversion Systems (Textbook). Tehnica Info, Chişinău, 595 p. ISBN 978-995-63-076-4.

Article and conference participations:

Bostan I., Dulgheru V., Sochireanu A., Ciobanu O., Ciupercă R. Floatable micro-hydro-power station with hydro-dynamic profile blades. International Conference on Technological Transfer in Electronic Engineering, Multifunctional Materials and Fine mechanics", 17 September 2005, p. 188–192.

Bostan I., Dulgheru V., Sochireanu A., Ciobanu O., Ciupercă R. Floatable micro-hydro-power station with adjustable hydro-dynamic profile blades/ Conference "Energetica Moldovei – 2005", 21-24 September 2005, Chişinău, p.604–608.

Bostan I., Bostan V., Dulgheru V. Numerical Modelling and Simulation of the Fluid Flow Action on Rotor Blades of the Micro-Hydropower Station. Ovidius University Annual Scientific Journal. Mechanical Engineering Series, Volume VIII, Nr. 1, 19–21 May 2006, Ovidius University Press, Constanţa, 2006. ISSN 1223-7221, p.70–78.

Bostan I., Dulgheru V., Ciobanu O. Some constructive-technological aspects regarding elaboration of mycro-hydro power station's multiblade rotor. Buletinul Institutului Politehnic din Iaşi, Volume LII(LVI) Fasc. 5D. Secţia Machine building section, Iaşi, May 25-27, 2006, p. 1275-1279. ISSN: 1011–285.

Bostan I., Ţopa M., Dulgheru V., Oprea A., Ciupercă R., Ciobanu O. Helicoid wind turbine. The second meeting of the National Club "Science & Business" "Energy-saving Technologies and Alternative Energy Sources", May 19, 2006. Catalogue of Innovation in Energetics,Vol.2, Chişinău, 2006 p.22–23.

Bostan I., Dulgheru V., Bostan V., Ciobanu O., Sochireanu A., Trifan N. Hydroelectric station. The second meeting of the National Club "Science & Business" "Energy-saving Technologies and Alternative Energy Sources", May 19, 2006. Catalogue of Innovation in Energetics,Vol.2, Chişinău, 2006 p.32–33.

Bostan I., Cebotari D., Donţu V., Dulgheru V., Ciobanu O. Elaboration of the low speed centrifugal pump. Journal "Meridian Ingineresc",Nr. 2, 2006, p. 11–15.

Bostan I., Cebotari D., Donţu V., Dulgheru V., Ciobanu O., Kokidico A. Low speed centrifugal pump. Nasosy & Oborudovanie, VNIIAEN, Ukrain, Nr.2, 2006. p. 32–34.

Bostan I., Bostan V., Dulgheru V., Ciobanu O. Elaboration of design of the minihydrostation multiblade rotor. National Seminar on Parts of Machine "Ioan Drăghici", University "Petrol şi Gaze", Ploieşti, Romania, 13-14.07.2006. p. 67-70. ISBN (10) 973-719-110-0.

Bostan I., Dulgheru V., Bostan V., Sochireanu A., Ciupercă R., Ciobanu O., Ciobanu R., Trifan N. Micro-Hydropower Station with multiblades vertical rotor and hydrodynamic profile of the blades. A XVIIIth International Conference on Inventics "Researches and advanced innovative technologies", Iaşi, 5-9.07.2006, p.60–64.

Jula A., Mogan Gh., Bostan I., Dulgheru V. et al. ECOMECA – ECO- mechanical engineering (monograph). Braşov, Publ. House of "Transilvania" University, Braşov, pp. 324.

Bostan I., Dulgheru V., Bostan V., Ciobanu O. Design of the construction for the micro-hydro-power station multi-blade rotor. The 8th National Symposium with international participation "Computer Assisted Design PRASIC'06". BRAŞOV, November 9-10, 2006, Proceedings, Vol. Mechanisms. Tribology, p. 67-70.

Bostan I., Dulgheru V., Bostan V. Micro-hydro-power station for flowing river water kinetic energy conversion without dam construction: an ecologically clean method for the satisfaction of energy needs of rural consumers. International Symposium on Socio-Economic Entanglements after Romania accession to the European Union. November 3-4 2006, Braşov, Romania. Proceedings, p. 34-38.

Bostan I., Bostan V., Dulgheru V. Numerical modelling of the hydrodynamic profile blades and simulation of the fluid flow action on rotor blades of the micro-hydropower station/ Meridian Ingineresc, Nr. 4, 2006, pp. 16-22.

Bostan I., Bostan V., Ciobanu O. Aspects concerning the constructive and operational optimisation of the multiblade rotors with hydrodynamic profile blades for energy conversion from the high power station //Acta Technica Napocensis. Series: Applied Mathematics and Mechanics 50, vol. II. ISSN 1221-5872. P. 255-258.

Bostan I., Bostan V. Conceptual design of the electrical micro-hydro-power station for conversion of flowing water kinetic energy into mechanical and electrical energy // Journal of the University "Dunărea de Jos" of Galaţi, Brochure XIV. Mechanical Engineering. Year XII. ISSN 1224–5615, 2006. p. 34-38 p.32-33.

Numerical simulation of the fluid flow interaction with hydrodynamic profile blades of the rotor of micro-hydropower station for river kinetic energy conversion // Pro-Activ Partnership in Creativity for the next Generation / Proceedings / 31st Annual Congress of the American Romanian Academy of Arts and Science / ARA Doval E. (Coord.) – Quebec Canada. Presses Internationales Polytechnique, 2007, pp. 59-62. ISBN 978-2-553-01412-3.

Bostan I., Dulgheru V., Bostan V., Sochireanu A., Ciobanu O., Dicusară I., Ciobanu R., Trifan N. Minigidroczentral' dlya konversii kineticheskoj energii tecushchei vody rek / Catalog III Mezhdunarodnogo Salona Izobretenii i Novyh Tehnologii "Novoe Vremea", 26-28 senteabrya 2007g, p. 35.

Bostan I., Tighineanu I., Dulgheru V., Sobor I. The aspects regarding the energetic factor in the durable rural development. International Symposium "The Business Environment in the context of Romania joining the European Union", Braşov, 1-2.11.2007.

Bostan I., Dulgheru V., Bostan V. The aspects regarding the use of microhydropowers in the irrigation works on agricultural lands. International Symposium "The Business Environment in the context of Romania joining the European Union",. Braşov, 1-2.11.2007.

Bostan I., Dorogan V., Dulgheru V., Vieru T. The aspects regarding the use of micro-hydropowers for illumination the streets of the communes situates along the rivers" International Symposium "The Business Environment in the context of Romania joining the European Union",Braşov, 1-2.11.2007.

Bostan I., Dulgheru V., Ciupercă R. Computational modeling of the helical wind rotor's functional parameters in ANSYS CFX 5.7 software's. The 30th Annual Congress of the American Romanian Academy of Arts and Sciences. Central Publishing House, Chişinău, Republic of Moldova, 2005, pag. 531–534.

Bostan I., Bostan V., Dulgheru V. Numerical Modelling and Simulation of the Fluid Flow Action on Rotor Blades of the Micro-Hydropower Station. Annals of the Ovidius University. Section: Mechanical Engineering, Vol. VIII, Nr. 1, 2006, Ovidius University Press, Constanţa, 2006. ISSN 1223-7221, p.70-78.

Bostan I., Dulgheru V., Ciobanu O. Design and technological aspects of the Micro-Hydropower Station Rotor multiblades elaboration. Bulletin of the Polytechnic Institute from Iaşi, Romania, Vol. LII(LVI) Fasc. 5D. Section. Machine building, Iaşi, 2006, pp. 1275-1279. ISSN: 1011-285.

Bostan I., Dulgheru V., Bostan V., Ciobanu O., Sochireanu A., Ciobanu R., Dicusară. I. Elaboration of the flotable micro-hydropower station with multiblade vertical axis / CSE. The 2nd Conference on Sustainable Energy. 3-5 July 2008, Braşov. Proceedings, Ed. Transilvania, p. 23-29. ISBN 978-973-598-316-1

Bostan I., Bostan V., Dulgheru V. Numerical modelling of the interaction between fluid flow and working elements/ CSE The 2nd Conference on Sustainable Energy. 3-5 July 2008, Braşov. Proceedings, ISBN 978-973-598-316-1, p. 381–386.

I. Bostan, V. Bostan, V. Dulgheru, O. Ciobanu. Study of the microhydrostation's hidrodynamic blades and water flow interaction. Scientific Bulletin, Serie C, Volume XXII. Fascicle: Mechanics, Tribology, Machine Manufacturing Technology, Part 2. Ed. Universitatea de Nord, Baia Mare, 2008. ISSN 1224-3264, P. 47–52. ¶

Patents:

Bostan I., Dulgheru V., Ciupercă R., Ciobanu O. Helical Aeolian turbine. Patent nr. 2995MD, 2006.

Bostan I., Dulgheru V., Bostan V., Sochireanu A. Trifan N. Hydraulic turbine. Patent nr. 2993MD, 2006.

Bostan I., Dulgheru V., Bostan V., Sochireanu A., Ciobanu O., Ciobanu R. Hydraulic station. Patent nr. 2992MD, 2006.

Bostan I., Dulgheru V., Bostan V., Ciobanu O., Sochireanu A. Hydroelectric station. Patent nr. 2991MD, 2006.

Bostan I., Dulgheru V., Bostan V., Ciupercă R. Hydraulic station. Patent nr. 2981MD, 2006.

Bostan I., Dulgheru V., Ciupercă R. Hydraulic station. Patent no. 2288, 2004.

Bostan I., Dulgheru V., Cartofeanu V., Ciupercă R., Ciobanu O. Floating hydrolectric station. Patent no. 2916, B.nr.11, 2005.

Bostan I., Dulgheru V.,Ciupercă R., Ciobanu O. Helical eolian rotor. Patent MD Nr. 2994, BOPI nr. 2/2006.

Bostan I., Dulgheru V., Bostan V. Sochireanu A., Ciobanu O., Ciobanu R. Dicusară I. Hydraulic station. Patent MD Nr 3104. BOPI nr. 7/2006.

International Exhibitions:

Bostan I., Dulgheru V., Bostan V., Sochireanu A., Ciobanu O., Ciobanu R., Dicusară I., Trifan N. Floatable Micro-hydropower Station with Adjustable Hydrodynamic Blades. Geneve'2006. 4–8 April. Official Catalogue. 2005, P. 47. (*Gold medal awarded*).

Bostan I., Dulgheru V., Bostan V., Sochireanu A., Ciobanu O., Trifan N. Mycro-hydro power station with multiblade rotor Inventions and Innovations – the 2nd International Trade Fair of Innovations, Inventions, Useful Models, Ideas in Various Spheres of Scientific and Engineering. Vinahodi+Innovacii", April, 10–13, 2006, Kiev. Official Catalogue, p. 68.

Bostan I., Dulgheru V., Sochireanu A., Ciobanu O., Ciobanu R., Ciupercă R., Dicusară I. Floatable micro-hydro-power station with adjustable hydro-dynamic profile blades: Poster, industrial prototype of the hydraulic pump and generator. International Exhibition INFOINVENT 2005, 9–12 November 2005, Chişinău (*Gold Medal awarded*).

International Exhibition of Invention, Research and Technology Transfer INVENTICA 2006, Iaşi, Romania, July 5–9, 2006. *Bostan I., Dulgheru V., Bostan V., Sochireanu A., Ciupercă R., Ciobanu O., Ciobanu R., Trifan N. Mycro-hydro power station with vertical multiblade rotor and hydrodynamic profile of the blades.* (*Award of the National Institute on Inventics, Romania*).

Exhibition "60 Years of the Moldovan Academy of Sciences, Chişinău, 11.06-14.06.2006. *Bostan I., Dulgheru V., Bostan V., Sochireanu A., Ciupercă R., Ciobanu O., Ciobanu R., Trifan N. Mycro-hydro power station with vertical multiblade rotor and hydrodynamic profile of the blades.*

International Exhibition of Innovation and new Technologies "Novii ceas", Sevastopol, Ukraine, 2006, September 27-29, 2006. *Bostan I., Dulgheru V., Bostan V., Sochireanu A., Ciupercă R.,*

Ciobanu O., Ciobanu R., Trifan N. Mycro-hydro power station with vertical multiblade rotor and hydrodynamic profile of the blades. (***Gold Medal awarded***).

International Exhibition of Invention, Research and Technology Transfer INVENTIKA 2006, Bucureşti, Romania, October 2-6, 2006. *Bostan I., Dulgheru V., Bostan V., Sochireanu A., Ciupercă R., Ciobanu O., Ciobanu R., Trifan N. Mycro-hydro power station with vertical multiblade rotor and hydrodynamic profile of the blades.* (***Gold Medal awarded***).

Bostan I., Dulgheru V., Bostan V., Sochireanu A., Ciobanu O., Ciobanu R., Dicusară I., Trifan N. Floatable Micro-hydropower Station with Adjustable Hydrodynamic Blades. Brussels EUREKA'2006. 23–27 November. Official Catalogue. 2006 (***Diploma and Gold medal***).

Bostan I., Dulgheru V., Bostan V., Sochireanu A., Ciobanu O., Dicusară I., Trifan N. Micro-hydro-power station for flowing river water kinetic energy conversion without dam construction/The tenth Moscow International Exhibitions of the Industrial Property "ARHIMED", 27–30 March 2007 (***Diploma and Gold Medal***).

Bostan I., Dulgheru V., Toca A., Ciupercă R. *"Helical blade wind turbine with vertical axis"* // International Exhibitions of Inventions, Research and Technological Transfer ECOINVENT 2007. 30.05-02.06.2007. Official Catalogue, p. 28, 2007 (***Gold Medal with the jury special prize***).

Bostan I., Dulgheru V., Bostan V., Sochireanu A., Ciobanu O. *"Micro-hydro-power station for flowing river water kinetic energy conversion"* // International Exhibitions of Inventions, Research and Technological Transfer ECOINVENT 2007. 30.05-02.06.2007. Official Catalogue, p. 27–28, 2007.

Bostan I., Dulgheru V., Bostan V., Sochireanu A., Ciobanu O., Dicusară I., Ciobanu R., Trifan N. Minigidroczentral' dlya konversii kineticheskoj energii tecushchei vody rek/III Mizhdunarodnii Salon Vinahidiv ta Novyh Tehnologii "Nivii Cheas", 26-28 veresnya 2007p. (***Diploma and Gold medal***).

Bostan I., Dulgheru V., Bostan V., Sochireanu A., Ciobanu O., Dicusară I., Ciobanu R., Trifan N. Minigidroczentral' dlya konversii kineticheskoj energii tecushchei vody rek/III Mizhdunarodnii Salon Vinahidiv ta Novyh Tehnologii "Nivii Cheas", 26–28 veresnya 2007p. (***Special Prize of the enterprise "Raduga 57", or. Moscova, Rusia***).

Bostan I., Dulgheru V., Bostan V., Sochireanu A., Ciobanu O., Ciobanu R. Energy cinversion of the waves into electric power. Brussels Innova-Energy. The Belgian and International Trade Fait for Technological Innovation, 21–25 November 2007 (***Gold Medal***).

Bostan I., Dulgheru V., Bostan V., Sochireanu A., Ciobanu O., Ciobanu R. Flotable Micro-hydropower Station with Adjustable Hydrodynamic Blades/The 6th International Exhibition (SuZhou) of Inventions, China, 9–11 octombrie 2008 (***Diploma and Gold medal***).

ON THE MODELLING AND EVALUATION OF SECURITY OF ENERGY SUPPLY IN REGIONAL ENVIRONMENTS

DAN SERBANESCU[*]
DG ENER D2 Unit Nuclear Safety, Decommissioning and Transport, Luxembourg

Abstract. This paper presents some approaches used in the modelling of the Security of Energy Supply for the case of regional environments, with some specific results obtained for specific cases from the European environment.

Energy is one of the most important driving forces of the modern world and it is critical that it be developed further. One of the most important objectives for all countries involved is to assure the necessary energy for the support and moving ahead of society, while acknowledging step-by-step the high diversity of the aspects involved as well as the complexity of the energy systems.

For various reasons, access and/or availability of energy for a certain period or certain area of the planet are becoming very important aspects of the energy systems, which require deeper understanding and more careful consideration.

In this context, Security of Energy Supply (SES) has become a priority issue for political decision-makers. To assist them in this task, models and systematic methods of energy systems are needed in order to assess vulnerabilities to a system of such complexity as is with energy supply networks.

There were five general aspects, which were used in the work presented at this chapter as a guiding strategy in the development of the tools for SES evaluation for some European regional environments. These aspects were considered to be relevant for the description of any approach attempting to describe the issue of security of energy supply and to propose strategies to cope with specific situations. The five guiding aspects of the strategy for a search and use of tools for the SES evaluation are as follows (Serbanescu, 2008a):

1. A certain model of the SES has to be developed as a basis for any study. During this phase it should be noted that the topic has a set of diverse approaches from various levels (academic and research organization, various governmental and nongovernmental organizations at international and/or national levels, etc.). The development of most of those models is based on notions

---

[*] To whom correspondence should be addressed: Dan Serbanescu, *DG ENER D2 Unit Nuclear Safety, Decommissioning and Transport, Luxembourg,* e-mail: Dan.SERBANESCU@ec.europa.eu

A. Gheorghe and L. Muresan (eds.), *Energy Security: International and Local Issues, Theoretical Perspectives, and Critical Energy Infrastructures,*
DOI 10.1007/978-94-007-0719-1_11, © Springer Science+Business Media B.V. 2011

reflected in keywords, such as: energy security, comprehensive approach, interface sustainability, governance, climate change, international level, regional level (Europe, North America, Black Sea, Baltic Sea), etc. It is also important to mention that in specific cases notions from theories already developed related to complex systems, such as information entropy, are used.

2. A set of methods has to be developed in order to obtain qualitative and quantitative descriptions of the objective functions defined in the models of SES. The type of methods could be diverse and are based in most cases on well known theories (game theory, risk analysis, reliability analysis, holistic/ expert opinion type approach, etc). However in each application the success of any of the methods is judged based on the results obtained in the verification and validation of results (benchmarking process V&V, etc.) for "normal cases" and "extreme" situations.

3. It is necessary to evaluate interfaces of other topics of the same complex systems with the models created for SES and the methods associated to them, which is an unavoidable step. Therefore, evaluation of the interface between the models and methods adopted for the evaluation of objectives in a SES system with other topics also associated with complex systems, such as IT, security, terrorism, and regional specifics is very important and has to be addressed.

4. Examples of applicability of the method are required to test the robustness of the adopted approach to diverse infrastructures, such as the interface between various types of energy sources, including the consideration of the placement of renewable energy sources (photovoltaic, wind, hydro, etc…), as well as their connectors to users and what lies between them (grids, various types of plant operation, various lifecycles of a given source of energy, etc.).

5. Lessons learned need to be evaluated for all performance of the steps of SES cases. This includes perspectives on future developments, intensive testing, sensitivity and robustness checks, as well as feedback from specific cases.

Since there are a multitude of approaches capable of solving the task of a SES evaluation a significant amount of work was dedicated to the evaluation of the applicable approaches (models, methods, case studies and expected results and feedback process) (Serbanescu 2008a).

The SES models and the methods used to evaluate the objective functions should be able to capture the complexities of such a system and to provide information needed to the decision-makers. The decision making process on SES objective functions requires basic information which allows the definition of the prioritized action options. This can be achieved by indicating and ranking both risks and uncertainties associated with the analysis.

The approach 'SES-RISK' (Security of Energy Supply Risk model) (Serbanescu, 2008a) was proposed to support such decision-making processes. The main features of SES RISK are as follows:

- It uses probabilistic risk assessment in order to assess the impact of threats to the energy supply system.
- The energy system is composed of energy sources (nuclear, fossil, and renewable energy) and input (e.g. primary energy supply network) and output grids (the electricity network).
- It is modelled as subsystems able to cope with technical, economical, terrorist, and socio-political challenges.
- Dependencies between subsystems are explicitly taken into account.
- The model is capable of incorporating quantitative and qualitative expert view data.
- The results are presented in the form of suggested actions based on the identified weaknesses of the network.

The goal of SES-RISK is to address the need of modelling complex systems using the probabilistic risk assessment approach. The model was built to be able to *(Serbanescu, 2005a):*

- Evaluate complex systems and systems of systems with interdependencies;
- Derive results and answers to the decision-makers' questions in qualitative and/or quantitative format(s), coupled with detailed scenarios for further actions;
- Perform systematic and traceable analyses, which can be easily modified if assumptions are changed;
- Evaluate the uncertainty of the analysis, including the uncertainty induced by various assumptions;
- Perform extended, fast, and user friendly sensitivity analyses; and
- Perform analyses for various levels of information available and with diverse sources (from numerical to expert judgment).

Nevertheless during all the related activities for the evaluation of SES objectives, which were performed based on the guiding aspects mentioned above, it was not at all assumed that SES-RISK is the only tool possible to address the modelling of complex systems, such as the security of energy supply.

SES-RISK was created to ensure interoperability with other tools. Tools' applicability to this task was evaluated based on three main criteria:

- The degree of credibility of the results obtained using the tool;
- The degree of conformity of the model developed for SES with the real system; and
- The degree of complexity of the tool.

Based on results obtained for a similar task (Serbanescu, 2007b) the expected positioning of some tools is shown as a sample in Figure 1.

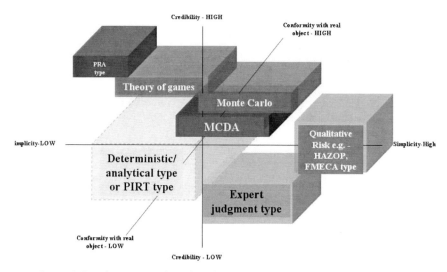

Figure 1. Sample representation of possible toolbox composition for SES evaluation.

The positioning of SES RISK reflects the initial expectations based on (Serbanescu, 2008a), i.e to use an approach able to be modeled in a systemic and systematic manner with integrated results for complex systems and to evaluate the credibility of the results based on extensive sensitivity analyses and a rigorous feedback process, even if the complexity of the tool itself is quite high.

The results obtained to date and the present status of verifications performed under the analysed case studies (as part of the Verification and Validation (V&V) process) confirm that this approach provides stability and robustness to the conclusions it is based on (Serbanescu, 2008a). These results also indicate that SES RISK is able to provide decision-makers with a prioritized set of actions indicating also the range of uncertainty to each of the recommendations.

Models

Development of SES RISK was based on the need to have a model to fulfil some of the needs mentioned in the first chapter. Due to the complexity of SES modelling there is a diversity of models, as partially reflected in Figure 1. This diversity is further increased by various media involved in the development of the models – universities and national and international organizations focused on various specific cases (general cases, regional cases, etc.)

Essentially, SES-RISK describes how well the energy supply network reacts to challenges it is facing. SES-RISK uses probabilistic risk assessment (PRA) based on previous work measuring vulnerabilities of complex systems (Serbanescu, 2005a; Serbanescu et al. 2007a; Serbanescu et al. 2008a). PRA uses event and fault trees to evaluate the resilience of a system to a particular challenge. This model tries to address the following requirements:

- Evaluate the energy supply network as a complex system with inter-dependencies;
- Present answers to the decision-makers' questions (qualitative or quantitative), coupled with detailed scenarios for further actions;
- Indicate the uncertainty of the analysis and sensitivity analyses;
- Cater for the use of information available from diverse sources, from quantitative data to qualitative expert judgment; and
- The model first considers the challenges presented, and then the energy system itself as a system of barriers to these challenges. Finally, results (risk levels and recommended actions) are presented to the decision-maker.

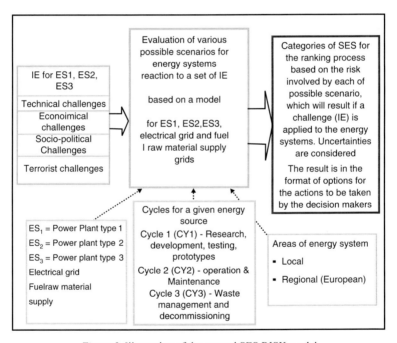

Figure 2. Illustration of the general SES RISK model.

The Energy System as a System of Barriers

The model of energy supply system consists of three parts: a primary energy supply (the '**input grid**', e.g. gas pipeline system), an output infrastructure to the energy customers' grid (the '**output grid**', e.g. electricity network), and the **energy sources** themselves.

In this example, three energy sources (nuclear, fossil and renewable) are considered. The first two are assumed to cover the base load (ES1 and ES2 in Figure 3, while the latter serves local needs, e.g. a windmill (ES3). Each sub system can be modelled to the required level of detail.

For each energy source, three stages in its life cycle are modelled: research, testing, and prototype (CY1), operation and maintenance (CY2), waste management and decommissioning (CY3).

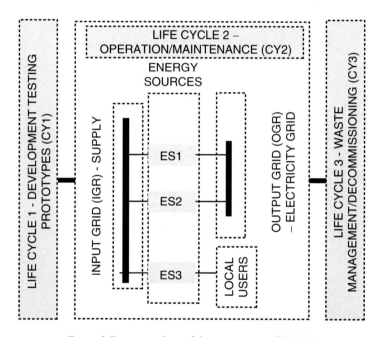

Figure 3. Representations of the components of SES pilot case.

SES risk models the energy supply system as subsystems that act as technical, economic, socio-political, and counterterrorist **barriers** to security challenges. Connections between challenges and barriers are an important feature of the model.

The barriers have four levels of robustness: barrier level 1 (able to cope with a low degree of challenge); barrier level 2 (able to cope with a medium degree of challenge); barrier level 3 (able to cope with a high degree of challenge). If the challenge exceeds level 3, then the energy system is assumed to collapse.

Challenges

A challenge to the energy system consists of two parts: first, the external threats, e.g. a terrorist attack, and second, the weaknesses of the energy supply system.

(a) Types of challenges

Similar to barriers, challenges (known in PRA as Initiating Events) also come in four types: technical, economic, socio-political and terrorist. Each of these threat types has one of three associated degrees of severity (1-low, 2-medium, 3-high).

(b) Initial conditions of the energy system

Initial conditions of the energy system may add to the challenge. The degree of resilience of a particular part of the energy supply system is expressed in one of four levels:

o   normal initial state (GC0)
o   low disturbance of the initial state (GC1)
o   medium disturbance of the initial state (GC2)
o   high disturbance of the initial state (GC3)

These initial conditions are derived first via a PRA by developing event and fault trees considering scenarios. This is shown in a simple example in Figure 4.

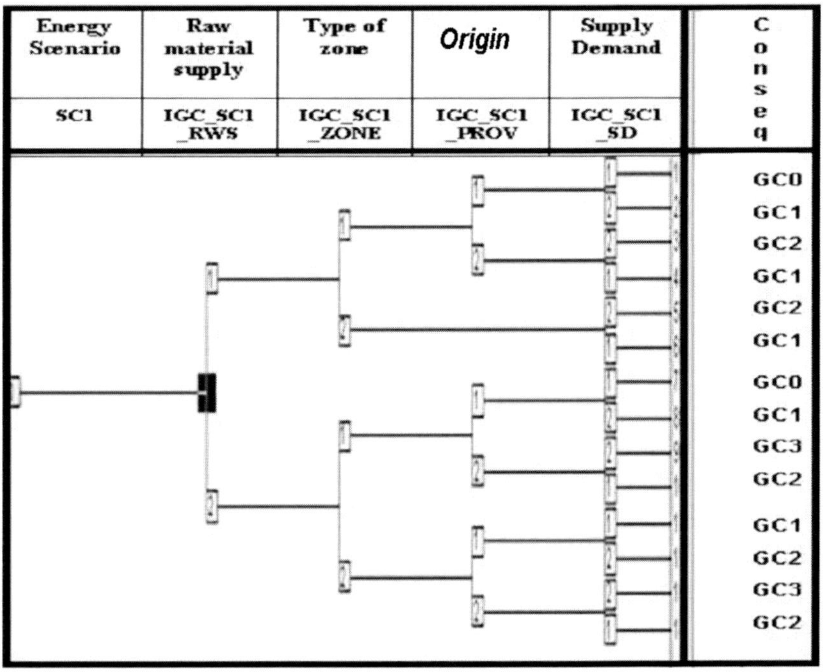

Energy Scenario	Raw material supply	Type of zone	Origin	Supply Demand	C o n s e q
SC1	IGC_SC1 _RWS	IGC_SC1 _ZONE	IGC_SC1 _PROV	IGC_SC1 _SD	
					GC0
					GC1
					GC2
					GC1
					GC2
					GC1
					GC0
					GC1
					GC3
					GC2
					GC1
					GC2
					GC3
					GC2

Figure 4. Schematic representation of the evaluation of initial conditions.

For each scenario (SC), the system is expected to activate various barriers to cope with the potential disturbances, e.g. by triggering:

- o   mechanisms built in to cope with the primary energy supply vulnerabilities, for example to pipelines taking different routes;
- o   mechanisms built in to cope with zone vulnerabilities, e.g. is the energy produced locally or imported;
- o   mechanisms built in to cope with vulnerabilities due to potential unbalances between supply and demand of energy.

Assumptions

After the energy system has been defined, a number of assumptions have to be made. A sample of assumptions adopted for case studies includes, but it is not limited, to issues as follows:

1. How the types of energy sources and the grids are connected (e.g. which energy source feeds into the electrical grid)?
2. How the barriers are dependent on each other? For example a major failure of the electricity barriers (switchers, control systems, and dispatcher) is likely to have economic effects, thus weakening the robustness of the economic barrier. The interconnections between the barriers of the energy system illustrated in Figure 5 are chosen by the modeller in a binary matrix format where "1" (logic true) means that barrier elements are dependant) and "0" indicates independence (Table 1).
3. What are the initial conditions of the system (see above)?

TABLE 1. Sample of the method implementation in Risk Spectrum

	T 1	T 2	T 3	E 1	E 2	E 3	S P1	S P2	S P3
T 1	1	0	0	1	0	0	0	0	0
T 2	0	1	0	1	1	1	1	1	1
T 3	0	0	1	1	1	1	1	1	1
E 1	1	1	1	1	0	0	0	1	1
E 2	0	1	1	0	1	0	1	1	1
E 3	0	1	1	0	0	1	1	1	1
S P1	0	1	1	0	1	1	1	0	0
S P2	0	1	1	1	1	1	0	1	0
S P3	1	1	1	1	1	1	0	0	1

4. What are the acceptability criteria (end states), i.e. how robust does a system need to be?
5. How should results be grouped in order of importance and relevance to a decision?

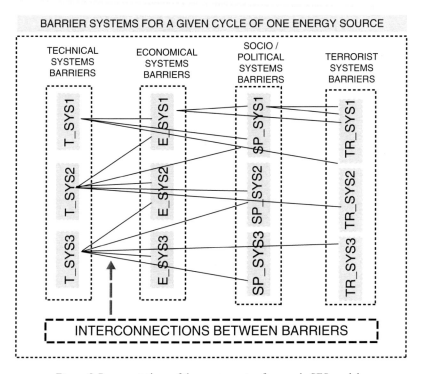

Figure 5. Representations of the components of a sample SES model.

6. How to evaluate the barriers? In this context each of the barriers are evaluated using a fault tree approach as per (USNRC 1983; USNRC 1998; Serbanescu et al. 2008a), connected in scenarios, called event trees. A more detailed presentation of this method is presented in the next chapter. However at this stage of presentation of the assumptions it is important to mention that the scenarios are considered as leading to some end states, which have to be evaluated and defined. These resulting end states are grouped into categories (GC) based on assumptions, which reflect expert opinion(s) and existing documents. The high subjectivity possibility in this process is compensated by the possibility of a set of sensitivity calculations, which can be done with the use of integrative risk models as tested in other applications (Serbanescu 2005a). The initiating events (IE), end states, systems, etc. are coded so that the use of specialized computer codes (such as (© Relcon Scandpower AB 2008) would be possible. A representation of this scenarios is represented schematically in Figure 6, where the end states are noted as $ES^{(\alpha)}_i$.

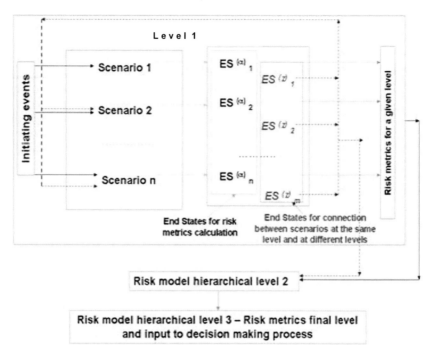

Figure 6. Representation of a CAS model to be implemented in PRA codes (e.g. RiskSpectrum).

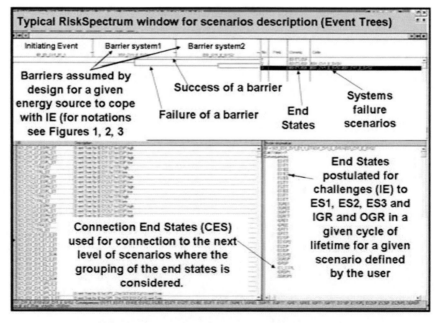

Figure 7. Sample of binning rules as used in ©RiskSpectrum.

As shown in Figure 7, the scenarios used are based on Event Trees, which are similar to the Event Trees used in PRA level 2 as defined in (USNRC, 1983), e.g. they consider and evaluate both failure and success branches – for more details on this issue in (Serbanescu et al. 2008a).

Based on list of challenges and end states possible for installation, customers, environment and workers a set of scenarios can be built for each IE. The scenarios start with the challenge (IE) and then takes various pathways depending on success or failure of the barriers/layers of protection assumed by design for each of them (as illustrated in Figure 9). Each scenario ends with an ES for a given part (installation, customers and environment, workers) then it continues to another set of scenarios describing success or failures of the buildings, zoning and distances to cope with the negative effects of the challenges (IE). Then the scenarios are grouped by impact (IMP1, IMP2, IMPI, etc, as represented in Figure 9 by the notations $ES^{(z)}_n$); between them weather impact scenarios are modelled. The final results of the scenarios, after weather is also considered, indicate the final risk level for the model. The scenarios for the model without buildings/zoning/distances are called PRA Nuc level 1, those which include buildings/zoning/distances are called PRA Nuc level 2 and the final scenarios including weather conditions are called PRA Nuc level 3, in analogy with PRA Nuc standards. Each scenario for a given IE (i) consists of a set of logic combination of the following types:

$$Scenario\ \boldsymbol{l} = \boldsymbol{IE}\ \boldsymbol{(i)}\ *\ barrier\ \boldsymbol{1}\ failed\ *\ barrier\ \boldsymbol{2}\ successful\ .....+ \qquad (1)$$
where
$$Barrier\ \boldsymbol{w}\ failed = component\ \boldsymbol{1}\ failed\ *\ component\ \boldsymbol{2}\ failed\ ....\ +$$
$$+\ component\ i\ failed\ *\ component\ k\ failed\ + \qquad (2)$$

If (2) is included in (1) and the laws of Boolean logic are applied then one can obtain a set of equations of the type (3) for each scenario:
$$Scenario\ 1 = \boldsymbol{IE(i)}*\{\ component\ \boldsymbol{k}\ *\ component\ \boldsymbol{l} + component\ \boldsymbol{j}\ *\ component\ \boldsymbol{n}$$
$$+\ component\ \boldsymbol{t}\ *\ component\ u\ *\ component\ v+\ ....) \qquad (3)$$

In all the equations mentioned above, * and + are logical operators meaning AND, respectively OR, in a Boolean logic sense. Summarizing this short clarification: the main aspects that have to be considered for the scenarios calculations are related to the fact that PRA Nuc is giving a set of Boolean logic combinations of failures and successes of various components of the model, leading to a given ES, if the system is challenged by a set of IEs. That is why even if failures are represented by probabilities and by using specific methods to interpret such results, (Serbanescu, 2007; Serbanescu et al. 2007a; Serbanescu, et a.l 2007b; Serbanescu, 2008a) the main message contained in the list of failures leading to given ES, is mainly in their logical mainly and may be a second aspect in the absolute value of their probabilities. The use of the results for ranking purposes considering the relative probabilities is therefore a very important application of PRA Nuc for early phases of risk analyses of a given system. In each scenario a set of barriers is assumed to be available to cope with the challenges. These barriers are defined in detail for each analysis as in the specific case evaluated in (Serbanescu et al. 2010a; Serbanescu, 2010b).

Methods

The next guiding aspect, which had to be considered in the work performed for the evaluation of a SES of the chosen case studies (Serbanescu, 2008a; Serbanescu, 2008b) was to decide and define the method to be used for the adopted SES model described in the previous subchapter. While performing this task, one has to be aware of both the diversity of existing tools for the evaluation of complex systems, based on game theory, risk and/or reliability analyses, etc. and on the need for validation of results and benchmarking for "normal cases" and "extreme" situations.

The method chosen for the SES model evaluation was called SES RISK and it consists of five steps (Serbanescu, 2008):

STEP1: Define the initial conditions and assumptions.

STEP2: Calculate main flow of the end states.

STEP3: Obtain from computer code (© Relcon Scandpower AB 2008) the following:

- o   combination of events, leading to various end states, which are different for each energy system and grids;
- o   transition end states used to connect subsystems between them and groups of end states (Serbanescu, 2008);
- o   final grouped states, called survivability states;
- o   results for the end states parameters.

STEP4: Post-processing of (© Relcon Scandpower AB 2008) model results, which are introduced and analyzed in decision-making type tables and where the end states calculated in the previous step are inserted.

STEP5: Interface with users with a view to reformulate results and ranking. Restart new iterations, if needed, and perform extensive sensitivity analyses.

In the description of the SES RISK the fact is underlined that it is based on risk analysis methods as used in the nuclear field and other technical and non-technical evaluations of non-nuclear complex systems as communicated in (Serbanescu, 2008).

In these approaches, the starting point for the development of the tool was the Probabilistic Risk Analysis for Nuclear Applications (PRA Nuc) as defined in the existing standards (USNRC, 1983), which assumes performance of a series of tasks for a given model of the installation.

The process of using PRA Nuc starts with the definition of the limits and details of the installation to be evaluated. A series of assumptions are defined in order to set the limits of the model and the targets of the analysis. The next step consists of defining the possible challenges to the installation, environment and workers; these challenges are called in PRA Nuc Initiating Events (IE).

The existing information from deterministic analyses, previous records of operation of the installation, as well as previous risk analyses provide a list of possible scenarios to follow if the installation is challenged by a given IE are defined. In order to describe the scenarios a previous definition of the results of the IE (end states ES) and the consequences (CSQ) are needed. Some ES are just an intermediate phase of the installation and are connected to the reaction of the building (confinement) in which the installation is located and further in the environmental conditions (weather, surroundings, etc). Based on this information, a PRA Nuc model is built consisting of a series of scenarios (called Event Trees (ET)). Event Trees assume that the installation has, by design, a set of protecting layers (consisting of technical systems, human actions, etc.) to cope with undesirable events (IE). The possible failure of these layers is also modelled in a specific set of failure scenarios called Fault Trees (FT). Then the scenarios of ET are using the combination of failures of the barriers as defined in the FT in a process called "integration of FT into ET". The result of this integration is a set of combination of failures of various components of the installation and other barriers leading to a given set of end states. The end states are grouped in levels of risk based on deterministic calculations and/or previous probabilistic analyses of PRA type, called risk categories. These risk categories could also be used from existing requirements, regulations, etc. However in the case of new installations, such as hydrogen ones, the analyses of PRA type are meant by themselves to help in defining such standards.

The results are then grouped and ranked, and for the important contributors to risk, specialized analyses are done: uncertainty evaluations for the results and sensitivity of the results to various assumptions made during the modelling. The inputs of the ranking by risk for the contributors and whole set of scenarios are then reiterated for new set of analyses and also to the deterministic list of aspects needing in-depth review (detailed evaluation in deterministic terms of the installations in a set of specific conditions, which results in high impacts from PRA Nuc results). The process is represented in a simplified manner in Figure 7, described in detail in (Serbanescu, 1991; Serbanescu, 2005a; Serbanescu, 2005b; Serbanescu, 2007; Serbanescu et al. 2007a; Serbanescu et al. 2007b; Serbanescu et al. 2008a).

SES Risk is based on definition of the CS and risk as presented in the previous paragraphs. The model is based on a generalized format of PRA (Figure 8), as already implemented in some applications. This approach allows:

i.   To model complex systems and systems of systems with interdependencies.
ii. To derive results and answers to the decision-makers questions in qualitative and/or quantitative format(s) and with specific details, not only with suggestions for further actions.
iii. To perform systematic traceable analyses, able to be modified if assumptions are changed.
iv. To evaluate the uncertainty of the analysis, including the modelling uncertainty induced by various assumptions.
v.   To perform extended fast and user friendly sensitivity analyses.
vi. To perform analyses for various levels of information available and with diverse databases (from numerical to expert judgment).

vii. If risk is used in the sense of a parameter combining both probability of
occurrence of an event and the impact produced by it, then this approach
can provide the results in both quantitative and qualitative manner with
the evaluation of uncertainty.

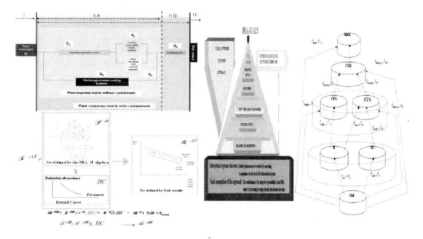

Figure 8. General representation of a PRA model and its metrics.

In analogy to PRA used in technical risk assessment based on PRA Nuc, the
PRA analysis used in SES-RISK has three levels. At each level, the subjective
appreciation of the risk increases. End states at each level are combined according
to 'binning rules' to yield end states of the next level.

PRA Level 1 (The Physical Level)
The model is first given initial start values. Then challenges are presented.

Each of the barriers is evaluated using an event and fault tree approach
(USNRC, 1983; Serbanescu et al. 2008a). These scenarios are solved through
converging iterative calculations.

A scenario finishes in an end state. The end state indicates the energy source,
the type of failure (i.e. technical, economical, socio-political or terrorist protection),
and its level of gravity (low, medium, high).

Examples of end states are:

o   economic disturbance in energy source 1 (fossil) with low impact, short
term oil price increase due to market disturbances.
o   disturbance in ES2 on technical level of low gravity magnitude, e.g.
failure of the first safety system assumed to cope with technical failures
(involving malfunction of ES1 without risk impact on environment).

End states can also be combined, e.g. a transient increase in oil price leading
to high impact on trucking and fishing industry and subsequent strikes.

PRA level 2 (survivability criteria):

The end states of PRA level 1 are grouped according to predefined rules, **'survivability criteria'**, (Tables 2 and 3) which reflect the impact of end states and their combinations on the survivability of the system. These are the end states of PRA level 2. The subjectivity comes through the definition of survivability.

TABLE 2. Sample of survivability categories for each level SES

SES–Specific[1] Survivability Subgroup (SES-SSS)	SES-SSS short code	Short description of specific survivability group
SURV-T0	**T0**	Survivability with Technical aspects challenged – Low impact leading to minor follow up corrective actions needed
SURV-T1	**T1**	Survivability with Technical aspects challenged – Medium impact leading to some important follow up corrective actions needed
SURV-T2	**T2**	Survivability with Technical aspects challenged – High impact leading to major follow up corrective actions needed
SURV-ESP0	**E0**	Survivability with Socio Economical or Public Interface aspects challenged – Low impact leading to minor follow up corrective actions needed
SURV-ESP1	**E1**	Survivability with Socio Economical or Public Interface aspects challenged – Medium impact leading to some important follow up corrective actions needed
SURV-ESP2	**E2**	Survivability with Socio-Economical or Public Interface aspects challenged – High impact leading to major follow up corrective actions needed

For example, one of the scenarios leading to a final end state in the highest risk category (survivability level 3; 'SURV 3') is shown in Table 4 (in the row shaded in *yellow*):

"The worst survivability category (3) could be reached if, for an IE in the group SC1 of initial assumptions (IE_SC1), consisting of a medium impact terrorist attack (which assumes that attack barriers are in place by design, no extended impact happens, but requires significant damage repairs) on the ES2 during operation (ES2_CY2) by the time the initial conditions of ES and grids were in a medium affected stage and, subsequent to the moment of the attack, an independent disturbance of low magnitude of the barrier is designed to cope with economic minor challenges for ES2 during the performance by the energy source of environmental protection tasks. (ES2_CY3_E_SYS1) – such as the high price increase of environmental protection actions."

---

[1] Specific implies that the survivability states are applicable only to specific and very detailed situations, and not to the whole system (e.g. technical, economic, socio-political, terrorist attack) at various magnitudes.

TABLE 3. Sample of general survivability categories for SES

t3.74 SES – General[2] Survivability Group (SES – GSG)	Short description of general survivability group	SES – Specific survivability subgroup combination of categories defined in Table 2 (Binning)
SURV 0	Very Low Impact leading to no significant follow up corrective actions needed (NCO)	T0 E0 E1 T1 E2 T2
SURV 1	Small Impact leading to Minor follow up corrective actions needed (CO1)	E0T1 E0E1 E0T1 E0T0
SURV 2	Medium Impact leading to Some Important follow up corrective actions needed (CO2)	E0T1 E1T0 E1E2 E1T1 E1T2 E2T1
SURV 3	High Impact leading to Major follow up corrective actions needed (TCO)	E2T2

TABLE 4. Sample of results for survivability category 3 for the scenarios contributing mostly to it

SURV3–MCS		
ES1_CY3_T_SYS1	ES1_CY3_T_SYS2	IE_SC1_ES2_CY3_T1
ES1_CY3_E_SYS1	GC2_SF	IE_SC1_ES2_CY3_TR1
ES3_CY3_SP_SYS1	GC2_SF	IE_SC1_ES2_CY3_T2
ES1_CY3_E_SYS1	GC1_SF	IE_SC1_ES2_CY3_TR1
ES2_CY3_E_SYS1	Gc2_SF	IE_SC1_ES2_CY2_TR2

---

[2] General implies that the survivability states are applicable to any of the energy sources or grids modelled.

PRA level 3 (recommended actions):

At PRA level 3, the end states of level 2 (the combination of scenarios leading to a specific survivability level) are ranked by **impact** and **uncertainty** associated with the result (Serbanescu et al 2008; Smithson, 2000; Howard 1984).

Table 5 shows an example of the risk ranking for all final states reaching survivability level 3. These survivability groups are now associated to four categories of **corrective action** required:

- no change of objectives,
- mild change of objectives,
- important change of objectives,
- total/fundamental change of objectives.

Thus, the optimal actions are shown to the decision-makers in order to decrease the risks in a SES evaluation case for a given challenge.

TABLE 5. Example of the risk ranking for all final states reaching survivability level 3

	Components and their occurrence / failure impact for a SES state of type SURV3 (as defined in Table 2)	Rank of Impact	Confidence in results	Group of impact
A	Failure of the barrier defined by System 2 of TR type, for ES2 in cycle CY1	H	L	I = HL
	Failure of the barrier defined by System 2 of SP type, for ES2 in cycle CY2	H	L	
	Failure of the barrier defined by System 2 of T type, for ES1 in cycle CY2	H	L	
B	Failure of the barrier defined by System 2 of E type, for ES1 in cycle CY3	H	M	II = HM
	Occurrence of an IE (Challenge) to OGR of TR2 type in cycle CY2	H	M	
	Failure of the barrier defined by System 1 of TR type, for ES1 in cycle CY1	H	M	
C	Occurrence of an IE (Challenge) to OGR of T2 type in cycle CY2	M	L	III = ML
	Failure of the barrier defined by System 1 of T type, for OGR in cycle CY2	M	L	
	Failure of the barrier defined by System 2 of T type, for ES3 in cycle CY1	M	L	
D	Initial condition of worst type (GC3)	H	H	IV = HH
	Failure of the barrier defined by System 2 of SP type, for ES1 in cycle CY2	H	H	

There are two ranking criteria: one is the risk impact and the second is the uncertainty evaluation – considering (Serbanescu et al., 2008a) approach based on (Smithson, 2000 and Howard, 1984). To further assist the decision-maker, Table 6 is a sample of results presented in text format, with more details, that could be useful to provide further support to decision-making.

TABLE 6. Text formulation for selected failures (in yellow) which will have an impact on SURV 3, including comment on actions needed

No	Sample of presenting results from a chosen challenge (Initiating Event) from the list, which is ranked as a high risk, accompanied by suggested actions to be considered by decision-makers
1	This challenge consists in a technical failure of barrier 2 of ES1 (e.g. failure of safety systems discovered during an operation on a nuclear power plant, i.e. CY2) leading to the preventive shutdown of the plant. This then results in the disruption of electricity supply to the grid, even if there is no impact on environment, workers, and the public.
	The decision-maker could consider as a priority simply to improve the safety systems. However, this may not be the best course action from the perspective of assuring survivability of category 3 of the entire energy system, as this scenario has a high-risk impact but low confidence (see Table 5, row **A**). Thus, based on boundary conditions, the decision-maker may decide to choose another scenario.
2	This challenge consists of a terrorist attack of average magnitude on a high voltage switchyard within the EU electrical grid. This leads to an unstable electrical grid and a possible blackout across Europe, with a high impact on the base load sources (nuclear [ES1] or fossil [ES2]). This, in turn, could lead to failures of technical, political, and economic barriers for these energy sources.
	In this scenario, the decision-maker could consider, as a priority, improving the security (including preventive measures) of the key infrastructures of the grid. This could be a better course of action because this scenario has a high-risk impact with a medium confidence in the results (see Table 5, row **B**).
3	This challenge consists of a technical failure of barrier 1 of the electricity grid (e.g. technical failure of a switchyard due to severe weather conditions). The system will switch to the next level of protection (barrier 2) but may still have minor impact on users and other energy sources connected to the grid.
	In this scenario, the decision-maker could prioritise fixing barrier 1 of the electrical grid. However, this may not be the best course action, as this scenario has a medium-risk impact and low confidence (see Table 5, row **C**). Thus, based on boundary conditions, the decision-maker may decide to choose another scenario.
4	This challenge consists of a failure of barrier 2 of socio-political type for ES1, i.e. nuclear (e.g. failure of reaching consensus between government, industry, and the public regarding the continuation of nuclear power plant production (cycle 2).
	In this scenario, the decision-maker could shut down the nuclear plant, but this could have serious repercussions to the entire survivability of the whole energy system, as nuclear is one of the important sources of the energy mix. However, this could lead to even worse public reaction when they will realize that their everyday lives may be drastically changed due to lack of electricity. Thus, as this scenario is of high-risk and high confidence, the decision-maker may have no other choice but to quickly reopen dialogue with the public to seek consensus on the best course of action (see Table 5, row **D**).

In this example, iterative calculations were performed for survivability level 3 and ranking is convergent after three iterations. Furthermore, the initial conditions converge to a stable set of combinations of end states.

Sensitivity analysis:
In SES Risk, a decision maker can perform a sensitivity analysis by changing the assumptions and characteristics of the challenges.

INTERFACES OF SES EVALUATION WITH OTHER TOPICS

The third guiding aspect, which had to be considered in the work performed for the evaluation of a SES of chosen case studies, was to evaluate and define the interface with other topics, which could have a significant impact on the evaluation itself, such as topics related to:

- The decision-making process and the use of the SES RISK results in this process.
- Specifics of modelling infrastructures which are part of a SES model (general modelling and method related aspects, as well as aspects related to some infrastructures for which the evaluation of risks is more or less in the beginning: renewable energy, hydrogen installations, etc).

Risk Informed Decision Making (RIDM)
Development of tools like SES RISK was done while there is an increasing need for the use of integrative approaches to the risk governance (RG).
In this light there are many other approaches for the evaluation of risk to be used in a RIDM process. These approaches could focus on other aspects than risk analyses as known in technical systems such as the implementation of the precautionary principles or deliberative techniques in risk governance (Serbanescu, 2008c).
The main steps of the risk governance have to give answers to issues as follows:

- Identify risk, which is of concern.
- Plan the process of risk evaluation and management.
- Define the scope, the parts involved in the process, the objectives of the evaluations and actions involved by them.
- Define the uncertainty and boundaries in the decision process framework for each of the activities.
- Perform a risk analysis.
- Define the risk mitigation actions based on the risk analyses results.

Three approaches to support risk governance were compared in (Serbanescu, 2008c) with a focus on RIDM:

- Risk informed decision-making (RIDM).
- Precaution and risk reduction principle (PRECP).
- Risk deliberation (DELIB).

While there are general features to all those approaches the RIDM is more based on quantitative assessments than the other two. Decision-making (DM) is considered to be a cognitive process leading to the selection of a course of action among variations. Every decision-making process produces a final choice. It can be an action or an opinion. It begins when *we need to do* something but we do not know yet what exactly. Decision-making is therefore considered to be a reasoning process, which can be rational or irrational and can be based on explicit assumptions or tacit assumptions. A decision process can be represented as shown in Figure 9.

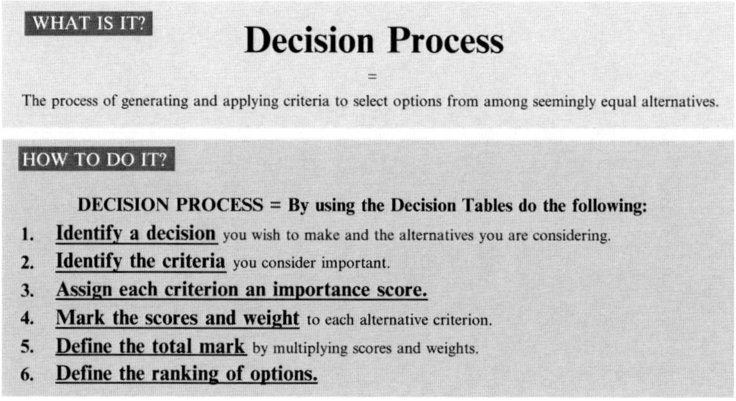

WHAT IS IT?

Decision Process

=

The process of generating and applying criteria to select options from among seemingly equal alternatives.

HOW TO DO IT?

DECISION PROCESS = By using the Decision Tables do the following:

1. **Identify a decision** you wish to make and the alternatives you are considering.
2. **Identify the criteria** you consider important.
3. **Assign each criterion an importance score.**
4. **Mark the scores and weight** to each alternative criterion.
5. **Define the total mark** by multiplying scores and weights.
6. **Define the ranking of options.**

Figure 9. Representation of the decision process *http://deepimpact.jpl.nasa.gov/ collaborative_ed_ module/7Appendix/AppendxFDecisionMaking.pdf.*

The methods mentioned above can be used with strengths and weaknesses for various types of risk analyses. However the use of risk methods based on PRA Nuc for RIDM specific tasks was considered to have important strengths. There are some reasons for which this type of methods can be preferred, of which the most important are as follows:

1. The method may be easily used as a support for decision-making process for CS with interdependencies.
2. The expected answers can be evaluated in both in qualitative and quantitative format and with more details, not only as a general recommendation/direction.
3. The method is a systematic tool, which is suitable for complex systems and with a very well defined and verified procedural set of principles and methods in some complex technical applications.
4. The method has the potential to adapt to more tasks than performed now, by including more decision-making areas into the existing framework.
5. The method can be verified and validated and used with high confidence for a given set of boundary conditions and very well defined and known limitations.
6. The apparent complicated aspect can be managed rather easily and is actually dependent on the resolution of the model (subsystems, components modules, etc).

7. The solutions can be provided in the format from Figure 3, which makes them an important tool for the support of decision-makers. The results are able, as can be seen from Figure 3, to support answers and suggestions to decision-makers on various types of questions, related not only to risk ranking, but also to the credibility of the results and limits of the evaluations.

The DM process, which considers the aspects related to risk impact of a certain option chosen as an information source, is a RIDM process. Based on the features described above the method PRA Nuc was considered a suitable tool for the evaluation of risks and support to the decision-making process PRA Nuc is considered a triple S method (Serbanescu, 1991; Serbanescu, 2005a; Serbanescu, 2005b), i.e. it is:

- Systematic
- Structured
- Systemic

It was also shown that the use of any method could generate some level of agreement on safety risk aspects between various actors in the DM process on risk for complex systems (industry, regulators, or the public) but there are important expected problems, too.

In this sense, the results of the evaluation of some cases showed that by using RIDM, PRECP, and DELIB, the following main conclusions can be drawn:

1. For a given combination of actors in the evaluation (Public-P, Industry-I, and Government-G) the applicability of an approach depends in a high degree on the CAS (Complex Apoietic Systems as defined in (Serbanescu, 2007a; Serbanescu, 2007b; Serbanescu et al. 2008a) evaluated in (Serbanescu et al., 2008c) three cases A, B, or C were chosen for the evaluation). If an approach is used for a CAS is considered of a lower ranking from the point of view of applicability of the method then one should expect systematic biased results and conclusions/recommendations to be included in RG should consider this aspect. The conclusions on applicability could be made based on a process considering as many and diverse opinions as considered to be relevant for a statistics – and/or decision-theory based trusted statement. However the evaluation should also consider that for different combinations of actors the ranking on importance from a risk point of view of various cases is different due to an apparent bias on the method induced by the combination of actors using it. This confirms and refines the "user effect" of a given approach. Applicability of various approaches to the cases in Figure 10.

2. The use of various tools of PRA Nuc type, which are amended for the interface with DM process is considered highly desirable from the perspective of having a set of tools and results presented with the consideration of their uncertainty (credibility) and a clear traceable and easy-to-modify list of assumptions. The recommendation is based on the experience for using SES RISK as a PRA Nuc method to perform RIDM tasks.

3. The goal of using the tool in order to derive recommendations to support RG process by adequate analyses and also by adequate communication to the decision-makers can be fully supported.
4. The object/situation modelled (the type of CAS analysed) – in this case it is SES – is actually imposing the type of tools to be used. The tool has to be adequate to the CAS modelled. SES RISK, which is PRA Nuc type of approaches is adequate for SES modelling.
5. There are specific benefits possible for evaluating not only risk ranking but also uncertainty calculations and periodical fast and traceable review for many sensitivity calculations given by the use of PRA Nuc type of calculations for specific CAS consisting of components modelling technical and scientific facets/relationships, as well as economic and social aspects. PRA Nuc type of analyses can comply with the requirement imposed for CAS systems of complex type during their use as RIDM tools.

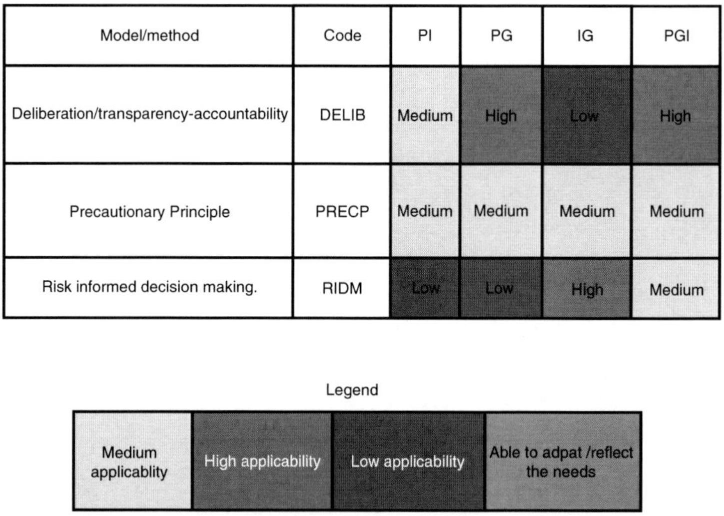

Model/method	Code	PI	PG	IG	PGI
Deliberation/transparency-accountability	DELIB	Medium	High	Low	High
Precautionary Principle	PRECP	Medium	Medium	Medium	Medium
Risk informed decision making.	RIDM	Low	Low	High	Medium

Legend

Medium applicablity	High applicability	Low applicability	Able to adpat /reflect the needs

Figure 10. Sample illustration of some conclusions on the use of various methods for risk governance processes.

Examples of Specific Infrastructures

The development of the SES model and the implementation of SES RISK method for its evaluation require pre-existing models and tools for infrastructures, which are included in a SES. It is therefore considered important to illustrate some of the aspects encountered related to the modelling and evaluation of various infrastructures as part of the SES tasks.

	CRITERIA/OPTIONS RANKING	Option A	Option B	Option C
1	Criterion 1 = Cost	Medium	High	Medium
2	Criterion 2 = Uncertainty	High	Medium	Low
3	Criterion 3 = Probability	Medium	Low	Low
4	Criterion 4 = Risk	Low	High	High
	TOTAL	Low	High	High

Figure 11. PRA Nuc type supported inputs for the decision-makers for ranking options.

Some General Common Aspects Related to Specific Infrastructures

The work performed for various infrastructures confirmed both the specifics of each and the common features of modelling and evaluation.

The main topics presented in the paper are aimed at describing the experiences gained and lessons learned from four applications. An important experience gained in this research is the discovery that the problems encountered in the use of this approach in nuclear and non-nuclear applications have common problems and solutions that can be grouped into a set of paradigms, which are presented in a set of seven "problem-solution" paradigms in the last subchapter. For the risk defined as a norm for a given CAS mentioned above a risk metric as per formula (4) can be used for any infrastructure:

$$\text{Risk} = f\,(P_{IE} * P_{PR} * P_d) \tag{4}$$

Where

P_{IE} is the probability of the IE

P_{PR} is a probability representing the plant reaction pattern for each IE challenge

P_d is a normalized probability representing the damage produced by a given IE

The risk model is defined by the contribution of various minimal cut sets (MCS) to the global risk frequency. It shows that there is the following type of dependency on a certain probability of a given event:

$$\text{MCS} = p_1 * (\,1 - g(p_i)) + h(p_i) \tag{5}$$

Where:

p_1 is the probability of the event for which various changes and impacts are being evaluated during a specific analysis $g(pi)$; $h(pi)$) are functions of the probabilities of basic events other than p_1

The CAS model for any infrastructure is structured on three levels (in analogy with the PRA Nuc three levels). More details on CAS and the related aspects are presented later in this paper.

The first level of CAS, the physical one, may be considered as a complex hierarchical system (CHS). A CHS may be represented as a structured system. Safety of such a system is considered a structural function of CHS, which has risk as a measure (Serbanescu, 1991). A sample of such a level for a CAS is represented in Figures 12–14 (Serbanescu, 2005a; Serbanescu 2005b).

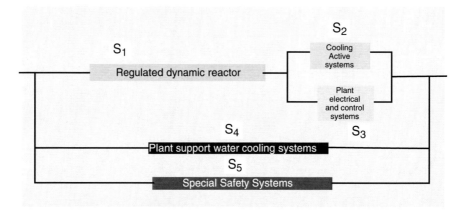

Figure 12. CAS level 1 sample model.

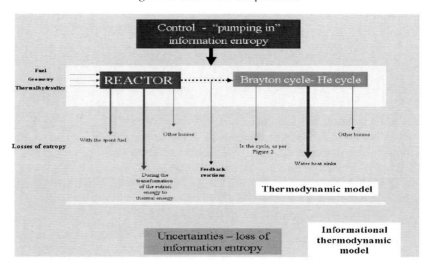

Figure 13. CAS level 1 sample model from energy and information exchange view point.

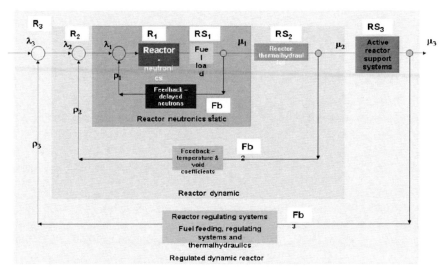

Figure 14. CAS level 1 sample model from cybernetic approach view point.

Figures 12–14 show three different possible models adopted for a CAS level 1:

- "classical" technological approach used as a basis for both reliability and risk analyses of a system
- an approach considering CAS from the point of view of information exchange
- a cybernetic approach

The development of PRA Nuc approaches for other non-nuclear or new nuclear applications was based on the use of the "classical" modelling of the systems as per Figure 12. However for the adaptation to the needs to model subjective probabilities approaches of the type represented in Figure 13 were used and for the features related to modelling of interdependence between subsystems of various type the Figure 14 approach was used.

With respect to these new approaches some previous results were used, as follows:

- For the modelling of subjective aspects a methodology for considering information processes as part of any safety model and included in the definition of its measure, the risk, was used as per (Serbanescu, 1991; Serbanescu, 2001; Serbanescu 2005a). The synergy was calculated as a function called Lagrangean, which is a function of the risk, considering as boundary condition the condition of independence between the basic events. The Lagrangean and the limiting conditions are as per (6) and (7).

$$L = R(x, p_i) - \lambda_i * B_i \tag{6}$$

$$B_i = \Sigma\, p_i - 1 = 0 \qquad\qquad i = 1, 2, 3, \ldots n \tag{7}$$

where:

      B   is the boundary condition function

  λ_i is calculated from (6), (7) and the condition to have an extremum value for L in the point

where $p_i=0$ (i=1,2,...n) and p_i (i=1,2, ...n) are given by the model B values.

     X   is the variable of the problem (for which the extremum of L is searched for, for instance one of the basic events probabilities, which is of interest for a specific application)

- For the modelling of a CAS as a complex hierarchical system the adopted approach was based on the results of considering the whole PRA model as? A hierarchical system and to model its tasks and components accordingly. A representation of this approach is shown in Figure 15 and the results were searched using approaches communicated in (Serbanescu, 1991; Serbanescu, 2001; Serbanescu, 2005a) and are schematically represented in Figures 17 and 18.

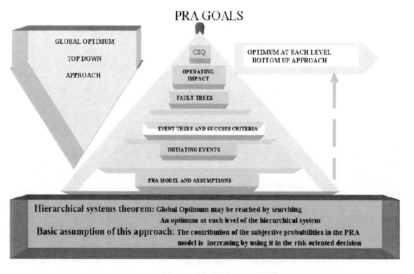

Figure 15. PRA as a CHS.

The objective function for a system modelling a given infrastructure as represented in Figure 15 is calculated in accordance with Formula (8):

$$R_j = Q * F_{IE} * P_{system,j} * P_{structures,j} * r_{fuel,j} * r_{HPB,j} * r_{conf,j} \tag{8}$$

Where

R_j	Expected quantity of radioactive material released per year
Q	Quantity of radionuclide in the reactor core inventory
F_{IE}	Frequency of the initiating event associated with sequence "j"
$P_{system,j}$	Probability of the system successes and failures for sequence "j"
$P_{structures,j}$	Probability of the passive structure successes and failures along sequence "j"
$r_{fuel,j}$	Release fraction from the fuel, given system and structure response for sequence "j"
$r_{HPB,j}$	Release fraction from the Primary Release Source Boundary, given system and response for sequence "j"
$r_{conf,j}$	Release fraction from the confinement, given system and structure response for sequence "j"

- During the process of adaptation and/or building new capabilities for PRA Nuc to be able to model new infrastructures, there was an extensive use of perturbation theory approaches. Starting from the "classical" basic PRA Nuc it was considered that a PRA-like model for new a CAS and/or new features of already modelled CAS assumes that the changes to the model impact only on the linear part of the model. The basic reason for this assumption was that as far as the CAS model for any new model is in compliance with the requirements for a CAS a modelling is possible. In subchapter 2.4 a more detailed presentation of the CAS and related requirements is presented. The results of the models for a CAS have to demonstrate whether the Pareto Set (Figure 16) of the solutions comply with the boundaries as defined by the licensing/societal requirements. It is assumed that linearity is within the perturbed limits to a given start point for a CAS model. This is based on the assumptions that the CAS model is linear in the logarithmic space is defined by risk as a norm of sigma algebras built to represent it (details in (Serbanescu 2005a)). It uses a combined set of PRA and other risk analysis methods as well as qualitative and quantitative pre- and post-screening criteria. The method is therefore considered an adaptation of the pre- and post-processing of the baseline PRA data for a given design change.

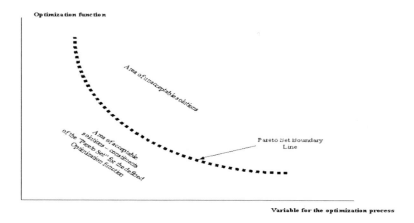

Figure 16. CAS level 1 Pareto set for risk objectives.

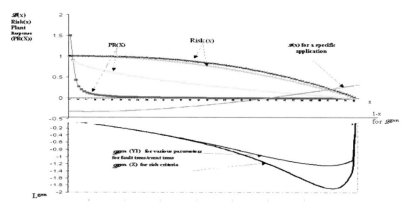

Figure 17. Sample objective functions for optimization of CAS level 1.

Examples of Specific Infrastructures

This paragraph shows some new developments on modelling new infrastructures for SES purposes. While modelling using PRA Nuc type of approaches of nuclear and fossil plant-related infrastructures as well as the electricity, gas, and steam grids, is very well represented in the literature, the risk modelling of other infra-structures such as renewable energy, or hydrogen installations and/or networks is in the beginning phases. This paragraph presents some of the results of the work done for the modelling of such new systems.

Item code	Model components	CAS Level 1 Physical level risk metrics	CAS Level 2 Society Decision Objectives	CAS Objectives definition Level 3			
				No Change Objective NCO	Small Change of Objectives CO1 (Note 1)	Major Change of Objectives CO2 (Note 2)	Total Change of Objectives TCO (Note 3)
PUBL1 *	Public challenges manageable by the existing system of public interface	NA	H	H	H	H	H
PUBL2 *	Public challenges NOT manageable by the existing system of public interface	NA	L	H	H	H	H
POL1 *	Political challenges manageable by the existing political system features	NA	H	H	H	H	H
POL2 *	Political challenges NOT manageable by the existing political system features	NA	L	H	H	H	H
ECON1 *	Economical challenges manageable by the existing economical system	NA	H	H	H	H	H
ECON2 *	Economical challenges NOT manageable by the existing economical system	NA	L	H	H	H	H
SEC1 *	Security challenges manageable by existing security systems		H	H	M	M	M
SEC2 *	Security challenges NOT manageable by existing security systems		H	M	M	M	M
EXTEV ***	External events challenges		H	H	M	M	M
BR **	System breaks leading to possible contamination	M	H	L	L	L	M
COOL **	Support cooling water systems failures	M	L	VL	VL	VL	L
HE **	Support high energy CAS part failures	M	H	L	L	VL	L
CONTR **	CAS control and operation	M	L	VL	VL	VL	L
CONT **	Last barrier to the containment release - containment	M	H	M	M	M	M

Notes

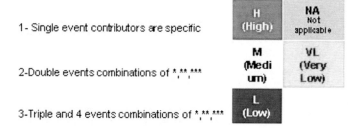

1- Single event contributors are specific

2- Double events combinations of *,**,***

3-Triple and 4 events combinations of *,**,***

H (High) NA Not applicable M (Medium) VL (Very Low) L (Low)

Figure 18. Sample results for a CAS levels 1–3 (Serbanescu et al. 2007a; Serbanescu et al. 2007b).

Renewable Energy Systems

PRA Nuc development was done in order to model renewable systems aspects to integrate them into a SES model. (Serbanescu, 2009; Serbanescu et al., 2010a; Serbanescu et al., 2010b)

Renewable energy systems (RES) have some specific features. Some of them is related to their dependence on geographical position of the areas modelled, as represented in Figure 19.

The PRA Nuc type of analyses performed so far for some RES (photovoltaic manufacturing, hydrogen refueling station) confirm some important features for the use of risk analysis for the evaluation of these complex systems The evaluation of risk of RES and the systems supporting their lifecycle related activities are part

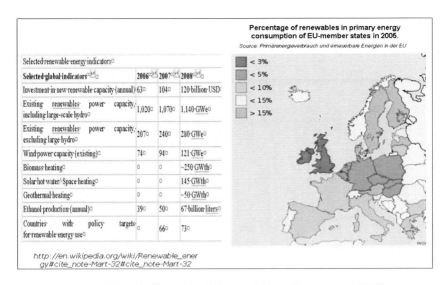

Figure 19. Sample illustration of the magnitude and importance of RES.
(Source: Primärenergieverbrauch und erneuerbare Energien in der EU; http://en. wikipedia.
org/wiki/Renewable_energy#cite_note-Mart-32#cite_note-Mart-32)

of a very dynamic trend of recent years to develop risk tools for new complex systems. The main new aspects which were considered as having impact on the RES type models were as follows:

- Definition of the objects to be studied has to consider carefully
  o defining the systems
  o lifetime cycle
  o type of subsystems
- The models to be used for risk analysis have to
  o define the potential risks from the object the type of model
  o consider applicability and/or need for changing of existing risk analysis tools
  o evaluate the detailed tasks, which need to be performed on a certain set of methods
- The framework in which the results of those risk analyses are to be used is defined by the success of evaluating their limits and feedback control.

The performance of risk analyses for RES highlights a series of issues which have to be considered in the author's opinion (described under the term paradigms and presented in detail for all SES models and SES RISK process in subchapter 2.4). The RES modelling for risk analyses purposes requires its representation of this complex system (CS) by using assumptions and increased design and operation information. In order to develop a computer usable model for RES, a consideration of its lifecycle phases, components, and interfaces between them and with other systems and the environment, public, and workers has to be done. RES models for risk analysis purposes describe scenarios of its behaviour if challenged by various disturbances. The description of the scenarios includes reaction of the model and its barriers to various challenges, as well as detailed descriptions of the barriers designated to cope with challenges. The risk metrics used for various levels (up to description of risk categories) was done based on damage states of RES in cases where various undesired challenges take place. Analogue results are expected for various criteria to describe survivability of RES in case its reaction to challenges is important from the perspective of interface with other energy systems. For all criteria the significance of the results resides so far in their relative values (between them) and not in the absolute values. The risk metrics and survivability criteria impose a certain ranking by importance of various parts of RES and the ranked contributors and the evaluation of the uncertainties in their calculations are provided as an input to the decision-making process targeting improvement of RES from risk perspective.

A feedback reaction for restarting the risk analysis process can be performed by using systematic tools to identify areas for the next iteration. Iterations are performed so that a convergence of the results related to the ranking of contributors to various risk levels is achieved Further reviews are expected to identify areas of strategic concern in RES risk analyses. Some specific features of the RES having impact on the PRA Nuc like modelling are presented in Table 7.

TABLE 7. Conclusions on renewable

Item	Feature	Expected positive aspect	Possible negative aspect	Possible impact on overall energy system risk process
1	Large scale or regional availability	Close to the user	Distribution and storage, smart grids needed	Medium
		Availability in various forms almost everywhere	Need to be identified and studied – their magnitude, availability, costs, social impact	High
		Diversity of renewable energies	Priority of their availability and coordinated use needed	
2	Renewable energy systems – features	Possible basis to building revolution, energy security supply strategies	They have features of complex systems: complex economical geographical engineering etc systems having a high uncertainty in modelling and management	High
		End use oriented energy systems	Complex uncertain systems	
		Lifecycle of the system makes it attractive and usable	Detailed evaluations of risks and problems in various phases of the lifecycle needed	

More details on the model are presented in subchapter 2.4 under the description of the paradigm 3 and in (Serbanescu, 2009; Serbanescu et al., 2010a; Serbanescu et al., 2010b).

Hydrogen Installations Systems

Risk evaluation of a hydrogen installation by using the methodology from (Serbanescu, 2008; Serbanescu, 2009; Serbanescu et al., 2010a; Serbanescu ct al., 2010b) also starts with the development of the model. The model is based on the description of the components of the installation, their interconnecttions, and the installation operating requirements. A set of assumptions and the consideration have to be considered in developing the model.

A schematic representation of the case study modelled so far is presented in Figure 20. As was mentioned in (Serbanescu, 2008) the main groups of components represented in Figure 20 of a hydrogen refilling station are:

- Hydrogen inlet pipe (HYIP) for periodic refilling of the storage tanks with hydrogen
- Purification and drying part (PD), consisting of a set of components for hydrogen purification and drying in order to maintain it at the required purity for the use in cars
- Compressor unit (CU), consisting of a set of two groups of compressors (low and high pressure) to assure the pumping and storage of hydrogen
- Storage Cabinets (SC) consisting of a group of tanks used for storage, called cylinders (CYL) connected in groups to assure the designed storage capacity and redundancy of the minimum required capacity if some of the cylinders cannot provide hydrogen
- Underground connecting pipe from SC to the parts where hydrogen is delivered to consumers
- Disposal part (DISP) where the hydrogen is directed to independent connections to each consumer (DISPU).

Various components and groups of components are located in different buildings. On the other hand it should also be mentioned that each component complies with some rules of protection by distance to possible explosions and relief valves to protect an installation from over-pressurization. The operational connections, location, and barriers are included in the model and this is reflected in the following steps of the methodology. Their representation as per the needs of the risk analysis code are presented in (Serbanescu et al., 2010a; Serbanescu et al., 2010b). More details on the model are presented in subchapter 2.4 under the description of the paradigms 2 and 3 and in (Serbanescu, 2008; Serbanescu, 2009; Serbanescu et al., 2010a; Serbanescu et al., 2010b).

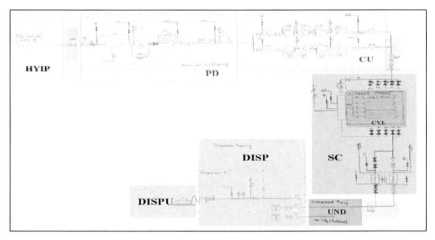

Figure 20. Hydrogen refilling station. (Serbanescu 2008).

LESSONS LEARNED FROM THE PERFORMANCE OF SES EVALUATIONS

The last guiding aspect was to perform an evaluation of the lessons learned after each iteration of the SES RISK process and to define actions for the next iterations.

Returning to the issue of why to use a SES RISK method after all, one has to mention that currently there are several decision-support methods available to be used for the assessment of the multiple risks faced by a complex industrial-based society. However there are some important aspects to be mentioned about the work performed for SES modelling as presented in this chapter:

- It was done by using risk type approaches for the reasons mentioned in the previous subchapters.
- It was also based on extensive results on the application of PRA Nuc type of analyses for many areas (USNRC, 1998; Haimes, 2004; USNRC, 1983; Serbanescu, 2001; Serbanescu, 2005a; Serbanescu, 2007; Serbanescu et al., 2007a; Serbanecsu et al., 2007b).
- It had the goal of showing how to develop a SES model using risk approaches based on PRA Nuc experience and its developments to various other areas (Serbanescu et al., 2008b).

These developments have been shown to trigger two main methodological novelties:

- The extensive use of subjective probabilities evaluations, and
- The use of hierarchical systems theory in the modelling.

Some of the most novel features had to be added to the PRA Nuc approaches in order to be able to use them for SES risk-like models are as follows:

- Combination of the risk modelling with decision theory and energy technology insights in order to deal with the complex issues of technical and non-technical character.
- Extensive use of new developments of the use of PRA Nuc like methods for new – non-nuclear infrastructures.

The most important experience gained in this work is the conclusion from the problems encountered in the use of risk approaches for SES evaluations a set of common problems and solutions appear. These problems are called paradigms of the approach and are listed below. They embed the most important set of conclusions on the lessons learned and the actions taken to solve them. These paradigms are the following:

1. A suitable system for PRA modelling SES and its infrastructures (various energy sources lifecycle etc) as a complex system (CS) has to be built.
2. Detailed SES description as a CS has to be developed for which risk analyses are to be performed.
3. The SES and infrastructures included in it, as well as their interdependencies model description has to consider the challenges to it, for various lifecycles stages and a description of mitigating actions (barriers etc.) designed to cope with these challenges.
4. SES risk analyses require definition of objective functions and risk metrics.
5. Risk calculation results have to be provided in a format usable for the decision-making processes.
6. The validity of risk results and soundness of the decisions based on them is highly dependent on the understanding of errors and systematic biases.
7. The risk analyses are an iterative process in which a systematic search for solutions to paradoxes appeared during the process is the basis for the next iterations.

Paradigm 1. Suitable Systems for PRA Modelling

A modelling system has to be built in order to represent RES as a complex system for further risk analyses. However building the model requires making assumptions and this induces the need for careful consideration of the results and a solution to prove by iterations, benchmarking against other methods, etc their convergence to a stable unique real solution.

The systems that could be modelled based on PRA-Nuc approaches should be a special type of complex systems, i.e. complex, self-regulating, self-generating, and hierarchical – Complex Apoietic Systems-CAS-(Serbanescu, 2007; Serbanescu et al., 2007a; Serbanescu et al., 2007b). Some clarifications on the adopted definitions on complex systems as adopted in the work performed are presented below.

Complex Systems (CS) are systems with specific features from which the most important could be considered as following:

1. The system is characterized by the fact that the interrelations between its elements/components create a synergetic effect, which actually makes the system unique. The interactions are of short distance between elements or of long distance (in space and time).
2. The existing relationships in a CS is defined by the existence of feedback loops, which results in a dynamic system structure, with constant nonlinear interfaces.
3. CS' are interacting with the environment, being open to the exchange of energy and information.
4. No part of a CS can contain the whole and therefore no part can assure the control of the whole CS. This could lead to specific tools for controlling it, such as distributed control, hierarchical control, and/or external to the CS unitary control.
5. CS elements are CS' themselves and are highly adaptive, which introduces another difficulty in the choice of the modelling tools. CS' have a historical hysteresis characteristic, in the sense that they have a history and that leads to an effect known as "the butterfly effect", i.e. a small change in time and space will impact the future history of the whole CS or to a completely different component that the one that had the initial change. However, even if the interrelations of the components are nonlinear they comply with the cause and effects law in this specific manner. The concept of linearity and nonlinearity apply to the models defined for CS and it is one of the important modelling issues. CS has boundaries that are difficult to define, which has a high impact on the content of their models and their own changing dynamics and flexibility.

Complex Apoietic Systems (CAS) were introduced by the works of (Maturana et al., 1980) and have the specific features additionally to the generic CS ones, sometimes refining the generic CS features. CAS are systems, for which an autopoietic mechanism can be defined, leading to the system possibility not only to self regulate, but also to recreate itself, as follows:

1. The system boundaries have to be clearly defined at any moment in time
2. The system has to have components, being themselves a CS
3. The cause – effect law interactions have to be operable
4. The system boundaries have to be self-produced by the system, as well as
5. The system components
6. The rest of the components should be also be able for most of them to be self-produced by the system

In other words, autopoiesis means "self-production" (self-creation or production) and expresses a fundamental complementarity between structure and function. An autopoietic machine is a machine organized (defined as a unity) as a network of processes of production (transformation and destruction) of components which:

- through their interactions and transformations continuously regenerate and realize the network of processes (relations) that produced them; and
- constitute it (the machine) as a concrete unity in space in which they (the components) exist by specifying the topological domain of its realization as such a network.

The space defined by an autopoietic system is self-contained and cannot be described by using dimensions that define another space. When we refer to our interactions with a concrete autopoietic system, however, we project this system on the space of our manipulations and make a description of this projection. One can recognize in this problem similarity with the definitions and methods in integrated risk models and their connection to the Gödel's Theorem and hence with the issue of control from inside (by a given component of the system) of that system.

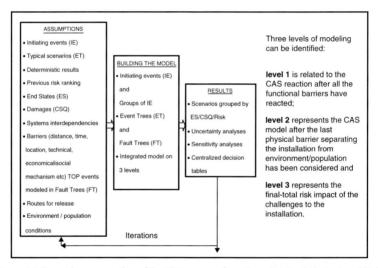

Figure 21. General representation of the PRA process for a Complex Apoietic System (CAS).

Representation of the complex system – the nuclear power plant for the purpose of the evaluation of their safety was referenced in (Serbanescu, 2005a). The complex system was represented as a hierarchical structured one on three levels, of which the first one is related to the generic societal targets and objects. The next levels go to subsystems related to the regulatory, design, and concept subsystems and to the operation and detailed evaluation. The system is represented with a feedback connection to the previous levels, assuring self-regulation. Each level is defined by elements of the subsystems composing the whole. The elements of the sets at each level are connected on vertical (from level to level and on horizontal (at a given level). The vertical structure is defined by a relationship of a type "Element X is requiring element(s) Y with the degree of certainty W". At each level there are uncertainties connected to the rules environment, which indicates the high importance of the type of inferences to be used, usually grouped in a conventional manner into deterministic and probabilistic ones. The three levels, with a defined relation between levels, generate a sigma algebra on the created structure, for which a norm can be used. For the third level of detail, risk can be used as a norm for the sigma algebras defined as per (Serbanescu, 2005a).

The model to be built and the method to be used for the evaluation of the risk induced by CAS have to comply with seven requirements:

(i)   To be systemic (i.e. use of systems theory in modelling), systematic (i.e. use the same approach throughout the entire PRA process) and structured (i.e. consider the model hierarchical and evaluate each of the levels one by one – either by using top-down or bottom-up approach;

(ii) To be able to define structures of dynamic cybernetic interrelations between components with both random and intentional types of challenges to CAS, and to solve nonlinear dynamic models by defining what linearity means for a CAS;

(iii) To define the system as a whole, as being a result of the synergetic interfaces of its components and to define also the CAS interface with the environment;

(iv) To have a solution for the system control (like for instance distributed control, hierarchical control and/or external to the CAS unitary control);

(v)   To have a system management based on predefined objectives, such as energy/substance balance or risk impact, including validation and verification processes;

(vi) To solve the specifics of the cause – effect issue for a CAS, which are connected in turn to other issues such as definition of linearity, uncertainty, and system structure modelling;

(vii) To be dynamic and highly flexible in defining initial and boundary conditions.

Feasibility of extension of PRA-Nuc application was previously demonstrated for other CAS-like systems (Serbanescu, 2005a; Serbanescu 2005b) and an integrated risk model of random technical and intentional man-made events for a nuclear power plant was described in (Serbanescu, 2007a; Serbanescu, 2007b). The same approach was adopted for SES models. Figures 22–25 illustrate the approach adopted for the modelling of CAS systems as shown in (Serbanescu, 2005a; Serbanescu, 2007a; Serbanescu 2007b).

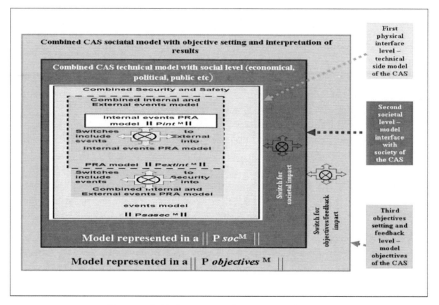

Figure. 22. PRA representation as a generic CAS.

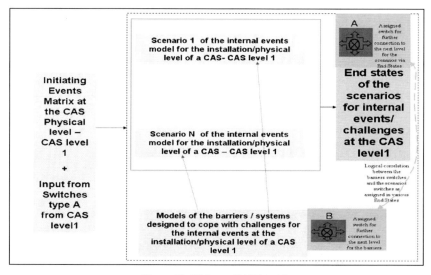

Figure 23. PRA as a CAS level 1.

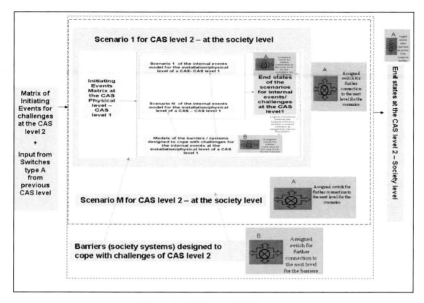

Figure 24. PRA as a CAS level 2.

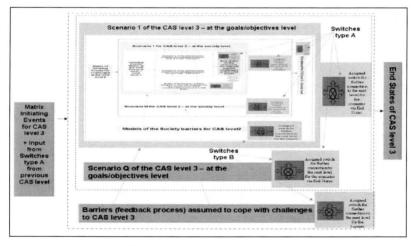

Figure 25. PRA as a CAS level 3.

Paradigm 2. System Description

A more detailed description of the system built to represent SES for risk analysis purposes was done by considering its components, lifecycle phases, and their interfaces. For a given model of SES there is a need to have information on its components, lifecycle, and their interdependencies. The model built for SES has to be of a specific type.

The general process of PRA Nuc approach analysis for a CAS system was represented in subchapter 2.3. Reiterating the main flow of the process, it has to be mentioned that for all the Initiating Events (IE), the reaction of CAS is also modelled considering the expected End States (ES) of each scenario. A PRA Nuc like model for a CAS was represented in Figures 22–25 from Paradigm 1. There are three levels of modelling, which can be identified:

- level 1 is related to the CAS reaction after all the functional barriers have reacted;
- level 2 represents the CAS model after the last physical barrier separating the installation from environment/population has been considered and
- level 3 represents the final-total risk impact of the challenges to the installation.

However the details of the model require that the system has to be built in order to represent SES for risk analysis purposes, i.e. it has to consider its components, lifecycle phases, and their interfaces. Therefore for a given model of SES there is a need to have information on its components, lifecycle, and their interdependencies, all the time considering that the model built for SES has to be of a specific type. This paradigm addresses the need for accurate details on the systems and traceable assumptions on their modelling.

As the modelling requires also a representation of the interconnections between the systems, a set of "Connecting End States" (CES) is defined. CES are used to assure both "horizontal" (at the same level of the CAS model) and "vertical" (between levels of CAS) connections between scenarios.

As an example, Figure 26 illustrates a CAS model being developed for modelling of the hydrogen distribution station (presented under subchapter 2.3 and with more details under next paradigm). Similar models were also built for security of all infrastructures part of the SES model and the SES model itself.

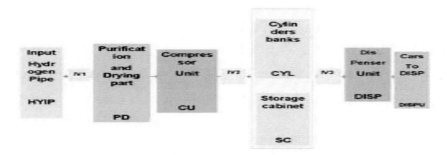

Figure 26. Sample representation of the main systems of a hydrogen distribution installation.

This illustrates the so-called Reliability Equivalent Diagram (RED), the basis for any reliability and/or risk analyses as per PRA Nuc approaches.

> **_Paradigm 3. Scenarios and Faults (Barriers) Description_**
>
> _The SES model has to consider the challenges to it and a description of mitigating actions (barriers, etc.) designed to cope with these challenges for each SES model including a set of potential challenges and the barriers designated to cope with them are defined. For each challenge a scenario of the SES model reaction to the challenge and the possible end situations of the system defined under SES is defined. The scenarios and end states are grouped based on criteria adopted for degree of damage of the model, based on acceptance criteria defined a priori._

After REDs were built for every barrier of every infrastructure, the scenarios of SES mode reaction to each of the challenges have to be built and combined with each other.

Figure 27 illustrates the general process of generating ES by combining the scenarios defined in the Event Trees with the Fault Trees of the systems, which are assumed to operate as barriers against various challenges. Such a process is known in PRA-Nuc as "integration of Fault Trees into Event Trees".

In this figure, the representation technique shown is similar to the graph theory approach and is in accordance with the existing standards *(USNRC 1983)*. Nodes and basic events (i.e. final failures of a given chain of failures) represent failures.

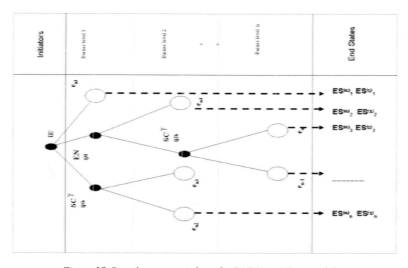

Figure 27. Sample representation of a CAS Event Tree model.

Once the ET and FT are built then the next phase is the "integration of FT into ET". The principle of this phase was presented actually in the formulas (1) and (2) in subchapter 2 and they are represented in Figure 29. Therefore the result of the integration process is a set of combinations of failures for a given IE leading to a given ES. Further on the ES can be grouped and a set of corresponding failures leading to a given risk category is obtained.

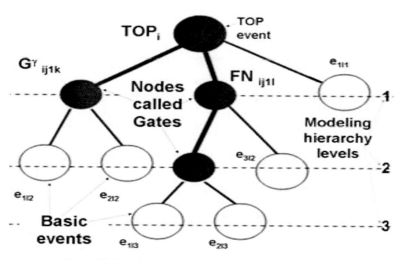

Figure 28. Sample representation of a CAS Fault Tree model.

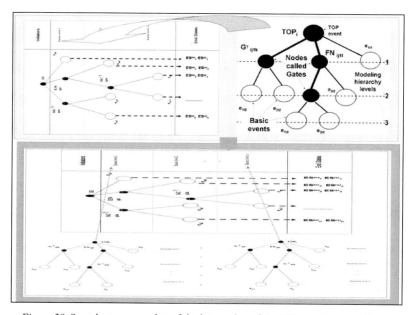

Figure 29. Sample representation of the integration of Fault Trees into Event Trees.

However while in Figure 27 the nodes show the possibility of failures for entire systems of the CAS, which were designed to act as barriers, in Figure 28 the nodes show the possibility of failure of given components in a given system.

In Figure 29, $ES^{a}_{(1-n)}$ represent end states of a given scenario, which indicate certain levels of damage after the scenario took place, while $ES^{a}_{(1-n)}$ are used as

CES. The TOP nodes from Figure 28 are connected with the SC and EN type nodes in Figure 27. The calculation of scenarios leading to an end state in Figure 29 which includes combination of failures at the level of components of the whole CAS. Please note that the notations shown in these figures follow the PRA-Nuc principles and are used during the implementation of the model into the computer codes.

As it was mentioned in (Serbanescu, 2005a; Serbanescu et al., 2007a; Serbanescu et al., 2007b) a CAS model of challenges is generating a σ-algebra over the set of all possible ES. In such a structure "risk" is defined as a norm in the measurable vector space of the sets of scenarios. The risk is defined as the distance between normal state and an altered end state. This distance is usually calculated, as a product between the probability of a given sequence of scenarios and the damages associated with the end states after the scenario will take place. The scenarios resulting from Boolean combinations of failures differ in many aspects, such as the combination of failed barriers or the ES (final conditions) after the scenario comes to an end for the given level of PRA Nuc type of model.

Figure 30 shows an example of a CAS scenario definition. End states for levels 1 and 2 are also known as Release Categories (RC) while ES of level 3 are known as Risk Categories (RK). More details of these and representative examples of these entities are presented in (Colli et al., 2008). Riskspectrum program provides two type of results of risk – so called risk metrics (Serbanescu et al., 2008a):

- the sum of all the components failures leading to a given risk level (called Minimal Cut Sets MCS)
- the relative contribution of a given component to all the sum of combination (a weighting factor (called "importance" in PRA Nuc – Imp FV or simply FV) and for which, in some papers, a modified indicator is used to manage large databases with results for a faster processing and ranking, as shown in (Serbanescu et al., 2008a).

In any analysed case the ES identification is subject to continuous iterations and sensitivity analyses following the process previously represented in Figure 30. A sample of a typical IE list for the hydrogen installation presented in Figure 30 is:

(i) *Breaks* (e.g. break of vessels/cylinders in storage cabinet unit);
(ii) *Leaks* (e.g. leak from the hydrogen inlet ESD);
(iii) *Overpressurization* (e.g., overpressurization of 10 mbar in PD Unit);
(iv) *Thermal challenges to installation* (e.g., thermal impact of 3 kW/m^2 in PD Unit);
(v) *Fire, explosions, missiles* (e.g., fire generated by installation failures in PD unit);
(vi) *Support systems failure* (e.g. loss of central control);
(vii)*External event* (e.g., external Flood in UND);
(viii)   *Security threats* (e.g., security challenge in area close to units).

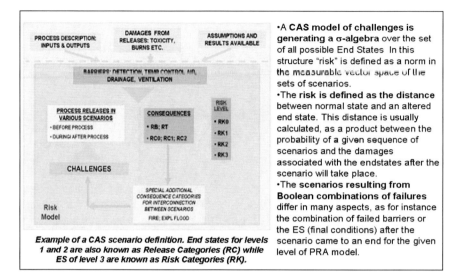

Figure 30. Example of end states for hydrogen installation.

It is very important to mention that a specific feature of a CAS model level 3 for SES is that ES are SES risk-related criteria. In this case, scenarios leading to various SES categories are ranked on the basis of their risk and importance, in accordance with PRA-Nuc methodology. It is also important to note that for the SES case, a set of economical (e.g. energy price variations) and socio-political initiators (e.g. result of referendum for a given energy source) were also defined (Serbanescu et al., 2008a).

Paradigm 4. Objective Function and Risk Metrics

SES evaluation using the method SES RISK requires definition of objective functions and risk metrics. Objective functions for the risk analyses of the SES model were adopted using risk metrics (risk levels 1–3) based on various type of damage states of SES model at various challenges and general SES system survivability categories for SES model as a complex system if challenged by various factors.

After the SES model is prepared and the method SES RISK is defined, an evaluation of SES can be done, with the goal to identify the weak points of the system if challenged by external undesired factors. In order to evaluate this aspect, as in any system modelling and evaluation, it is necessary to define the objective functions and risk metrics. There is a set of objective functions for the risk analyses of the SES model, which were adopted using risk metrics (risk levels 1–3) and which were based on various types of damage states of SES model as a whole and its component infrastructures. Therefore damage end states and general SES system survivability categories for SES model were defined in subchapter 1.

However under this paradigm a very important specific aspect of any CAS risk evaluation is presented; it is related to the fact that for any type of CAS, constraints exist in terms of risk. These constraints are usually represented in the form of a linear dependency between the acceptable threshold for risk and the frequency of a given event (IE) to happen.

From a topological point of view, this threshold in a 3-dimensional subset (defined by the risk, probability of an IE and the dominant parameter of the CAS) splits the space of possible solutions into two areas separating the acceptable solutions from the unacceptable ones. The threshold line is linear if the axes of risks and event probability/frequency are represented in the logarithmic scale. This line is another representation (in 3D form) of Pareto sets of solutions as presented in (Serbanescu, 2005a; Serbanescu et al., 2007a), and for this reason, PRA Nuc-like analyses can be also used to define the Pareto set of acceptable solutions for a CAS.

More detail is shown in (Serbanescu, 2005a) – one can consider a system as a cybernetic hierarchical one (e.g. of a CAS type as explained before). As it was shown in (Serbanescu, 2005a) for such systems the distance from the best solution is quantitatively described by a specific type of risk measure calculated as a loss of information entropy (Jaynes, 2003; Smithson, 2000). Consequently, the optimum for a certain problem is achieved when the Shannon Information Entropy[3] reaches a minimum, given the variables constraints. In mathematical terms, this leads to the task of finding the optimum for (9) with the limit conditions (10).

$$S_{inf} = X_I * \ln X_I , \qquad (9)$$

$$\sum X_i \leq \alpha , \qquad (10)$$

In Formulas (9) and (10), X stands for any risk metric of a given PRA level (e.g. probability of failure or risk) and α is used for the limitations imposed to those parameters (e.g. optimization target for acceptable probability of a given ES or risk values). The solution of equations (9) and (10) are represented in the form of a Lagrangean Function (LF). The whole process is schematically represented in Figure 31.

In order to analyse possible changes to the risk metrics, a tool was needed to evaluate the resulting impacts so that decisions can be taken on what course of action to follow and it is on Perturbation Theory (Kato, 1995). This approach is used for matrices built with event trees and fault trees in PRA Nuc specialized computer codes. Therefore, in order to evaluate the impact of challenges on the CAS models and/or modifications performed mainly under the sensitivity analyses for SES model, linearity is assumed within the "perturbed" limits of the CAS model.

---

[3] Shannon information entropy is defined for the set of CAS risk metrics solution.

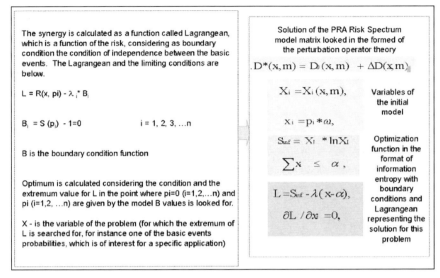

The synergy is calculated as a function called Lagrangean, which is a function of the risk, considering as boundary condition the condition of independence between the basic events. The Lagrangean and the limiting conditions are below.

$$L = R(x, pi) - \lambda_i {}^* B_i$$

$$B_i = S(p_i) - 1 = 0 \qquad i = 1, 2, 3, \ldots n$$

B is the boundary condition function

Optimum is calculated considering the condition and the extremum value for L in the point where pi=0 (i=1,2,…n) and pi (i=1,2, …n) are given by the model B values is looked for.

X - is the variable of the problem (for which the extremum of L is searched for, for instance one of the basic events probabilities, which is of interest for a specific application)

Solution of the PRA Risk Spectrum model matrix looked in the formed of the perturbation operator theory

$$D^*(x, m) = D_0(x, m) + \Delta D(x, m)$$

$$X_i = X_i(x, m), \qquad \text{Variables of the initial model}$$

$$X_i = p_i {}^* \omega,$$

$$S_{nf} = X_I {}^* \ln X_I \qquad \text{Optimization function in the format of information entropy with boundary conditions and Lagrangean representing the solution for this problem}$$

$$\sum x \le \alpha,$$

$$L = S_{nf} - \lambda(x - \alpha),$$

$$\partial L / \partial x = 0,$$

Figure 31. Illustration of the process of calculation of the risk Lagrangean.

This linearity is related to the logarithmic risk metrics set of results. The verification of the impact of this assumption is one of the main targets of the sensitivity cases.

As a consequence of such an assumption the building of the model and the calculation of solutions is guided by formulas of type (11), in which "R" represent the risk metrics function, "NL" index is related to "non-perturbed" initial model/solution, "HC" is related to the induced perturbation in a format of a Hazard Curve, "R⁰" is the non perturbed solution for risk metrics (reference solution), ΔR the modification induced in the results by a given assumption and ε_{error} is the tolerance.

$$R^{NL} = F^{NL}(P^M, HC) = F^0 \otimes HC = R^0 + \Delta R + \varepsilon_{error} \qquad (11)$$

As it is shown in Figure 32 and in Formula (11), F is the function that builds the solutions of the PRA-Nuc model and it is defined based on the sets of end states P^M and the hazard curve of a given new phenomenon to be modelled (HC), and with values in the sets of values obtained for risk calculations of all ES as defined by the set R^{NL} (Serbanescu, 2007).

An example of risk metrics results for development of new applications for PRA Nuc has been explained in more detail in (Serbanescu, 2005a), where different approaches to evaluate risk metrics were discussed regarding their use in decision-making related to optimized solutions from a risk perspective.

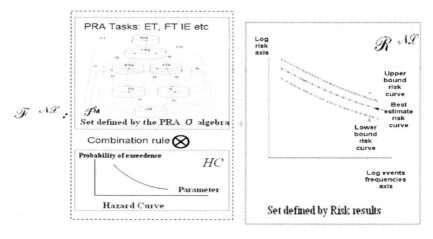

Figure 32. Sample representation of the definition of the solutions for PRA-Nuc in new applications based on the old ones.

Figure 33 shows two approaches in presenting risk metric results:

- A2 is the classical approach, as defined by (USNRC, 1983).
- B2 is the approach based on the Lagrangean Function (LF) approach mentioned above.

If one would like to use these approaches in decision-making, it is important to note the subtle difference between A2 and B2:

- In the classical approach (A2), the user has to bear in mind both the space of acceptable solutions from a risk perspective (i.e. paraboloid inside the cone) and the limitations (the cone).
- In the LF approach (B2), the limitations are already embedded in the space of acceptable solutions defined by the internal paraboloid (i.e. the cone is already embedded within the paraboloid). The external paraboloid in B2 indicates the external upper bound of the solutions and not the limits defined by the cone in A2.

In both cases the acceptable solutions from a risk perspective indicate scenarios for acceptable design solutions with various degree of feasibility, providing also important information for the CAS risk optimization process.

As shown in Figure 33 there are 10 types of solutions identified in the solutions space. For A2 these are:

1. low risk impact – not difficult design solutions;
2. low risk impact – solutions defined mainly by active systems features;
3. medium risk impact – achievable design solutions;
4. medium to high risk impact – solutions supported by passive features;
5. high, but still acceptable risk impact – difficult design solutions

For B2 these are *6–10*, where the note 6 in Figure 33 corresponds to group *5* mentioned above, i.e. high but still acceptable risk impact-difficult design solutions, and *10* corresponds to *1* (i.e. low risk impact-not difficult design solutions). It is also worth noticing that results shown in Figure 33 are typical for all the CAS mentioned in this chapter -examples also in (Colli et al., 2008; Serbanescu et al., 2008a; Serbanescu et al., 2008b).

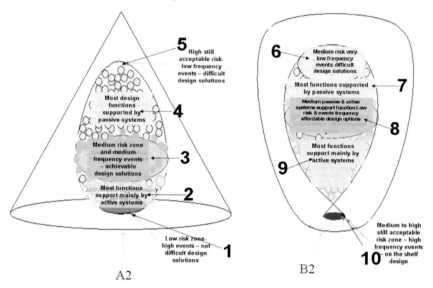

Figure 33. 3-dimensional representation of the risk metrics results for CAS level 3.

Paradigm 5. Risk Calculation and Decision-Making Processes

SES RISK calculation results have to be provided in a format usable for the decision-making process. Risk analyses results for SES are provided in a form of: a list of contributors to various risk and damage levels ranked in the order of importance – given the whole process and assumptions of the model the results have significance if considered in their relative values and not in absolute one; a similar set of results for survivability criteria adopted for SES can be also obtained and in both cases an indication on the expected error magnitude is given and results are then used for the risk-based decision-making process.

The main goal of the SES evaluation is to be able to support the DM process. Therefore SES RISK calculation results have to be provided in a format usable for the DM process. Risk analyses results for SES are provided in a form of: a list of contributors to various risk and damage level ranked in the order of importance – given the whole process and assumptions of the model the results have significance if considered in their relative values and not in absolute one; a similar set of results for survivability criteria adopted for SES can be also obtained and in both cases an indication on the expected error magnitude is given and results are then used for risk based decision-making. One of the main uses of the risk analysis is to support the DM process in various areas. The graphical representation of the commonly used decision analysis problems methods was introduced by (Howard, 1984).

PRA–Nuc has been used to support the decision-making process both as guiding information (RIDM) or to assess risk-based criteria (Risk Based Decision Making – RBDM). The experience gained in its application to the nuclear system led to the identification of a set of areas of applicability for both the deterministic and probabilistic tools, as shown in Figure 34.

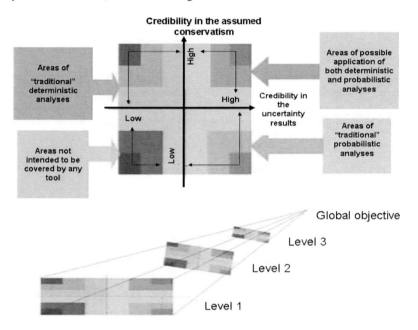

Figure 34. Sample representation of areas of applicability for decision-making of deterministic and probabilistic approaches. (Serbanescu et al., 2007b).

In this classification the area of applicability is defined by two criteria:

(i) credibility in the assumed degree of conservatism of the model, and
(ii) credibility in the uncertainty level of the built model and applied method to evaluate risk.

$$\mathbf{O} = (\mathscr{P} \otimes U_{(P)}) \otimes (\mathscr{D} \otimes U_{(D)}) \otimes (\mathscr{F} \otimes U_{(F)}) \tag{12}$$
$$\quad R_P \qquad R_{G1} \qquad R_D \qquad R_{G2} \qquad R_F$$

Formula (12) can be applied in order to find out how certain could one be on a risk analysis based on probabilistic and deterministic set of results and/or how to "combine" them (if possible). The function O (objective of the decision process) is a result of a combination using a series of logic operators:

- ○ $\mathbf{R_{P,D}}$ for reasoning on credibility of probabilistic and respectively deterministic results, and $\mathbf{R_F}$ for the reasoning on the credibility of reasoning based on feedback from experiments/real cases;
- ○ $\mathbf{R_{G1}}$ and $\mathbf{R_{G2}}$ for connecting results on reasoning based on:

  - • probabilistic evaluations (for the terms noted with $\mathbf{P}$ – probabilistic statements and $\mathbf{U_p}$ – probabilistic statements uncertainties),
  - • deterministic evaluations (for the terms noted with $\mathbf{D}$ –deterministic statements and $\mathbf{U_d}$ – uncertainties of deterministic statements),
  - • feedback review statements (for the terms noted with $\mathbf{F}$ – statements based on feedback review and $\mathbf{U_f}$ – uncertainty of the statements from feedback review).

It is also important to note that risk analysis results are usually divided in "deterministic"-oriented statements and "probabilistic"-oriented statements.

For deterministic judgments the result is composed of the criteria value $\mathbf{D}$ and the level of uncertainty in this values ($\mathbf{U_d}$); for the probabilistic results the components of the results are $\mathbf{P}$ and $\mathbf{U_p}$. There is also a component of results given by feedback from real object while compared to the model ($\mathbf{F}$ set of statement).

Paradigm 6. Understanding Errors and Judgment Biases

The validity of risk results and soundness of the decisions based on them is highly dependent on the understanding of errors and systematic biases. Results provided to the decision-making process contain information on the uncertainties encountered due to the modelling process, as mentioned in the previous paradigm and with details in the annex and references and information on the systematic biases induced by the process due to specific "myths" used explicitly or tacitly in risk analyses as used now.

SES results on the risk impact of various challenges are delivered as an input for the DM process. However the issue of credibility of the results, transposed for the decision-makers in the format of degree of certainty of recommended actions is crucial for any method. Therefore understanding of errors and systematic biases in the model is essential.

In the search for solutions on the task of self-assessing the credibility of risk analysis results, two groups of systematic biases are identified.

The first is related to the generic scientific method and its drawback, for which a series of systematic errors (called "scientific myths") exist as defined in (Mc Comas, 1996):

(i)   Hypotheses become theories which become laws;
(ii) Hypothesis is an educated guess;
(iii) A general and universal scientific method exists;
(iv) Evidence accumulated carefully will result in sure knowledge;
(v)   Science and its methods provide absolute proof;
(vi) Science is procedural more than creative;

(vii) Science and methods can answer all questions;
(viii)   Scientists are particularly objective;
(ix) Experiments are the principle route to scientific knowledge.

TABLE 8. Sample representation of the reasoning operators from Formula (4) used in decision-making statements

Impact function / cases	Case1 Optimistic trust in risk results	Case 2 Pessimistic trust in risk results	Case 3 Neutral trust attitude on risk results	Case 4 Overoptimistic trust of risk results	Case 5 Over pessimistic trust attitude in risk results
P	L	L	M	L	M
R_P	L	L	M	L	H
U(P)	H	H	H	H	H
TOTAL p	L	L	M	L	H
D	H	H	M	H	L
R_P	H	L	L	H	L
U(D)	L	L	L	L	L
TOTAL D	M	M	L	M	L
F	H	H	M	H	L
R_F	H	L	L	H	L
U(F)	L	L	L	L	L
TOTAL F	H	M	L	M	L
R^{G1}	M	M	L	L	L
R^{G2}	H	M	M	H	M
O Total objective function	M	M	H	L	L

The second group of biases is related to the "myths" generated by risk analyses for CAS, for which a list as defined by (Hansson, 2000):

(i)   "Risk" must have a single, well-defined meaning;
(ii) The severity of risks should be judged according to probability weighted averages of the severity of their outcomes;
(iii) Decisions on risk should be made by weighing total risks against total benefits;
(iv) Decisions on risk should be taken by experts rather than by laymen;
(v)   Risk-reducing measures in all different sectors of society should be decided according to the same standards;
(vi) Risk assessments should be based only on well-established scientific facts;
(vii) If there is a serious risk, then scientists will find it if they look for it.

Paradigm 7. Solutions to Paradoxes

Feedback from each SES RISK iteration was provided. In order to perform this task, the lessons learned were considered for the next iterations. A risk analysis was considered as an iterative process in which a <u>systematic search for solutions to paradoxes</u> appeared during the process that was the basis for the next iterations. A systematic review of <u>the sources of uncertainties and biases</u> identified in the previous phases was performed. Based on the findings and specific features of each phase a set of <u>feedback actions</u> for next iterations were defined.

Identification of the existence of possible systematic errors as described in the previous paragraph on the 6th paradigm is very important. However understanding what are the biases and limitations is only part of the feedback problem. It is very important to be able to understand the sources of those limitations and to define actions and strategies for further reviews of the SES model and the improvements of the results in the sense of getting results with better confidence.

In order to search for possible solutions for the systematic errors in the models and in the methods some criteria were identified. One of them is related to the knowledge acquisition process. If the theory of triadic approach as defined in (Peirce, 1931) is applied as one possible interpretation of the process, then the process of developing CAS models could be represented as in Figure 35.

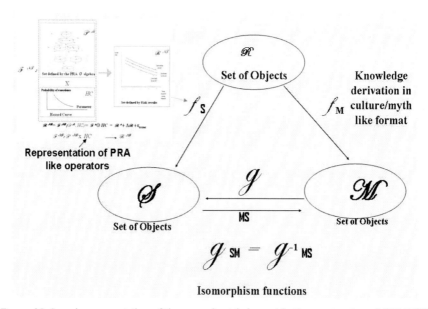

Figure 35. Sample representation of the reasoning triads used in the construction of SES RISK tools.

The process of finding solutions for the CAS model using an adequate method can then be represented as in Figure 35, where the areas of acceptable solutions for the CAS model is defined by the intersection between the areas of the set of model representation and the set of solutions obtainable with the CAS method described before. The model is therefore defined by a set of three relations between the set of real objects (**R**), for which the set of CAS (**S**) models are built, while generating the set of systematic biases (**M**).

As it is shown in Figure 36 the *triad of models* is represented by KNW-SOC-ENV (being equivalent to the triad in Figure 17) and the *triad of methods* is represented by RSK-DEC-CAS, which is built based on PRA-Nuc with the specifics presented previously in (Serbanescu et al., 2008).

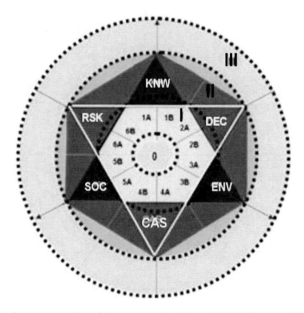

Figure 36. Sample representation of the process of building SES RISK methods in order to find solutions to the SES models.

The intersection of the two triads (models and methods) generates four areas of solutions represented by the concentric circles (0, I, II and III) shown in Figure 36. These circles indicate, in increasing order, the level of uncertainty and knowledge gained by a certain risk evaluation process.

Another important aspect is related to the search of systematic errors in CAS risk analysis. This search is done using a Cartesian approach in each triad of the *model* or *method*. This search, which is shown in Table 9, is used for all steps in CAS modelling and solution-finding processes, independent of the risk evaluation approach type, i.e. "deterministic" or "probabilistic".

TABLE 9. Representation of the reasoning process as per (Descartes, 1637)

	GOALS AND CRITERIA
R1	Goal of knowledge is to formulate judgments (true and fundamental) on objects
R2	Objects are targets, for which knowledge is indisputable and real
R3	The main criterion for sound knowledge is to get intuition from the objects
	THE TOOL IS DEFINED BY THE RULES
R4	The tool is the method
R5	The tool features are to aim at Organizing and Developing Gradually Increased Complexity Hierarchy of the Objects – **GICHO**
R6	In order to fulfil rule R5 a Complexity Level Criteria (CLC) derivation process is used so that the following are performed: To define what is simple and hence the reference CL_0 To define the departure from CL_0 of other objects $\square C = CL - CL_0$
	EVALUATE COMPLETENESS OF KNOWLEDGE
R7	Evaluate of the objects in the interaction *(synergetic mode).* The condition for a successful method is to consider knowledge **(KNW)** as a process **(Process)** – **KNWP**
R8	The application of the method has a necessary and sufficient condition, i.e. the existence of an intuitive understanding of each object at all levels
R9	The **KNWP** has to start from CL_0 and to get confirmation at each step that the subject can understand the object.
R10	Subject improves himself by practicing on CL_0
	RESULTS OF KNWP
R11	Subject has to use all mind tools to achieve **KNW**: (i) Reasoning, (ii) Imagination, (iii) Senses and (iv) Memory In order to: Perceive CL_0 by intuition Build **KNW** starting from CL_0

In (Serbanescu, 2005a) a suggested series of steps was identified in all the modelling processes to address the issue of search for biases and solutions to cope with them in a systematic and traceable manner.

A similar approach was adopted for the SES model and SES RISK process. In order to define the specific actions needed to avoid loss of control over the process of credibility self-assessment of results, the feedback process was performed in three major systematic steps (Serbanescu et al., 2007b; Serbanescu et al., 2008b).

***First step:** Identification of the detailed **governing principle** of each of the phases described below*:

P1   Unique source modelled with integrated method leads to the need to diversify definitions of objective functions.

P2   Deterministic versus probabilistic approaches as complementary approaches seem to cope with the dual features of SES risk models, but their use creates paradoxes if the applicability for each of them is not defined.

P3 Combine deterministic and probabilistic by using weights in judgments on random variables generating numerical results for which credibility depends on a solution not at the level of a simple combination between the two types of data (deterministic and probabilistic).

P4 Test the model in order to check its stability for paradoxes identified up to this phase, which results in a closed logic loop that defines the reference case while remaining in the limits of the model itself.

P5 A SES model could be in one of nine stages. They follow each other and their duration, intensity, and degree of challenge of paradoxes at each phase differ from one SES type to another. Furthermore, the causal approach of identification of beliefs behind the paradoxes for each phase results in challenges to the cause-effect approach.

P6 The SES model is built for a given set of objective functions and the metrics of this model is expected to include risk, which usually is not one of the intended goals for the user.

P7 Merging deterministic and probabilistic models forces the modeller and user to have common statements about combinations of probabilistic (which are by default uncertain) and deterministic events. However, this can sometimes lead to conflicting conclusions with problems of logic inconsistence.

P8 Management of risk model leads to managerial/procedural control in order to limit the uncertainty in the real process of SES evaluation. However this action is in itself creating new systematic assumptions and errors and is shadowing the ones accumulated up to this phase.

P9 The completion of a nine-cycle phase of a given SES model while using a SES RISK method. The completion of this process reveals the need to restart the process after each iteration in order to improve the model and to make corrections to the method SES RISK.

*__Second step__: Definition of the **beliefs** identified to be the main features of each of the steps:*

P1 It is assumed that there is a unique definition for risk and the risk science has a unique and unitary approach to give all the answers.

P2 It is assumed that the well-established scientific facts showing both random and deterministic data used in a scientific manner could provide support for certitude by using risk assessments.

P3 It is assumed that in the case of risk analyses a scientific method of universal use exists to evaluate severity of risks by judging them according to their probability and the outcomes/damages produced.

P4 It is assumed that by using carefully chosen experience and model results one can derive objective results proving the validity of results for the given SES model.

P5 It is assumed that by using educated guesses and experiments scientists can find and evaluate any significant risk due to the objectivity and other specific features of science.

P6   It is assumed that based on the objectivity of science and the approach in risk analyses to evaluate risks against benefits; the results could be used as such in decision-making process.

P7   It is assumed that by using a scientific method, which is honest and objective and by using risk reducing measures in all sectors of society (any type of SES model) the combined use of well proven tools in all science of analysis/deterministic and synthesis/probabilistic approaches assures success in SES modelling.

P8   It is assumed that science is more procedural than creative (at least for this type of activity) and the decisions themselves have to be made by trained staff and scientists.

P9   It is assumed that science is evolving based on laws which appear by the transformation of hypotheses into theories, which become laws and for any SES, in this case, if there will be a real risk then the scientists will find it.

***Last step**: Definition of a set of **actions** to prevent generation of paradoxes:*

P1   Model diversity of objective functions for SES metrics and use hierarchy for its structure looking for optimum at each hierarchical level.

P2   Applicability areas of deterministic and probabilistic parts in the SES model and in the decision module have to be clearly defined and used.

P3   Use numerical and logical functions as switches and connectors between the deterministic and probabilistic part of SES models and their metrics.

P4   Use a special sensitivity analysis phase to define the sensitivity to unseen/not clearly formulated assumptions embedded in the model based on their reflection in the paradoxes. In the screening process of those issues, use diverse methods not included so far in SES model.

P5   Perform a full inventory of identified paradoxes in SES model and the possible beliefs generating them in order to have a better understanding of the SES model bias and define the SES reference model for further analysis.

P6   Identify the rules for the post processing of risk analyses results in order to be used in the decision-making process by defining the place and desirability for the user of RIDM – a module to be added to the actual results from risk analyses for SES models.

P7   The merging action of deterministic and probabilistic approaches in SES risk models is accompanied by an improved set of logical construction of the formulations of results and modules merging the two approaches.

P8   Continuously check the objectivity of the SES model through risk managerial actions, including their potential distortion into the model.

P9   Restart the cycle of modelling even if there is no user request for it since there is always a need to have full coherent answers to all paradoxes encountered, even if they seem of no interest to the user and/or scientific community. The expected effect of the actions implemented in the SES risk analysis process was represented and discussed in *(Serbanescu*

2005a). The main conclusion of this representation showed the importance of the feedback modelling in a SES RISK analysis in order to assure a convergent set of solutions by identifying and managing mainly the systematic possible errors.

References

Colli et al., 2008, ColliA., Serbanescu D., PRA-Type Study Adapted to the Multi-crystalline Silicon Photovoltaic Cells Manufacture Process, ESREL 2008

Descartes R., 1637, Discours de la méthode, Paris, Garnier - Flammarion, 1966, edition *Discourse on Method* (1637).

Haimes, Yacov Y., 2004, Risk Modeling, Assessment and Management, 2nd Edition, Wiley & Sons, New Jersey.

Hansson S.O., 2000, Myths on Risk Talk at the conference Stockholm thirty years on. Progress achieved and challenges ahead in international environmental co-operation. Swedish Ministry of the Environment, June 17-18, 2000 Royal Institute of Technology, Stockholm

Howard et al, 1984, Howard R.A. and Matheson J.E., (editors), Readings on the Principles and Applications of Decision Analysis, 2 volumes (1984), Menlo Park CA: Strategic Decisions Group

Jaynes, E. T., 2003,Probability Theory - The Logic of Science, Cambridge University Press, Cambridge, UK

Kato, Tokyo, 1995. Perturbation Theory for Linear Operators", Springer Verlag, Germany, ISBN 3-540-58661

Mc Comas, W. 1996, Ten Myths of science: Re examining what we know, vol. 96, School Science & Mathematics, 01-01-1996, pp 10

Maturana et al. 1980, Maturana H.R.,Varela F.J. Autopoiésis y Cognición. Dordrecht, holanda: D. Reidel, 1980.

Peirce, C.S., 1931, *Collected Papers of Charles Sanders Peirce*, 8 vols. Edited by Charles Hartshorne, Paul Weiss, Arthur Burks (Harvard University Press, Cambridge, Massachusetts, 1931-1958, *http://www.hup.harvard.edu/catalog/PEICOA.html*

© Relcon Scandpower AB, 2008. RiskSpectrum® PSA Professional, developed and maintained by Relcon Scandpower AB in Sweden, http://www.riskspectrum.com/

Serbanescu D. 1991., Serbanescu D., A New Approach in the Decision Phases of the PSA Studies, PSA91, Vienna, 1991

Serbanescu D. 2001, Serbanescu D., The use of the decision theory and probabilistic analyses in the npp licensing decision process, IAEA-CN-82/28, 2001

Serbanescu 2005a. Some insights on issues related to specifics of the use of probability, risk, uncertainty and logic in PRA studies, Int. J. Critical Infrastructures, Vol. 1, Nos. 2/3, 2005

Serbanescu 2005b. Integrated Risk Assessment, ICRESH 2005, Bombay, India.

Serbanescu 2007, Serbanescu, D., Risk Informed Decision Making, Lecture presented at VALDOC Summer School on Risk Issues, Smoegen (Sweden), Vol. Karita Research Sweden (Organiser), JRC PB/2007/IE/5019

Serbanescu et al, 2007a, Serbanescu, D. & Kirchsteiger C., 2007a. Some methodological aspects on a risk informed support for decisions on specific complex systems objectives, ICAP 2007

Serbanescu et al, 2007b, Serbanescu D., Kircsteiger C., Some methodological aspects on a risk informed support for decisions on specific complex systems objectives, EC DG Joint Research Centre, Institute for Energy, presented at SRA Conference 2007, Netherlands

Serbanescu 2008, Serbanescu D., Some aspects, models and results on the use of PRA – Nuclear for the modeling of risks on hydrogen installations – preliminary results of the work performed in the HYSAFE/HYQRA project in 2008/ Feb. 2009 JRC Petten, Presented in HYSAFE workshop Tirennia, Italy 2008

Serbanescu et al 2008a, Serbanescu D. &Vetere Arellano A.L., SES RISK a new method to support decisions on energy supply, ESREL 2008

Serbanescu et al 2008b, Serbanescu D., Colli A., Vetere Arellano A.L., On some aspects related to the use of integrated risk analyses for the decision making process, including its use in the non-nuclear applications, ESREL 2008

Serbanescu et al 2008c, Serbanescu D. & Vetere Arellano A.L., Risk-Informed Decision Making (RIDM), in the report CARGO- sixth framework programme-Citizens and governance in a knowledge-based society, Proposal/Contract no.: FP6-036720 Comparison of Approaches to Risk Governance WP1 – 30 January 2008

Serbanescu 2009, On some aspects of performing probabilistic risk assessment for regional renewable energy systems (T2-G.2), SRA Conference 2009 Risk Analysis: The Evolution of a Science T2-G Symposium: Overcoming Risks Inherent to Renewable Energy Technologies and Systems, Baltimore, United States of America December 6–9, 2009

Serbanescu et al 2010a, Serbanescu D., Baraldi D., Vetere Arellano A.L. Some aspects of the practical use of probabilistic risk analyses for the evaluation of risks in hydrogen installations, EC DG Joint Research Centre, Institute for Energy, Petten, Netherlands, DGENER D2Nuclear safety, transport and decommissioning, Luxembourg, 2010

Serbanescu et al 2010b, Serbanescu D., Baraldi D., Vetere Arellano A.L, On some specific methodological aspects of using probabilistic risk analyses for the evaluation of risks in hydrogen, EC DG Joint Research Centre, Institute for Energy, Petten, Netherlands, DGENER D2Nuclear safety, transport and decommissioning, Luxembourg, 2010

Smithson, Michel J., 2000, Human judgment and imprecise probabilities, web site of the imprecise probabilities project http://ippserv.rug.ac.be 1997-2000 by Michel J. Smithson and the Imprecise Probabilities Project

USNRC 1983, PRA Procedures Guide. A guide for the performance of Probabilistic Risk Assessment for Nuclear power Plants (1983) USNRC, NUREG/CR-2300, February 1.

USNRC 1998, Regulatory Guide 1.174, An approach for using PRA in Risk Informed Decisions on plant specific changes to the licensing basis, July 1998.

ENERGY SECURITY: A PROBLEM OF COMPLEX SYSTEMS AND COMPLEX SITUATIONS

ERIC BOSWORTH*, ADRIAN GHEORGHE
Old Dominion University, Norfolk, VA, USA

Abstract. This paper addresses the following aspects: Identifies the domain of energy security, describes the energy security in relation to a concept of Homeland Security, identifies issues of energy security in relation to complex systems, such as the risk of nuclear installations and the availability of resources at national level, such as oil, and identifies the vulnerability of the national economy to a lack of energy security.

Problem Statement: Energy Security

Energy is the catalyst of progress for all human kind. Major nations have come to rely on this "drug", of sorts, to build their respective economies and futures. In doing so, these nations have created a severe dependence. Thus, as a result, a huge vulnerability is created that incurs tremendous risk to ensure its reliability. Vulnerability enters the equation due to the dependence, or the reliance, on the energy to keep respective economies and nations progressing. An example is of a nation, not too long ago, that entered a world war over their dependence on foreign oil. That nation was Japan. As a result, the United States and other countries placed embargos on the growing nation's ability to progress. As Japan had no means to generate their own oil, they looked to find and secure it. Therefore, Japan entered World War II. Ironically enough, Japan is now the world's leader in renewable energy.[1]

Japan, of course, is not alone making the mistake of becoming oil dependant. Many of the world's economies and nations are oil dependent, especially the United States (U.S.). What is troubling is that the U.S. did not learn from Japan's mistake or from Japan's example of becoming a leader in the renewable energy arena. One can argue that the U.S. was suckered into the cheap oil and gas of the 1980s and 1990s. Part of the complacency that took place during the 1980s and 1990s (when the U.S. truly let its habit get out of hand) was due to the simple fact

---

* To whom correspondence should be addressed: Eric Bosworth, Old Dominion University, Norfolk, VA, USA, e-mail: humbe001@odu.edu

[1] US Energy Information Administration Country Analysis Briefs, Japan: Environmental Issue, Jan 04. Retrieved October 10, 2009 from http://en.wikipedia.org/wiki/Energy_security.

A. Gheorghe and L. Muresan (eds.), *Energy Security: International and Local Issues,*   235
Theoretical Perspectives, and Critical Energy Infrastructures,
DOI 10.1007/978-94-007-0719-1_12, © Springer Science+Business Media B.V. 2011

that imported oil and gasoline were dirt cheap—too cheap to warrant serious efforts to develop alternatives.[2] Now, in order to protect its economy and the way of life of its citizens, the U.S. must protect the sources of its vulnerability. Along with this protection, of course, come many risks and complexities. Some countries, other than Japan, have vowed to be oil free by a certain timeframe. Increasing energy security is also one of the reasons behind plans for an oil phase-out in Sweden, together with a block on the development of natural gas imports.[3]

So, what is the domain of energy security? First, one must identify the various types of energy. These types of energy are: Petroleum (oil, gas & diesel), Natural Gas, Nuclear Power, and Renewable Energy (bio-fuels, wind, thermal, solar and water generation, including hydroelectric and tide turbines). Second, one must identify the security threats to present energy in the short and long-terms. Effective programs must address the physical, cyber, and human elements of CIKR (Critical Infrastructures and Key Resources), as appropriate, and consider long-term, short-term, and sustainable activities.[4]

Energy, the products themselves, and the infrastructure that ensure its exploration, production, refinement, storage, distribution, and consumption are in need of analysis. Energy drives economies. This drive is the major impact energy has, or the lack thereof, on the U.S. as a progressive civilization. One of the editors of Complex Interdependent Systems, Dr. Adrian Gheorghe, has identified an analysis that looks at infrastructures and their economic value as "infranomics." Infranomics is the understanding of the rules of interactions, dependability, and complexity, as well as their implications for complex critical infrastructure systems and their problem solving.[5] The U.S. energy infrastructure fuels the economy of the 21st century. Without a stable energy supply, health and welfare are threatened and the U.S. economy cannot function.[6] Implementing a framework to mitigate threats is complicated by the fact that more than 80% of the infrastructure is privately owned. The problem area, or risk, is being able to get each of the 80% of nations with privately owned infrastructures to work together and identify risks and vulner-abilities. Although, I have used the U.S. as an example, it must be mentioned that almost every country faces the same complexities and needs for risk identification.

Energy Security and Homeland Security

Homeland Security Presidential Directive 7 (HSPD-7) identified 17 critical infra-structure and key resources (CIKR) sectors and designated Federal Government Sector-Specific Agencies (SSAs) for each of the sectors.[7] Within this structure,

---

[2] Durante, D., & Sneller, T. (2005). Issue Brief: Energy Security. A publication of Ethanol across America. Retrieved October 10, 2009, from http://www.ethanol.org/pdf/contentmgmt/Energy_Security_Brief.pdf.

[3] Retrieved October 10, 2009 from http://en.wikipedia.org/wiki/Energy_security.

[4] U.S. Department of Homeland Security. (2009). National Infrastructure Protection Plan: Partnering to enhance protection and resiliency.

[5] Gheorghe, A. (2009, August), Complex Interdependent Systems, Module 1, Slide 3. Received August 28, 2009.

[6] Retrieved October 11, 2009 from http://www.dhs.gov/xlibrary/assets/nipp_snapshot_energy.pdf.

each of the identified sectors is responsible for the generation and employment of their individual plans. Additionally, these sectors have the responsibility of identifying gaps within their protection plans and with those dependent upon their sector and vice versa. The various sector specific plans are: Agriculture and Food; Banking and Finance; Communications; Defense Industrial Base; Energy; Information Technology; National Monuments and Icons; Transportation Systems; and Water Systems.

Specifically, this paper looks at the Energy aspect of the sectors. The Department of Homeland Security (DHS) has generated the Energy Sector Plan. This governmental organization recognizes the nation's reliance on energy. Specifically, the governmental organization recognizes the consequences the nation could face should energy be interrupted and the fact that threats exist to impair the progress, the health, and the well being of the U.S. The plan states, "The Energy Sector includes assets related to three key energy resources: electric power, petroleum, and natural gas. Each of these resources requires a unique set of supporting activities and assets".[8] DHS has also built a framework to assess the risks and vulnerabilities and to identify the critical infrastructures and their interdependencies. Figure 1 illustrates this framework. Most importantly, the framework establishes risk goals to be attained.

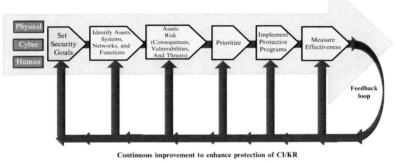

Continuous improvement to enhance protection of CI/KR

Figure 1. DHS NIPP Risk management framework[9].

The process depicted in Figure 1 clearly shows a systems engineering process incremented into steps, or phases, with a constant feedback loop. Step one is highlighted above as Set Security Goals. The steps continue from Identify Assets, Systems, Networks, and Functions through Measuring Effectiveness.

The Energy Sector Plan states, "A healthy energy infrastructure is one of the defining characteristics of a modern global economy. Any prolonged interruption of the supply of basic energy—electricity, petroleum, or natural gas—would do

---

[7] Retrieved October 11, 2009 from http://www.dhs.gov/xlibrary/assets/nipp_snapshot_energy.pdf.

[8] U.S. Department of Homeland Security (2007). Energy: Critical Infrastructure and Key Resources Sector-Specific Plan as input to the National Infrastructure Protection Plan (Redacted).

[9] U.S. Department of Homeland Security (2007). Energy: Critical Infrastructure and Key Resources Sector-Specific Plan as input to the National Infrastructure Protection Plan (Redacted).

considerable harm to the U.S. economy and the American people".[10] The use of energy is ubiquitous, for this aspect of infrastructure touches every aspect of the U.S. economy. In fact, one could argue, successfully, that energy affects all of the world's progressive economies.

DHS has broken energy down into three segments: Electricity, Petroleum, and Natural Gas. Within these segments, DHS has broken each down further into various segments within each segment. Figure 2 below illustrates this breakout.

Electricity	Petroleum	Natural Gas
• Generation • Fossil fuel power plants • Coal • Gas • Oil • Nuclear power plants* • Hydroelectric dams* • Renewable energy • Transmission • Substations • Lines • Control centers • Distribution • Substations • Lines • Control centers • Control Systems • Electricity Markets	• Petroleum • Onshore fields • Offshore fields • Processing • Transport (pipelines)* • Distribution (pipelines)* • Storage • Liquefied Natural Gas Facilities • Control systems • Gas Markets	• Crude Oil • Onshore fields • Offshore fields • Terminals • Transport (pipelines)* • Storage • Petroleum Processing Facilities • Refineries • Terminals • Transport (pipelines)* • Storage • Control Systems • Petroleum Markets

Figure 2. Segments within the energy sector[11].

Figure 2 identifies the intricacies of each energy domain, such as electricity. The figure details the generation, particularly sources of generation, such as transmission, distribution, controls systems, and markets. The figure does this for the two remaining domains of Petroleum and Natural Gas and represents a good start to the identification of infrastructures and their dependencies/interdependencies.

Interdependencies exist between the transfers of electricity to public domains. Figure 3 illustrates one such example of interdependencies: the communication system between electric infrastructures.

Interdependencies not only exist within the electric infrastructure, like communications as illustrated above, but also within the transportation system to transfers coal, oil, and natural gas to the generation plant. If some adverse affect that has an impact on the trucks, rail system, and waterways that haul the energy

---

[10] U.S. Department of Homeland Security (2007). Energy: Critical Infrastructure and Key Resources Sector-Specific Plan as input to the National Infrastructure Protection Plan (Redacted).

[11] U.S. Department of Homeland Security (2007). Energy: Critical Infrastructure and Key Resources Sector-Specific Plan as input to the National Infrastructure Protection Plan (Redacted).

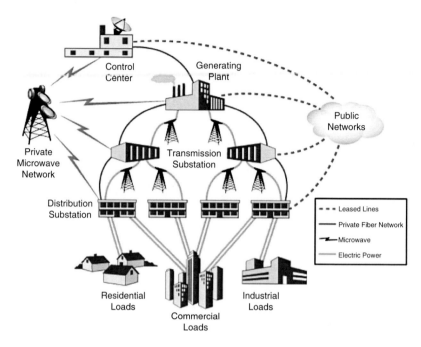

Figure 3. Electric power system and control communications[12].

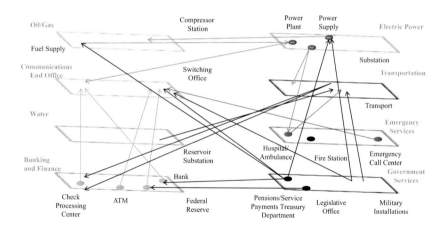

Figure 4. Interdependencies across the economy[13].

---

[12] U.S. Department of Homeland Security (2007). Energy: Critical Infrastructure and Key Resources Sector-Specific Plan as input to the National Infrastructure Protection Plan (Redacted).

source to the generation plant, then the ability to generate the required electricity will be affected. Additionally, there exist interdependencies with other providers, such as emergency response systems. Emergency services require electricity to power their buildings and to provide for the welfare of their employees or others who may be ill. Figure 4, below, illustrates, at a greater granularity, the details regarding other interdependencies.

Interdependencies also cross international borders. As stated by DHS, "Energy infrastructure interdependencies also cross international borders. Oil and natural gas pipelines and electric transmission lines have helped integrate the energy systems of North America. Moreover, increasing imports of petroleum products continue to highlight the dependence of the United States on foreign oil".[14] Again, one sees the reference of dependence by the U.S. on foreign oil. It appears that no matter how one slices and dices the framework, the analysis comes back to a single denominator – oil. Some argue, though, that the cross section of energy sources does not solely rely on oil. In fact, these individuals are correct, for power generation does come from coal, hydro electric, solar, wind, geothermal, and natural gas. However, even a slight interruption in oil for energy production could result in devastating consequences. Prior to solving the dilemma of oil dependence, one first needs to secure and then stabilize the current situation. Thus, one must perform risk analysis to determine where one's vulnerabilities are, to prioritize these risks, and to determine what mitigations need to employ. According to the Department of Energy (DOE) and DHS, the energy sector employs many methodologies for determining risk. The important idea to take away from this is that the energy sector works closely with most parties, such as other government agencies and private parties, to determine interdependencies. DHS states the following concerning risk assessments, "These criteria include both the analytic principles that are broadly applicable to all parts of a risk methodology and specific guidance regarding the information needed to understand and address each of the three components of the risk equation: consequence, vulnerability, and threat".[15] The process for developing and implementing effective protective measures for critical energy infrastructures has three phases: determining needs, developing programs, and finding long-term solutions.[16] DHS' process is better illustrated through Figure 5 below.

DHS' process, as illustrated above, collects data to determine needs, for example information sharing and building partnerships, developing programs by using effective practices already in use, and finding long term solutions, such as research and development needs. Additionally, DHS identifies long-term technological solutions for protecting physical assets, energy control, and cyber systems. Any

---

[13] U.S. Department of Homeland Security (2007). Energy: Critical Infrastructure and Key Resources Sector-Specific Plan as input to the National Infrastructure Protection Plan (Redacted).

[14] U.S. Department of Homeland Security (2007). Energy: Critical Infrastructure and Key Resources Sector-Specific Plan as input to the National Infrastructure Protection Plan (Redacted).

[15] U.S. Department of Homeland Security (2009). National Infrastructure Protection Plan: Partnering to enhance protection and resiliency. Appendix 3A: NIPP Core Criteria for Risk Assessments.

[16] U.S. Department of Homeland Security (2007). Energy: Critical Infrastructure and Key Resources Sector-Specific Plan as input to the National Infrastructure Protection Plan (Redacted).

risk management plan, regardless if it relates to this topic or not, should approach the problem(s) from a long-term perspective.

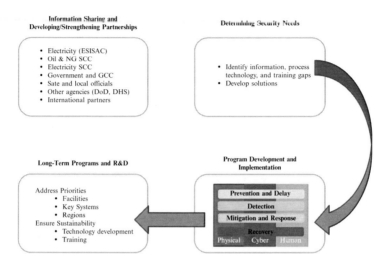

Figure 5. Evaluating and prioritizing needs, and implementing program.

Energy Security and Complex Systems

Protection and complexity are often used simultaneously when discussing critical infrastructures. Issues regarding the security, or insecurity, of energy are as follows:

- Oil and other fossil fuel depletion,
- Reliance on foreign sources of energy and geopolitics,
- Demands of growing countries (China and India, for example),
- Economic efficiency versus population growth, and
- Need to invest in alternatives to fossil fuels.[17]

These issues are very real and their solutions are extremely complex. Moreover, some of these issues are currently unknown. In fact, the U.S. is constantly looking for viable alternatives to resolve some of these issues. The U.S. DOE identifies energy issues on their web page. Their web page identifies seven categories of concern:

- Science & Technology – Biological Sciences, Carbon Sequestration, Chemical Science, Climate Change, Computing, etc.
- Energy Sources – Bio Energy, Coal, Electric Power, Fossil Fuels, Fusion, Geothermal, Hydrogen, etc.

---

[17] Shah, A. (2009). Energy Security. Global Issues: Social, Political, Economic, & Environmental Issues That Affect Us All. Retrieved October 12, 2009 from http://www.globalissues.org/article/595/energy-security.

- Energy Efficiency – Buildings, Energy Star, Financing, Homes, Industry, Power Utilities, etc.
- The Environment – Clean Air, Climate Change, Facilities, Oversight, Waste Management, and the National Environmental Protection Agency Act
- Prices & Trends – Annual Energy Outlook, Annual Energy Review, and Energy Statistics
- National Security – Cyber, Facility, Nuclear, Intelligence and Counter Intelligence, Weapons of Mass Destruction, Energy Responsibility, etc.
- Safety & Health – Chemical, Chronic Beryllium Disease, Facility, Nuclear, Waste Transportation, Worker Health, and Worker Safety.[18]

From the various categories above, DOE has dissected their concerns into many sections, not to mention complex areas of concerns. All of these categories have their own set of issues. One can argue that each of these energy related areas, or categories, have their own set of dependencies and interdependencies, including vulnerabilities and risks.

RISKS ASSOCIATED WITH NUCLEAR POWER PLANT INSTALLATIONS

Continuing the theme of complex systems, this document now addresses the risks associated with the installation of a nuclear power plant. By its very name, one can surmise that a nuclear power plant (NPP) is a complex system. Why? The term nuclear has a negative connotation, at least in the U.S., from historical accounts, such as the bombings of Nagasaki and Hiroshima, the events that ensued following the Three-Mile Island incident during the 1970s, and the 1980s Chernobyl fiasco. The infrastructure necessary to generate, cool, house, dispense and, most importantly, control the generation of power in a NPP is extremely complex. Frankly, radioactive by-products from the generation of power/energy are dangerous to the health of humans and most other living things; hence, the fear, the negative connotation, and the complexity surrounding its generation and disposal. If one decides to install a NPP, it is best to plan accordingly to minimize the risks. Additionally, it is necessary to consider nuclear safety in a community perspective. From the conception to the decommissioning, only a common approach can guarantee the maintenance of a high level of safety in nuclear installations in an enlarged EU.[19] Another aspect, according to the Commission of the European Communities, is that "In order to avoid risks to human health and to the environment it is necessary to guarantee, at Community level, that financial resources will be available for the completion of decommissioning work in conformity with safety standards. To this end, specific regulations must be put in place for the creation of decommissioning funds, to which the operators of nuclear installations will have to contribute throughout the active life of the installation." So, in short, funding can be a major risk for the installation of a NPP. Thus, prior to even breaking ground, one needs to acquire the security of such funding and the required needs. The program cannot "short [change]" funding, for, when this happens, neglect and adherence to safety standards can be affected.

Exposure to radioactive materials is always a risk to the communities in which they are built; although safety precautions are planned, enacted, implemented and audited. Although the radiation protection system created by the basic standards in force ensures a high level of protection for the health of the population based on current scientific knowledge on the subject, the protection system must be supplemented by strict application of safety standards designed to anticipate and control the risks of exposure for the population. In nuclear installations, in particular, keeping up high safety standards at all stages, from conception to decommissioning, is achieved by maintaining effective defenses against radiological risks and by preventing accidents that could have radiological consequences. This maintenance is a sine qua non in order fully to attain the objectives of health protection set out in Article 2(b) of the Treaty.[20] Risks to human and environmental health are also present during the decommissioning of a NPP. Decommissioning operations may also constitute potential threats to human health and to the environment, not only now, but also in the future, especially if necessary measures relating to the radiological risks of such operations are not taken in good time.[21] The best means of mitigating, or reducing, the risk to the environment or human health is to properly dispose of the waste. If the waste is disposed of properly, then reductions in harmful effects are greatly reduced.

So far, this document has mentioned safety and waste as two risks associated with NPPs. Two more are now in need of mentioning. First, nuclear power generation produces a by-product called plutonium-239. This by-product can be used to make weapons of mass destruction. Thus, installations lend themselves nicely and conveniently to terrorism and weapons proliferation. Secondly, and most obviously, NPP installations pose a health risk. For example if a disaster occurs, such as that in Chernobyl, the fallout is deadly. Exposure to radioactive fallout leads to an increased risk of genetic disorders, cancers, and leukemia. In some areas of Belarus, for example, national reports indicate that incidents of thyroid cancer in children have increased more than a hundred-fold compared to the period before the Chernobyl accident.[22] Also, according to this same source, working in a NPP can also be a risk.

It is important to know that another aspect of the complexity within NPPs is the fact that there are many redundant systems in place, in order to mitigate against the potential threats surrounding its existence. The nuclear power plant design strategy for preventing accidents and for mitigating their potential effects is "defense in depth" – if something fails, there is a back-up system to limit the harm done. If that system should also fail, there is another back-up system for it and so on.[23]

---

[20] Commission of the European Communities. Brussels, 30.1.2003.

[21] Commission of the European Communities. Brussels, 30.1.2003.

[22] (WISE/NIRS Nuclear Monitor, 2005) Retrieved October 14, 2009 from http://www10.antenna.nl/wise/index.html?http://www10.antenna.nl/wise/621-22/7.php.

[23] Cohen, B. L. Risks of Nuclear Power. Retrieved October 14, 2009 from http://www.physics.isu.edu/radinf/np-risk.htm.

Vulnerability of National Economy

If one was to say that vulnerability in the U.S. was solely a result of the country's dependence on oil, one would be wrong. If one continued by saying that the world's sole vulnerability was their dependence on oil, one would also be wrong. The fact of the matter is that world economies are far more complex than such a simplistic assertion. Through research, a rather old citation was discovered, though it was none the less accurate in the dissemination and explanation of such complexities. John Sterman asserts,

"Since 1973 there has been growing awareness that the energy transition will be more difficult, time consuming, and expensive than anticipated. There are already many signs of economic stress. During the 1970s, economic growth faltered from the 3.7%/year rate of the 1950s and 1960s to 2.7%/year. The nation experienced the two deepest recessions since the Great Depression, high unemployment, large trade deficits, slackened productivity growth, and the most severe peacetime inflation in U.S. history. While not all the nation's economic woes can be traced to energy, the impact of energy on the economic health of the nation is undeniable. The unemployment, factory shutdowns, hardship, and inconvenience caused by the OPEC embargo of 1973, natural gas shortages of 1976, coal strike of 1978, and gasoline shortages of 1979 all demonstrated the vulnerability of a modern industrial economy to disruptions in energy price and availability. But energy also creates vulnerability in more subtle ways: energy prices outpaced inflation for most of the decade, raising the real price of energy and adding to inflationary pressures; growing capital requirements for energy production threaten investment in other sectors of the economy; the costs of producing synthetic fuels and other alternative sources rise as OPEC prices rise; and high OPEC prices transfer income and wealth from oil consuming nations to oil producing nations".[24]

As the few statements above articulate, energy vulnerabilities are many and complex. Additionally, one should note that the statements are from an article that is over 27 years old. Does the statement ring familiar? It should. The world is currently undergoing a recession. Most blame this on the failures associated with the vulnerabilities of financial institutions, such as mortgage lenders and mortgage insurers. Perhaps, these individuals are correct. However, I postulate another scenario, or rather a triggering event, that merely brings to light shaky or vulnerable lending practices. My postulation is that I recall for several months leading up to this recession exponential energy prices (to the U.S. anyhow, but I am sure this was felt everywhere) where upon oil prices reached ~$160.00 per barrel of sweet crude. Until this experience, the economy was skyrocketing, which one can argue was fueled the shaky practices outlined by most media outlets. Was this mere coincidence? Perhaps, but there are uncanny resemblances to the 1982 quote above. I assert that the cost of energy just prior to the recession was a triggering event that exposed weaknesses in the economic infrastructure. Nothing exposes

---

[24] Sterman, J. (1982). Economic Vulnerability & The Energy Transition. Cambridge, MIT Alfred P. Sloan School of Management. Retrieved October 16, 2009 from http://dspace.mit.edu/bitstream/handle/1721.1/2016/SWP-1325-09065908.pdf?sequence=1

risks and vulnerabilities like a recession, or, even worse, a depression. The interdependencies become ever surreal at the expense of the middle and lower classes on the socioeconomic spectrum.

Now, let us diversify our vulnerability analysis and look at the vulnerabilities associated with a Middle Eastern country like Iran, for example. I know this sounds like an oxymoron, due to the fact that Iran has oil reserves and is a major exporter of energy to the world. However, the reality is that vulnerabilities within the context of energy do, in fact, exist to such countries. I bring to attention Iran's vulnerability that is due, largely, to its self-inflicted position on the world scene. The U.S. has negatively impacted Iran's economy by imposing sanctions on its ability to export its most precious commodity, oil, although, Iran, in turn, has threatened to reduce its production. Paul Rivlin in his article concerning Iran's energy vulnerability states, "Iran's warnings that it may stop oil exports are idle threats, because it cannot significantly reduce oil exports without inflicting massive damage on its own economy. By subsidizing all energy products, Iran has artificially boosted demand, while U.S. sanctions limit its ability to increase supply. As a result, Iran has become reliant on imports of gasoline and other products and so is exposed to potential international sanctions".[25] The quote makes one question if Iran is really seeking nuclear technology/capabilities for adversarial reasons, or, instead, seeking nuclear capability for energy usage. Rivlin goes on to say, "As the economy and the population have grown (in Iran), so has the demand for energy. Rapid urbanization has also contributed to the rise in demand, as has the system of massive energy subsidies. The amount of crude oil available for export has been constrained by a growing domestic demand."[26] In Iran's situation, with a growing population, progressive nation (economically at least), and with increasing demands, it only makes sense to find an alternative energy to consume in-house, while continuing to produce oil and gasoline for export at higher and higher prices. This sounds legitimate and makes economic sense. What is more is the by-product produced from NPPs, mentioned earlier in this paper, Plutonium-239. This by-product is icing on the cake, so to speak, for Iran's decision. Could not this convenient by-product be used to secure the nation's future on the world scene?

The most obvious vulnerability, not to mention complexity, is the emergence of new industrialized nations placing demands on energy and the world market. North America and Europe used to place high demands on energy. Now, India and China are coming on the energy scene placing higher and higher demands on the world's energy sources. India and China combined, as mentioned before, have the highest percentage of the world's population. The Laboratory of Energy Systems in Switzerland states,

"The emergence of new big consumer countries on the energy markets and the perspective of oil and gas depletion at the end of the current century raise the

[25] Rivlin, P. (2006 December). Iran's Energy Vulnerability. The Middle East Review of International Affairs, Volume 10, No. 4, Article 7/7. Retrieved October 16, 2009 from http://meria.idc.ac.il/journal/2006/issue4/jv10no4a7.html.

[26] Rivlin, P. (2006 December). Iran's Energy Vulnerability. The Middle East Review of International Affairs, Volume 10, No. 4, Article 7/7. Retrieved October 16, 2009 from http://meria.idc.ac.il/journal/2006/issue4/jv10no4a7.html.

concerns about fair distribution of the remaining resources for the common and sustainable well-being of the mankind. High volatility of energy prices discourages the investment and delays the energy technology transition. Voluntary measures are needed mainly in industrialised countries in order to develop alternative and sustainable energy sources, to enable technology transfer towards emerging and developing countries and to avoid struggle for energy procurement".[27]

Conclusions

Upon reading this paper, to emphasize the reality of energy diversification and development of energy alternatives is an extreme understatement, not to mention an overstatement of the obvious. This work exemplifies the understanding of the domain of energy security, including its relation to Homeland Security, NPPs, resource availability, and vulnerabilities to economies. There is, however, still one thought to ponder. Imagine the country that develops a renewable, alternative, and clean source of energy that can be produced economically and likewise distributed. The country, group or individual who solves this problem will be the next Rockefeller, Microsoft, Saudi Arabia, or United States on the world scene, simply as a result of the economic risks and vulnerabilities associated with energy. Holding this technological commodity is like having the world at one's finger tips. However, one also needs to imagine the next set of risks and vulnerabilities related to the new set of infrastructures.\

[27] Gnansounou, E. (2008 April 24). Assessing the energy vulnerability: Case of industrialized countries. Laboratory of Energy Systems, Ecole Polytechnique Federale de Lausanne, Switzerland.

CYBER SECURITY: THREAT IDENTIFICATION, RISK AND VULNERABILITY ASSESSMENT

HAROLD UMBERGER[*] AND ADRIAN GHEORGHE
Old Dominion University, Norfolk, VA, USA

Abstract. The task of preventing unauthorized users from compromising the confidentiality, the integrity, or the availability of sensitive information is increasingly difficult in the face of the growth in Internet use, the increasing skill level of attackers, and the technological advances in their tools and methods of attack. The threats and risks are real and increasingly frequent, with high potential to critical energy infrastructures. The Department of Homeland Security needs to have a clear plan in place to help better mitigate the problems faced with cyber and energy security.

Introduction

The U.S. faces vulnerabilities because its critical infrastructure systems are totally dependent on the Internet and IP-based technology for information and data exchange. The loss, or prolonged delay, of the information network would be catastrophic to the economy, and would cause many other failures with interdependent infrastructures. Power grids, water/dam control, transportation systems, health systems, finance systems, food, emergency services, telecommunication systems, nuclear power plants, water supplies, and others are attractive targets for the nation's enemies. Telephones and VOIP will continue to see an increase in attacks, as more data is sent over networks.

In the U.S., the Department of Homeland Security (DHS) is responsible for protecting the nation's critical infrastructure against attacks, both natural and malicious. The energy infrastructure is complex and each of the components has associated risks and vulnerabilities. The goal of DHS is to minimize those risks and to make the systems more resilient to attack. The security issues the nation faces apply to everyone. Internet hackers and attackers are responsible, as are individuals who leave sticky notes with passwords in plain sight, who leave sensitive information on their desktops, and who deposit sensitive information into the office or home trash receptacles.

The growing sophistication of hackers has increased the threat that the nation faces with regard to cyber security. Ironically, though, the same tools hackers use

---

[*] To whom correspondence should be addressed: Harold Umberger, Old Dominion University, Norfolk, VA, USA, e-mail: humbe001@odu.edu

A. Gheorghe and L. Muresan (eds.), *Energy Security: International and Local Issues,*   247
Theoretical Perspectives, and Critical Energy Infrastructures,
DOI 10.1007/978-94-007-0719-1_13, © Springer Science+Business Media B.V. 2011

can also be used to identify vulnerabilities and enhance the security of the system. One can view information warfare in three categories: Personal Information Warfare, Corporate Information Warfare, and Global Information Warfare. The intent of this paper is to evaluate information warfare by describing cyberspace, by discussing Homeland Security, by identifying threats to information networks, by assessing cyber risk and vulnerability, and by discussing cyber governance.

Cyberspace is a critical infrastructure for the U.S. and most of the world; all other critical infrastructures rely on the digital infrastructure for critical support. A U.S. National Plan for Critical Infrastructures (PDD-63) states that the "critical infrastructures as those systems and assets – both physical and cyber – [are] so vital to the Nation that their incapacity or destruction would have a debilitating impact on national security, and/or national public health and safety."[1] One can say about critical infrastructures that "As metasystems, critical infrastructures are characterized by a high degree of connectivity, complexity and relevance to society. A single failure can cascade, potentially affecting security and social welfare of any nation."[2]

Cyberspace is any process, program, or protocol relating to the use of the Internet, or an intranet, for data processing, transmission, or use in telecommunication. According to the Department of Homeland Security's February 2003 National Strategy to Secure Cyberspace,

"Our nation's critical infrastructures are composed of public and private institutions in the sectors of agriculture, food, water, public health, emergency services, government, banking finance, chemicals and hazardous materials, and postal and shipping. Cyberspace is their nervous system –the control system of our country, and that the cornerstone of America's cyberspace security strategy is and will remain a public-private partnership."[3]

National Security Presidential Directive 54/Homeland Security Presidential Directive 23 (NSPD-54/HSPD-23) defines cyberspace "as the interdependent network of information technology infrastructures, and includes the Internet, telecommunications networks, computer systems, and embedded processors and controllers in critical industries."[4]

Digitalization and the capability to handle multiple sources of information have enhanced telecommunications, radio, television, and the Internet. Information exchange is no longer limited to professionals due to the increase in technology, such as social websites, cameras, camcorders, iPods, video, and audiotapes. Almost 2 billion people around the world now use the Internet. From 2000 to 2007, Internet use jumped 245% globally, with the greatest percentage increases in Africa (875%)

---

[1] PDD-63 – Defining America's Cyberspace: National Plan for Information Systems Protection, Version 1.0, An Invitation to a Dialog, White House, 2000.

[2] Gheorghe, A.V. Critical Electricity Infrastructures at Risk: The Advent of the System of Systems Engineering Concept, Gridwise Architecture Council, Downloaded at www.gridwiseac.org/pdfs/ interop_papers_0407/.../gheorghe.pdf.

[3] The National Strategy to Secure Cyberspace, February 2003, downloaded at http://www.dhs.gov/ xlibrary/assets/National_Cyberspace_Strategy.pdf on 08 November, 2009.

[4] Cyberspace Policy Review, Assuring a Trusted and Resilient Information and Communication Infrastructure, Downloaded http://www.whitehouse.gov/assets/documents/ Cyberspace_Policy_ Review_final.pdf on 08 November, 2009.

and the Middle East (920%).[5] The world is full of communication for knowledge exchange, and the market is highly competitive and complicated. Figure 1 illustrates the current number of Internet users around the world by region.

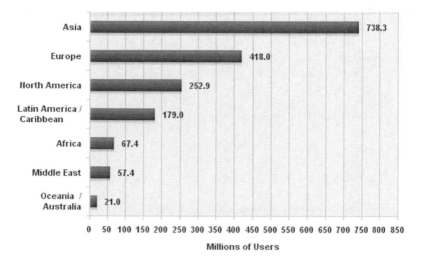

Figure 1. Internet users in the world (see footnote 5).

History of Cyberspace

The Internet is basically a giant network composed of thousands of smaller networks. The Internet includes all the computers, phone lines, cable lines, and other communication devices that hold the smaller networks together. It is a critical infrastructure that supports all other critical infrastructures by the transmission of electronic data. The U.S. first started the framework for the Internet in 1958 with the establishment of the Advanced Research Projects Agency (ARPA). In the early stages of Internet technology, most control systems were isolated, but that rapidly changed with the onset of the World Wide Web (WWW).

The WWW is a conglomerate of hyper connected multimedia data used to bring the Internet into the homes and offices of millions of people around the world. The Web was created in 1991 by Tim Berners-Lee while he was working at the Geneva, Switzerland-based European Particle Laboratory (CERN).[6] Berners-Lee began this proposal because he saw a need to create an interactive, inter-connected system that would let scientists and scholars share their ideas more efficiently. Figure 2, below, illustrates how rapidly the technology for information transfer has changed in the U.S., as well as some of the governance implemented to help with the decision-making.

---

[5] Internet World Stats, accessed at http://www.internetworldstats.htm, 10 November, 2009.
[6] http://www.w3.org/People/Berners-Lee/ShortHistory.html

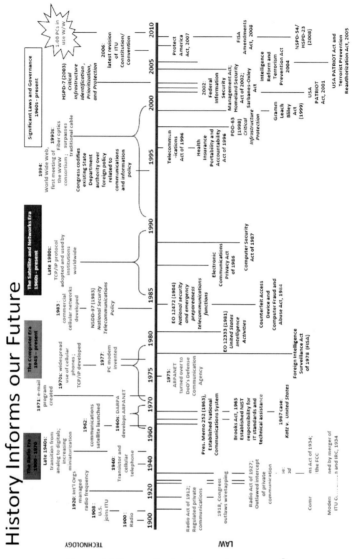

Figure 2. History of technology and governance[7].

[7] Cyberspace Policy Review, Assuring a Trusted and Resilient Information and Communication Infrastructure, Downloaded http://www.whitehouse.gov/assets/documents/Cyberspace_Policy_Review_final.pdf on 08 November, 2009.

In addition to the Web, email has become a dominant feature of the Internet. A handful of email applications currently dominate the market, but most individuals use a product from Microsoft, such as Outlook, a Netscape product, such as Netscape Messenger, Hotmail, Yahoo, or Gmail. Email has changed the way the world communicates, moving communication from written to electronic. The U.S. Postal Service has had a drastic drop in the amount of mail sent because of email and the ability to electronically pay bills on-line as opposed to mailing payments. Many people believe that email has made communication more effective in business, and has made the people more accountable. Email does bring the problems of phishing, the delivery method for many viruses, and it does overwhelm individuals with SPAM. The Internet is now generating lots of money with online commerce with Web sites, such as E-bay, that allow one to purchase anything electronically at online auctions.

The Internet is always becoming more convenient and user-friendly. Web portals, such as Yahoo, Google, or Info, seek to combine a number of popular Internet services, like viewing news headlines, Web searching capabilities, and finding information quickly and easily. The convergence of media activities is also becoming a reality with the ability the listen to radio, as well as watch television and movies online. As a result, many magazines and newspapers are now finding it hard to compete with the Internet and are going out of business. Libraries are another area that is changing, with many books and magazines becoming electronic.

Homeland Security

The Department of Homeland Security is responsible for assessing and mitigating potential vulnerabilities of the critical infrastructures of the United States, ensuring that the county is resilient to both actual and perceived threats. The Department is required to establish risk reduction for each sector and to work with other Federal, state, and local agencies to protect the country's assets. Vulnerabilities to disruptions of the information sector from either malicious acts or from natural causes could compromise the stability of the system as a whole and would result in a large economic loss.

This country has been thrown into a new sort of mindset concerning vulnerability. Prior to 9-11, there was a perception that the United States was a superpower that could not be affected by its enemies, both because of the geographic location of the United States and because of the nation's perceived invulnerability; however, being perceived as one of the biggest and wealthiest countries brings vulnerability. The United States, or even the world, could be thought of as a system of systems, with the subsystems being elements of the infrastructure, economy, culture, and politics. The mining of intelligence data can enable the assessment of the SOS and can determine vulnerability based on the observed indicators[8]

---

[8] Gheorghe, A.V. and Vamanu, D. (2008) 'Mining intelligence data in the benefit of critical infrastructures security: vulnerability modeling, simulation and assessment, system of systems engineering', *Int. J. System of Systems Engineering*, Vol. 1, Nos. 1/2, pp. 189–221.

Homeland Security is interrelated with different actors within the critical infrastructures (power plants, oil pipelines, electric grids), and one of the challenges these actors face is information sharing. The exchange of information about vulnerabilities, threats, and countermeasures is vital to protecting the critical infrastructures, since the different sectors are so interconnected. There are secure sharing networks available among the agents, which allow them to exchange sensitive information without fear of disclosure. This feature is important, since most infrastructures are privately controlled and hesitant to share information.[9]

Figure 3, below, illustrates how a cyber threat to one of the critical infrastructures can cause major problems for Homeland Security. Because of the interdependencies of the major systems, one issue can cause catastrophic effects to the country. This issue is hard to plan for and it is hard to have mitigations in place for every threat. For this reason, the systems need to be as resilient as possible. The ubiquity of digitalization is another concern of Homeland Security. Even though digitalization is good for performance and reliability, it brings new security concerns, as it opens the systems to hackers.

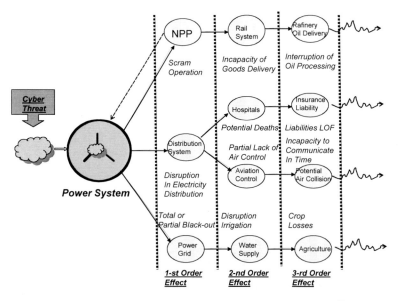

Figure 3. Interdependencies – A Homeland Security issue[10].

The National Cyber Security Division (NCSD) is an office in the Department of Homeland Security that is responsible for protecting the United States communications networks. The NCSD works with public and private organizations, nationally and internationally, to secure the country's cyber assets. They coordinate a

---

[9]   Braghn, S., Fovino, I.N., and Trombetta, A., Advanced Trust Negotiation in Critical Infrastructures,
      August 28, 2008.
[10] Gheorghe, A.V., Class Notes from ENMA 771, Module 1: Complex Interdependent Systems, 2009.

response team to respond to any problems that arise when dealing with the critical cyber infrastructure. The United States Computer Emergency Readiness Team (US-CERT) is a part of NCSD that is responsible for providing response support and defense against cyber attacks. The US-CERT provides a weekly Security Bulletin that gives a summary of new vulnerabilities. Figure 4 is one example of the very brief summary for the week of September 28, 2009. The entire summary has many products within each severity ranking. The ranking of vulnerability severity, or high, medium, and low, is based on calculations with a Common Vulnerability Scoring System11.

High Vulnerabilities				
Primary Vendor – Product	Description	Published	CVSS Score	Source & Patch Info
Apple – Safari	Apple Safari, possibly before 4.0.3, on Mac OS X does not properly handle a '\0' character in a domain name in the subject's Common Name (CN) field of an X.509 certificate, which allows man-in-the-middle attackers to spoof arbitrary	2009-09-29	7.5	CVE-2009-3455 BID
Medium Vulnerabilities				
Adobe -- Photoshop_Elements	Adobe Photoshop Elements 8.0 installs the Adobe Active File Monitor V8 service with an insecure security descriptor, which allows local users to (1) stop the service via the stop command, (2) execute arbitrary commands as SYSTEM by using the config command to modify the binPath.	2009-09-30	6.9	CVE-2009-3489 VUPEN SECTRACK BID BUGTRAQ SECUNIA MISC MISC
Low Vulnerabilities				
Juniper -- Junos	Multiple cross-site scripting (XSS) vulnerabilities in the J-Web interface in Juniper JUNOS 8.5R1.14 allow remote authenticated users to inject arbitrary web script or HTML via the host parameter to (1) the pinghost program, reachable through the diagnose program.	2009-09-30	3.5	CVE-2009-3486 VUPEN BID MISC SECUNIA

Figure 4. US-CERT vulnerability summary[11].

The Multi-State Information Sharing and Analysis Center (MS-ISAC) works with the NCSD of Homeland Security as a mechanism to gather and share information on cyber threats to critical infrastructures between the states and national levels. MS-ISAC promotes awareness of the interdependencies between cyber and physical critical infrastructures, providing the process for gathering and disseminating the threat information. The organization monitors the cyber activity and alerts the public with a Cyber Alert Level Indicators of Low, Guarded, Elevated, High, and Severe. There are different actions and notifications necessary,

---

[11] Cyber Security Bulletin SB09-278, Vulnerability Summary for the Week of September 28, 2009, Downloaded from http://www.us-cert.gov/cas/bulletins/SB09-278.html 09 Nov, 2009.

depending on the level that can be found on the MS-ISAC website at http://www.msisac.org/. The different levels show the current level of malicious cyber activity being monitored and that could cause potential damage to information systems. The current assessed alert level indicator is low.

LOW: Indicates a low risk. No unusual activity exists beyond the normal concern for known hacking activities, known viruses, or other malicious activity.

GUARDED: Indicates a general risk of increased hacking, virus, or other malicious activity. The potential exists for malicious cyber activities, but no known exploits have been identified and no significant impact has occurred.

ELEVATED: Indicates a significant risk due to increased hacking, virus, or other malicious activity that compromises systems or diminishes service. At this level, there are known vulnerabilities being exploited with a moderate level of damage or disruption, while the potential for significant damage or disruption is high.

HIGH: Indicates a high risk of increased hacking, virus, or other malicious cyber activity that targets, or compromises, core infrastructures, causes multiple service outages, causes multiple system compromises, or compromises critical infrastructures. At this level, vulnerabilities are exploited with a high level of damage, or disruption, or the potential for severe damage or disruption is high.

SEVERE: Indicates a severe risk of hacking, virus, or other malicious activity resulting in wide-spread outages and/or significantly destructive compromises to systems with no known remedy or the debilitation of one or more critical infrastructure sectors. At this level, vulnerabilities exploited with a severe level, or a wide spread level of damage or disruption of Critical Infrastructure Assets.

The formula used to calculate the severity, and thus the alert indicator level, is a formula that uses the criticality of the targeted system, the likelihood of perceivable damage, and the system, along with network counter measures that are in place to prevent the attack.[12]

Threats to Computer Systems/Information Networks

The recent cyber attacks by Russia against Georgia in August 2008, along with attacks on Estonia in 2007, Lithuania in 2008, and Kazakhstan in early 2009, have greatly increased the visibility of the international problem of cyber security. The cyber attacks were closely timed with actual Russian military operations against Georgia. Social networks operating on the Internet were used during the attacks and prior to the attacks for the recruiting of attackers. The cyber attacks appear to have been orchestrated by Russian organized crime working with the Russian government. The cyber attacks began with botnets, which consisted of denials of service and website displacement. Though they appeared minor in nature, the attacks significantly impeded the ability of the Georgian government to deal with the invasion. If more international cooperation had taken place, someone could have potentially warned Georgia of the pending attacks.[13]

---

[12] http://www.msisac.org, Downloaded on 09 November, 2009.
[13] Overview by the US-CCU of the Cyber Campaign Against Georgia in August of 2008, www.registan.net/.../US-CCU-Georgia-Cyber-Campaign-Overview.pdf

I have been a victim of a cyber crime, so I believe it is more frequent than the public realizes. I have an account on PayPal, which one normally thinks of as a secure website. Additionally, I have an EBay account that I sometimes use to buy things at auctions and I use Half.com to buy my college textbooks. PayPal gives one the option to have charges go towards a submitted credit card or have the amount withdrawn from a given bank account. I was browsing my bank statement online one day and observed a withdrawal of over $3,000. I did not purchase anything for that amount, so I dialed the phone number displayed on my account. The person answering the phone did not speak English, so I became even more concerned. I went into PayPal and it listed what it called my recent purchases as a website. I went to the website which was all in Chinese, but it was selling different things. After several phone calls and emails, I did get my money back, but I lost the money for about 2 weeks until the review was done. I immediately removed my bank account for automatic withdrawal and, as a result, I am more critical of my bank and credit card statements.

The economy, the infrastructure, the safety and the security of the United States, as well as much of the rest of the world, depend upon the Internet for the transfer of electronic data. The nation's critical infrastructures use the Internet as the primary means to interconnect. As the Director of National Intelligence (DNI) recently testified before Congress, "the growing connectivity between information systems, the Internet, and other infrastructures creates opportunities for attackers to disrupt telecommunications, electrical power, energy pipelines, refineries, financial networks, and other critical infrastructures".[14]

The American and international public, businesses, and governments have become increasingly dependent on automated information systems. These systems have also become extremely interconnected worldwide with the entire critical infrastructure. The cyber critical infrastructures are attractive targets for individuals and organizations that seek monetary gain, intelligence, or just the pleasure of doing damage to individuals, industry sectors, and even countries. Attackers use a variety of tools and techniques to identify and exploit system vulnerabilities and to collect information passing through the networks. As Figure 5 illustrates, as the sophistication of the attacker tools increase, the need for required complex knowledge is less important. As it becomes easier to attack network systems, the numbers of attacks increase, as does the potential damage increases.

The computing environment is transitioning to a globally integrated information structure and a worldwide effect will be felt across all sectors if a catastrophic event happens to the major networks. Many systems would feel the cascading effects as those effects ripples through the critical infrastructures of the developed countries. Attackers have stolen, modified, and destroyed both data and software. They have shut down entire systems and networks, thereby denying service to users who depend on automated systems to help meet critical needs.

---

[14] Director of National Intelligence, Annual Threat Assessment of the Intelligence Community for the Senate Armed Services Committee, March 10, 2009.

Some of the more common threats are listed and described below. Many are experienced at the same time. Zero day vulnerability is commonly observed, which is where a flaw in software is discovered, but not quickly corrected. Flaws in the system will continually be exploited until they are corrected or bypassed.

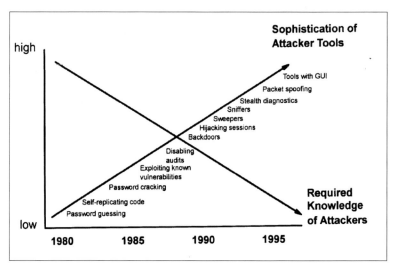

Figure 5. Cyber attacker knowledge required[15].

Malware	Malware is a general term for any malicious software designed to invade, spy on, or damage a computer or other programmable device. Malware is normally used as a single purpose attack.
Viruses	Viruses are software programs that are able to replicate their structure or effect by integrating themselves into existing files or structures on a computer. They usually are able to threaten, or modify, the actions or data of the host device or system without consent.
Trojans	Trojan horses are software programs that pretend to do one thing, but in reality are stealing and altering information, or cause other problems.
Spyware	Spyware includes programs that monitor keystrokes and report the information back to the originator. Spying and theft of data can be the focus of governments, industries, terrorists, and other criminals.
Voice Phishing	Using cell phones and VOIP for voice fraud, remote code execution, data theft, identity theft, and spam are common problems. The growth of the iPhone, iTunes, and mobile banking is an area of increased concern, as more payment infrastructure gets placed on these devices.
Financial Fraud	Credit card theft, identity theft, and corporate espionage are all growing concerns for cyber security.

---

[15] United States General Accounting Office Chapter Report, 05/22/96, GAO/AIMD-96-84, Information Security: Computer Attacks at Department of Defense Pose Increasing Risks.

Natural Disaster	Natural disasters, accidents, and unplanned errors are always a precursor for damage to systems without criminal intent.
Sabotage	Sabotage, theft, and other attacks by disgruntled employees is a hard threat to plan for, since the individuals possess the credentials, authentication, and knowledge of the system of interest.
Worms	Worms are programs that are able to replicate themselves over a computer network for the purpose of performing malicious attacks.
Bots	Bots are programs that take over and use the resources of a computer system over a network, reporting the results back to the originator. Bots can locate digital authentications, vouch for an identity, and purchase goods in electronic auctions. A Bot can remain on a machine, maintain a command, control mechanisms to communicate, and continue to communicate with the attacker while remaining undetected for long periods.
Password Cracking	Attackers try to guess and steal passwords with software that is able to try every possibility, like every word in the dictionary. Attackers can also gain access to a user's system privileges and enter a new password for themselves. Attackers can also use packet sniffing that monitors information packets as they are sent over the networks and records the beginning keystrokes, allowing attackers to learn user identifications and passwords.
Terrorist Attack	Terrorists use cyberspace for communication, coordination, fund raising, and recruitment. Electronic attacks can be launched from a mobile telephone or other type of communication device.
Cyber Warfare	Cyber Warfare is information warfare and can be used in a variety of ways, such as disrupting the military command and control. Cyber warfare is an international issue, as demonstrated by the warfare of Russia against Georgia in 2008. Internet traffic was blocked, routes were hijacked, and misinformation was posted on web sites.

Risk Assessment

All systems involve risk and will continue to have risk as they evolve. More complex technology will continue to bring more risk. As such, one needs to have plans to mitigate the risk to an acceptable level, or accept the risk, if it is deemed worthwhile. One of the major tasks of any systems assessment is to identify the hazards, assess them, and communicate the risk to management and the public. Risk assessment is a step in a risk management process: the determination of the quantitative, or the qualitative, value of risk related to a concrete situation and a recognized threat (also called a hazard). Quantitative risk assessment requires calculations of two components of risk: the magnitude of the potential loss and the probability that the loss will occur.[16]

The hazards are determined from the knowledge of past history and from the brainstorming of all potential hazards to the system. The knowledge of all risks and threats against the asset is an important step in risk management. The second step is to know what to do with the risks and linking the risks with the solutions.

---

[16] Wikipedia online dictionary, Downloaded, 9/13/2009.

Figure 6 is a tool to evaluate risk assessment that was adapted from MILSTD 882. Each potential mishap is evaluated for the severity and the likelihood of it occurring. The scale is determined for each system at the start of the risk management process. The severity can be based on factors, such as injury, damage, environmental factors, duration, and the number of people affected, as well as at different amounts. Some systems are classified catastrophic, if there is loss of one life, or thousands of dollars. Because of the scale difference, another system will only be catastrophic if it affects one million people, or $1 billion.

Mishap Severity: The severity categories provide a qualitative measure of the worst possible event that could result from a hazardous condition or event. These are as follows:		
1	Catastrophic	Death, permanent total disability, major equipment damage, severe financial losses, or irreversible severe environmental damage
2	Critical	Severe injury, permanent partial disability, system loss or damage, heavy financial damage, or reversible environmental damage
3	Marginal	Minor injury, minor equipment damage, financial damage, or mitigatible environmental damage
4	Negligible	Less than minor injury, less than minor equipment damage, or minor environmental damage
Mishap Probability: The levels of probability provide a similar qualitative measure of the likelihood of occurrence of the hazardous condition or event. These are as follows:		
		Specific Individual Item
A	Frequent (X > 10-1)	Will occur often in the quarter
B	Probable (10-1 > X > 10-2)	Will occur several times in the quarter
C	Occasional (10-2 > X > 10-3)	Likely to occur at some time in the quarter
D	Remote (10-3 > X > 10-6)	Unlikely, but possible to occur in the quarter
E	Improbable (10-6 > X)	So unlikely it is not expected to occur in the quarter

Figure 6. Risk assessment[17].

The two elements (probability and severity) are combined to form the matrix in Figure 7 to visually determine acceptable risk. The different levels (high, serious, medium, and low) have to either be reduced to lower risk level, or be accepted by the appropriate authority.

Figure 8 illustrates the risk assessment for cyber security. The list is a small part of the risks/threats that actually exist for the national/international community and is meant only as an example. The roll up risk for the global assessment is a 1D, which is a serious risk. The appropriate authority would then need to accept that risk, or require that further mitigations be instituted to lower the risk. The threat of Botnets is assessed at a Risk of 1C because of the possible severity and because the frequency is rapidly increasing. The Internet needs to use diversity, or redundancy, to increase options, reduce risks, and increase learning in to develop resilience into the network. Resilience is defined as the "capacity of a system to absorb disturbance, undergo change, and still remain essentially the same function, structure, identity, and feedbacks".[18]

---

[17] Adapted from MIL STD 882D.

[18] Managing and Regulating Networked Industries for Sustainability and Resilience, Longstaff, P.H., April 22, 2008.

FREQUENCY OF OCCURRENCE	MISHAP SEVERITY CATEGORIES			
	1 CATASTROPHIC	2 CRITICAL	3 MARGINAL	4 NEGLIGIBLE
A – FREQUENT	1A	2A	3A	4A
B – PROBABLE	1B	2B	3B	4B
C – OCCASIONAL	1C	2C	3C	4C
D – REMOTE	1D	2D	3D	4D
E – IMPROBABLE	1E	2E	3E	4E

Cells	Risk Level	Acceptance Authority:	
1A, 7C, 1C, 2A, 2B	HIGH		Federal Officials
1D, 2C, 3A, 3B	SERIOUS		State Officials
1E, 2D, 2E, 3C, 3D, 3E, 4A, 4B	MEDIUM		Program Manager/ Technical Director
4C, 4D, 4E	LOW		

Figure 7. Mishap risk index (see footnote 17).

Hazard	Risk	Possible Mitigation
Overall - 1D		
Malware	1D	Improved cyber security standards, personnel training, malicious code detectors, encrypted networks, Electronic Monitoring of traffic, firewalls, and email security
Viruses	3A	Improved cyber security standards, personnel training, malicious code detectors, audit trails, monitoring of traffic, firewalls, and email security
Trojans	3A	Improved cyber security standards, personnel training, malicious code detectors, monitoring of traffic, firewalls, and email security
Spyware	3A	Hardware based access control, secure domain name servers, firewalls, and email security
Voice Phishing	3C	Hardware based access control, digital authentication, and secure routing protocols
Financial Fraud	2C	Electronic monitoring of traffic, digital authentication, and audit trails
Natural Disaster	1D	Emergency management plan and resilient systems
Sabotage	1D	Maintain reliable systems, increase resiliency, smart grids electronic monitoring of traffic, emergency management plan, and resilient systems
Worms	3A	Improved cyber security standards, personnel training, malicious code detectors, audit trails, monitoring of traffic, and firewalls
Bots	1C	Improved cyber security standards, personnel training, malicious code detectors, audit trails, monitoring of traffic, firewalls, and buffer overflow protection
Password Cracking	2D	Personnel training, improved cyber security standards, digital authentication, and encryption
Cyber Warfare	1D	Emergency management plan, intelligence security, signal security, intelligence security, resilient systems, and backups
Terrorist Attack	1D	Emergency management plan, resilient systems, intelligence security, signal security, intelligence security, and backups

Figure 8. Risk assessment.

Figure 9 considers that risk and vulnerability can be viewed and assessed on the same graph. The idea is useful for times like when the risk of a threat is low, but the vulnerability to the threat is high. The items in the red area are unacceptable, the items in the green area are acceptable, and the items within the NEXUS are managed to achieve the best outcome.

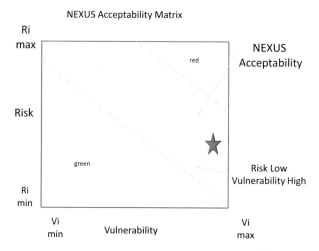

Figure 9. NEXUS acceptability matrix[19].

Cyberspace Vulnerability

There is no doubt that America and the rest of the developed world's critical infrastructure networks are under constant threat. Pervasive vulnerabilities of hardware and software and the connectivity of these mechanisms to the Internet make the multi-layered lines of defense, meaning anti-virus, firewall, and intrusion detection, relatively ineffective. The last 5 years have seen a dramatic increase in the number of Web application vulnerabilities. Both businesses and consumers are at risk; attackers target businesses with sensitive and valuable data, and consumers for their personal information, banking details, or simply their computer resources to create Botnets.

Many of the risks and vulnerabilities face in cyberspace seem to be because of a lack of planning, coordinating with different agencies, and being reactive instead of proactive. Networks are vulnerable because of inexperienced and untrained IT personnel and individuals using weak passwords. Moreover, identified weaknesses often go uncorrected for financial or time reasons. Another problem is that incidents are not reported to the appropriate authorities for dissemination to others. Part of this lack of reporting is because there is not a clear line of responsibility. This lack of a clear line of responsibility is discussed in the Section on governance. Lightening, power fluctuations, surges, blown fuses, and other power outages all disable computer systems, since they rely on an electrical source (Figure 10).

---

[19] Gheorghe, A.V., Risk and Vulnerability Management of Complex Interdependent Systems (ENMA 771/871) Class Notes 29 October, 2009.

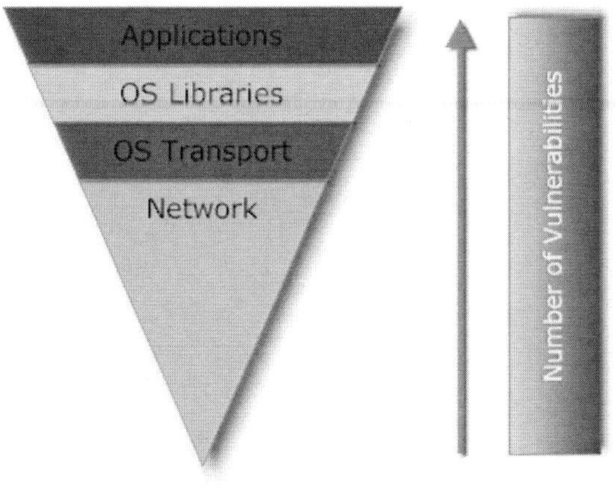

Figure 10. Vulnerability exploitation.

Developments in Information and Communication Technology have created a desire to have meters installed for household electricity monitoring and interfacing. There is now a capability to remotely monitor electric, water, and gas consumption for an individual's homes. There are a lot of benefits associated with the upgrade, such as improved efficiency, quicker response to problems, and more accurate billing. The data are sent over computer networks, however, this makes the system vulnerable to intercept and malicious attacks. The data could also be used by an intruder, to determine when the homeowner is home, based on electricity usage. With any new technology come new risks and vulnerabilities, which need to be identified and mitigated. The electric companies need to assure information security and prevent malicious attacks on the system.[20]

A recent trend in technology is the (E + I) paradigm, Electricity plus Information. Features, such as electricity transmission, distribution, and billing are all integrated with information networks. The technology has enhanced the performance part of the system with monitoring and controlling the system, but the openness of the networks creates threats and makes the system and the electrical system vulnerable to malicious attacks. The security of the system, along with the issue of maintaining the confidentiality and integrity of data on the networks, is a major concern. There are also failures caused by worms, or viruses, which can be attributed to terrorism, espionage, and other hacker activity. There are risks of computer and/or SCADA system failures. SCADA systems are used for process control in manufacturing and utilities. These systems are used to control refineries, power plants, factories, and other highly complex environments. SCADA faces vulnerabilities to security and stability issues. There needs to be better security technology with protocols, firewalls, and an audit/assessment of security administration.

---

[20] AlAbdulkarim, L., and Lukszo, Z. Information Security Assurance in Critical Infrastructures: Smart Metering Case, April 15, 2008.

According to the National Vulnerability Database and nCircle VERT (Vulnerability and Exposure Research Team), "Web application vulnerabilities have increased from 1.9% of all published vulnerabilities in 2006 to over 52% in 2009 (projected based on Q1 and Q2 growth rate). It is important to note that these figures only represent web application vulnerabilities in libraries, languages, frameworks and canned web applications; they do not account for the numerous custom Web Applications that contain their own web application vulnerabilities".[21] Applications seem to be more attractive, since they are not as effectively secured as operating systems. Specifically, there are fewer venders of operating systems, they pay more attention to the systems, and they send patches correcting identified problems quicker.

According to the Sans organization, during the last few years, the number of vulnerabilities being discovered in applications is far greater than the number of vulnerabilities discovered in operating systems. As a result, more exploitation attempts are recorded on application programs. The most "popular" applications for exploitation tend to change over time since the rationale for targeting a particular application often depends on factors like prevalence or the inability to effectively patch. Due to the current trend of converting trusted web sites into malicious servers, browsers and client-side applications that can be invoked by browsers seem to be consistently targeted.[22]

Information and Communication Technology (ICT) is so complex that the many different interdependent components make assessing the vulnerabilities difficult. There are so many possible unpredictable combinations that it makes security of the system difficult. There are available models that can generate an attack model graph as illustrated in Figure 11.

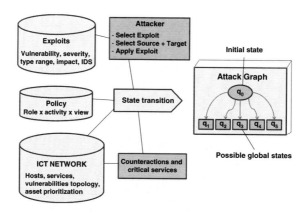

Figure 11. Computation of attack graph (see footnote 23).

---
[21] http://www.ncircle.com/index.php?s=solution_Web-Application-Vulnerability-Statistics
[22] http://www.sans.org/top-cyber-security-risks/

Variables, such as the network security policy, vulnerabilities of the system, and topology of the network, are input into the model. A cost benefit analysis can be conducted to assess the most likely behavior of the attacker. Cost ratings can also be based on the severity ratings given by CVSS or US-CERT. The Model can help to determine what processes, or components, are most in need of improvement.[23]

Governance

Governance is made of the laws, traditions, policies, and processes used to make effective decisions. "Risk governance refers to a decision-oriented process where joint solutions are defined by the involvement of all relevant stakeholders. The process should synthesize the multiple dimensions of the problem: the individual interests and concerns for each industrial company, the market and technical criteria for reliable operation, plus the objectives of different countries and European society as a whole".[24] A governance strategy develops procedures and policies for its activities, methods to monitor the status, and steps for corrective action to improve or correct any issues. According to the Cyberspace Policy Review ordered by the President to assess United States policies and structures for cyber security, "The Federal government is not organized to address this growing problem effectively now or in the future. Responsibilities for cyber security are distributed across a wide array of federal departments and agencies, many with overlapping authorities, and none with sufficient decision authority to direct actions that deal with often conflicting issues in a consistent way".[25]

Risk governance deals with many actors and governments when risk management within the individual organizations is insufficient. Critical Infrastructures, such as cyberspace, are in need of risk governance because a lot of the components are privately owned and controlled, though governments are responsible for the security of critical assets and society. We need to drive resiliency into processes, procedures, and technology to ensure survivability during times of stress. Weak governance, or the absence of control, increases the vulnerability to failure or attack. Making governance stronger makes the system less vulnerable and more resilient. The United States needs to work with the international community for the implementation of information security measures, the institution of computer crime laws, and the standards to pursue cyber criminals across international boundaries.

The National Institute of Standards and Technology (NIST) is an office within the United States Department of Commerce. The NIST establishes standards for all industries, including that of cyber security, for all Federal government and

---

[23] Rieke, R. (2008) Abstraction-Based Analysis of Known and Unknown Vulnerabilities of Critical Information Infrastructures, Int. J. System of Systems Engineering, Vol. 1, Nos. ½, pp. 59-77.

[24] Gheorghe, A.V., Masera, M., Weijnen M., and De Vries, L., Critical Infrastructures at Risk, Springer, 2006, pxx.

[25] Cyberspace Policy Review, Assuring a Trusted and Resilient Information and Communications Infrastructure, Downloaded at www.whitehouse.gov/assets/.../Cyberspace_Policy_Review_final.pdf on 10 November 2009.

government contractors. The NIST is responsible for the United States representation for all international cyber security standards development. They are in the process of coordinating a national licensing and certification for cyber security professionals. It will be unlawful to provide cyber security services without the appropriate certification, which should standardize and increase the education and training for IT professionals. The Computer Security Division is responsible for defining de minimis security requirements, for assessing the effectiveness of security require-ments, and for evaluating the security policies and technologies for the private sector and national security systems. The CSD's mission is to provide standards and technology to protect information systems against threats to the confidentiality of information, the integrity of information, and services to build trust and confidence in Information Technology systems.[26]

The International Organization for Standardization (ISO) is a network of 162 countries developed to establish a consensus on International standards. The ISO is a non-governmental organization that works with both private and public sectors. Standards, such as quality, safety, reliability, efficiency, and interchangeability are established. A standard must have global relevance to be implemented as an ISO. The ISO 27000 series are reserved for the family of information security management standards. There are ISOs for risk management standards, certification standards, auditing information security guidelines, information security governance, specific guidelines for cyber security, and many others covering specific information security areas. ISA/IEC 27014 specifically deals with the development of global govern-ance standard for information security. The standard is designed to cover risk management, management controls, and compliance and assurance of information security standards.

Emerging systemic risks are new risks faced due to technological advances, economy, and environmental issues that affect the major systems within critical infrastructures. Because of the new complex risks the world faces, a new international organization devoted to risk governance was launched. The International Risk Governance Council (IRGC) is an organization whose "purpose is to help the understanding and management of emerging global risks that have impacts on human health and safety, the environment, the economy, and society at large".[27] Based on the research, the IRGC has not been very involved with the global threat of cyber security, but that may be a growing area of interest for the organization. The United States needs to work towards an integrated global approach for unified policy guidance when it comes to cyber security.

The IRGC was legally established in Geneva, on the initiative of the Swiss government. Its board includes representatives from government, public research, private companies, and international organizations from various parts of the world and different professional backgrounds. Some of their specific projects include risk governance on critical infrastructures, such as the electric grid, nanotechnology,

---

[26] http://www.nist.gov, Downloaded 11 November, 2009.
[27] www.irgc.org

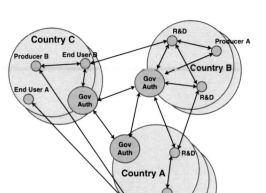

Figure 12. High level information exchange network scheme.

and the studies of risk governance strategies and policies. The IRGC Risk Governance challenge is: "The challenge of better risk governance lies in enabling societies to benefit from change while minimizing the negative consequences of the associated risks".[28]

Because of the sensitivity of the vulnerabilities of a system, many corporations and governments are hesitant about sharing information. The sharing of system threats and vulnerabilities makes one more vulnerable to attacks, if the obtained knowledge is in the wrong hands. Vulnerabilities that are not known by the attackers are less likely to be exploited. Figure 12 (see footnote 29) illustrates a high level information exchange network scheme for exchanging sensitive security information. There are already National initiatives in place, like the National Infrastructure Security Coordination Center in the U.K. This trust management framework allows security information and international threats to be passed among government authorities that are in agreement.[29]

Because the energy infrastructure is so important and problems with cyber security can be catastrophic to the energy infrastructure, we will look at its governance. The Federal Energy Regulatory Commission (FERC) is responsible for the energy power system reliability standards and the cyber security standards. The U.S. Congress has had several hearings on the topic of the power system and on the ability to adequately mitigate against the cyber threats the infrastructures face. It is believed that the FERC is not taking effective action to deal with emerging technology and threats. The FERC should be implementing infrastructure protection standards to ensure the electric grid remains available, even with cyber attacks. The Federal government may not be in a position to move quickly against

[28] Risk Governance: Trends and Challenges, Future Forum for Public Security, 15 Nov 2007.

[29] Advanced Trust Negotiation in Critical Infrastructures, Braghin, S., Fovino, I., Trombetta, A., August 28, 2008.

catastrophic hazards. The GAO has investigated one of the nation's largest power utility companies and found their security posture to be severely lacking.[30] The lack of cyber security actions, currently in place in the energy field, seems to be similar in other industries. Most private companies are more concerned with getting the technology on the market as soon and as cheap as possible, with little regard to the security of the system.

The United States does strengthen its stance on cyber security by implementing clear policies and guidance. Some experts have said that ownership of a computer system on a network should be like owning a car where one is required to have insurance. The requirement for a computing system could include features, such as certain firewalls and virus protection so individuals would not contribute as much to the Internet problems. Additionally, public awareness and education of the threats and vulnerabilities need to be the focus in correcting the problems with Internet security. Everyone should be held accountable for cyber security. If a person has a computer system connected to a network, they are contributing to the network. There needs to be clear guidance on who is responsible for what systems or networks those individuals need to be held accountable for their actions. US-CERT, as discussed in the section on Homeland Security, has partnerships with ISAC's, state and local governments, private sector cyber security venders, federal agencies, and other domestic and international organizations to coordinate the efforts to address key cyber security issues.

The National Science and Technology Council (NSTC) was established by Executive Order to coordinate science and technology across the different parts of federal research and development enterprise. An objective of the NSTC is to establish clear national goals for federal science and technology investments. Legislation, regulation, funding, intellectual property, Internet governance, IT workforce education, and training are all substantial roles in improving cyber security and information assurance.[31]

The National Science Foundation (NSF) is an independent Federal agency created by Congress to promote the progress of science. The NSF is the major source of funding for basic research conducted by American colleges and universities. The NSF builds the research capability for the country by providing the strategy and the scientific vision. Each year, the NSF supports an average of about 200,000 scientists, engineers, educators, and students. The organization does not perform its own scientific research, but funds groups that it approves.[32] Figure 13 displays an abbreviated list as an example of the NSF grants issued in FY04 towards cyber security.

[30] Implications of Cyber Vulnerabilities on the Resilience and Security of the Electric Grid, Subcommittee on emerging threats, cyber security, and science and Technology, May 21, 2008.
[31] Federal Plan for Cyber Security and Information Assurance Research and Development, National Science and Technology Council, April 2006, downloaded at www.nitrd.gov/Pubs/csia/csia_federal_plan.pdf on 11 Nov, 2009.
[32] National Science Foundation, Downloaded at http://www.nsf.gov on 11 November, 2009.

NSF Award #	Proposal Title	PI Linked to Abstract	Awd type	Solicitation Category	Security Lifecycle	Security Discipline	Lead Institution
0433540	CyberTrust Center: Security Through Interaction Modeling (STIM)	Reiter	CT-CS	CT-Sys	4	SM	Carnegie Mellon University
0433668	Collaborative Research: Cybertrust Center for Internet Epidemiology and Defenses	Savage	CT-C3	CT-Net	1	Net	U of Cal San Diego
0430378	Collaborative Reseach: Type Qualifiers for Software Security	Aiken	CT-T	CT-Fnd	5	T&E	Stanford University
0430510	Byzantine Replication for Trustworthy Systems	Alvisi	CT-T	CT-Fnd	7	Sarch	U of Texas Austin
0430271	Collaborative Research: A Survivable Information Infrastructure for National Civilian BioDefense	Amir	CT-T	CT-Net	3	Sarch	Johns Hopkins University
0430274	Collaborative Research: A Comprehensive Policy-Driven Framework for Online Privacy Protection: Integrating IT, Human, Legal and Economic Perspectives	Bertino	CT-T	CT-MI	7	SM	Purdue University
0430258	Cyber Trust - Restore the Trustworthiness of Digital Photographs: Blind Detection of Digital Photograph Tampering	Chang	CT-T	CT-App	1	App	Columbia University
0428422	An Economic Approach to Security	Feigenbau	CT-T	CT-MI	7	Net	Yale University

Figure 13. NSF cyber security grants FY04 (see footnote 32).

Figure 14 is a Governance Profiling Template that can be used as a model for determining the governance gaps with cyber security and for reducing the complexity in identifying governance problems.[33] To conduct a profile for the governance, the risk framework (scope, perception, and situational awareness) needs to be analyzed. The framework also characterizes the future implications, coordination and control, and risk characteristics. The profiling template shows a visual indication of the trends and shows areas of governance that need attention. A score to the left of the scale indicates good governance, while scores to the right show areas that need improved governance. The template can also show areas where governance may not be necessary and where risk management techniques are adequate.

Governance for information networks is not an easy process because of the high level of complexity and interdependencies of the critical infrastructures. Weaknesses in public and private governance hold back sustainable economic and social development.

System of Systems Approach

The nation's critical network and systems are vulnerable to persistent, evolving, and sophisticated cyber threats. The country and the international community must have a plan and be ready to deal with any situation as it occurs. The security of private sector information, systems, and networks is essential in all consumer activities. Because most critical infrastructures are operated by private organizations, they are also a vital concern for the nation's overall security.

Cyber security is not a single point system. The Department of Homeland Security is the hub activity surrounded by many public sector, private sector, multi-levels of local, national, and international governments. DHS needs to develop a system of systems (SOS) approach to the issue which includes people,

---

[33] Gheorghe, A.V., Masera, M., Weijnen M., and De Vries, L., Critical Infrastructures at Risk, Springer, 2006, p. 306.

activities, processes, and technologies. The government should ensure that it is making the most efficient use of its resources, reviewing and further defining the roles and responsibilities of the various advisory boards.

Characteristics	Scale of Governance Gap					
	Low					High
Risk scope	local	☐	☐	☐	☐	global
Risk perception	convergent	☐	☐	☐	☐	divergent
Public awareness	high	☐	☐	☐	☐	low
Probability	low	☐	☐	☐	☐	high
Damage potential	low	☐	☐	☐	☐	high
Ubiquity of damage	low	☐	☐	☐	☐	high
Persistence of damage	short	☐	☐	☐	☐	long
Reversibility of damage	high	☐	☐	☐	☐	low
Delay effects of damage	low	☐	☐	☐	☐	high
Level of complexity	low	☐	☐	☐	☐	high
Level of uncertainty	low	☐	☐	☐	☐	high
Level of ambiguity	low	☐	☐	☐	☐	high
Impact on equity	low	☐	☐	☐	☐	high
Public concern	low	☐	☐	☐	☐	high
Responsibilities pattern	clear	☐	☐	☐	☐	unclear
Regulatory basis	sub-national	☐	☐	☐	☐	international
Binding rules	binding for system	☐	☐	☐	☐	none
Level of compliance	high	☐	☐	☐	☐	low
Regulation adequacy	high	☐	☐	☐	☐	low
International cooperation	functional	☐	☐	☐	☐	not existing
Stakeholder participation	full eng	☐	☐	☐	☐	no engagement
Impact on global free trade	low	☐	☐	☐	☐	high
Impact on business	low	☐	☐	☐	☐	high
Impact on actors' power	low	☐	☐	☐	☐	high
Insurability	fully given	☐	☐	☐	☐	none
Technology change	incremental	☐	☐	☐	☐	break-through

Figure 14. Risk governance profiling template (see footnote 33).

A system of systems involves the integration of multiple independent systems to achieve a different desired outcome. Together, the systems are able to accomplish what one system cannot. A SOS requires multiple disciplines working together to design a system that achieves the overall goal, not necessarily that of the individual elements. Figure 15 illustrates the evolution of system engineering and demonstrates the importance of knowledge and research being fed back into the process to make a better product.[34]

Security must be designed from the beginning, not as an afterthought with patches. In the long run, it is more beneficial to make the systems more resilient and less vulnerable when designing them. There needs to be a focus on the overall system, like the vulnerabilities of wireless communications, the systems that control the energy generation and distribution, and the global positioning satellite systems on which we are becoming dependent.

---

[34] Sousa-Poza, A., Kovacic, S., and Keating, C. (2008) "System of Systems Engineering", Vol. 1, Nos. ½, pp. 1–17.

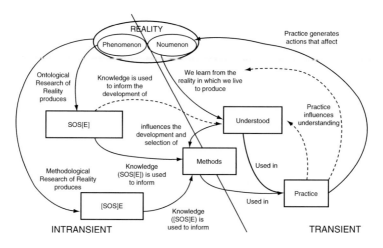

Figure 15. Foundation for the evolution of system of systems engineering (see footnote 34).

Governance is lacking in cyber security, both within the United States and internationally. There are many different agencies, as pointed out in the previous sections of this paper, working on different elements. But, there does not seem to be a coordinated effort with a clear line of responsibility. Cyberspace should be treated like a system of systems and should use a more top down approach, looking at the overall framework and not just separate elements. I believe that most of the elements are present, but there needs to be a clearer hierarchy of responsibility and accountability by the national and international communities.

Conclusion

The task of preventing unauthorized users from compromising the confidentiality, the integrity, or the availability of sensitive information is increasingly difficult in the face of the growth in Internet use, the increasing skill level of attackers, and the technological advances in their tools and methods of attack. The threats and risks are real and increasingly frequent, with high potential to critical energy infrastructures. The Department of Homeland Security needs to have a clear plan in place to help better mitigate the problems faced with cyber security. The Unites States needs to work with its international allies to help establish/improve the current governance for cyber security issues.

PART III: CRITICAL ENERGY INFRASTRUCTURES: OPERATIONAL EFFICIENCY, SECURITY, AND GOVERNANCE

ASSET CRITICALITY IN EUROPEAN GAS PIPELINE SYSTEMS – INCREASING CHALLENGES FOR NATO, ITS MEMBER STATES AND INDUSTRIAL PROTECTION OF CRITICAL ENERGY INFRASTRUCTURE

FRANK UMBACH*, UWE NERLICH
Center for European Security Strategy, Munich, Berlin, Germany

Abstract. The paper focuses on the critical infrastructure security of the European gas supply systems. First, it analyses the changing international security environment for CEIP (i.e. towards cyber attacks) and their critical assets, in particular energy control centers, of gas supply systems under varying conditions – i.e. specific conditions that may impact on the effectiveness and security of critical assets, such as gas control centers. Second, the paper also summarizes and highlights some results of the EU's OCTAVIO- and INSPIRE-research projects. Third, it also tries to identify new emerging risks in the European gas supply system along the EU's evolving common Energy Security Strategy, and subsequent measures, and initiatives on energy security adopted as of spring of 2010. Finally, against this background, some conclusions and recommendations are made for the future role of NATO for protecting CEI.

The Evolving Debate of NATO's Role in Energy Security

NATO has long recognized that energy security and, in particular, the disruption of the flow of vital resources like oil could affect Alliance security interests. The first Gulf War – albeit not a NATO operation – involved key NATO member states (U.S., France, Great Britain, Italy) that sought not only to liberate Kuwait but also to prevent Iraq from controlling Kuwaiti oil as well as threatening the oil fields in Saudi Arabia and other Gulf producers. Although both NATO's old Strategic Concept of 1991 and the present one of 1999 already stated the need to include global security challenges that can affect Alliance security interests such as "the disruption of the flow of vital resources",[1] energy security as a more

---

* To whom correspondence should be addressed: Frank Umbach, Center for European Security Strategy, Munich, Berlin, Germany, e-mail: FraUmbach@aol.com
[1] See The Alliance's New Strategic Concept agreed by the Heads of State and Government participating in the Meeting of the North Atlantic Council', London, 7-8 November 1991 (http://www.nato.int/cps/en/natolive/official_texts_23847.htm, downloaded 5 January 2010); and 'The

A. Gheorghe and L. Muresan (eds.), *Energy Security: International and Local Issues,*
Theoretical Perspectives, and Critical Energy Infrastructures,
DOI 10.1007/978-94-007-0719-1_14, © Springer Science+Business Media B.V. 2011

broadly defined security concept became a more important subject only after the first Russian-Ukrainian gas crisis in 2006 when a new cold war between the West and Russia had threatened the already fragile bilateral relationship between NATO and Russia that existed since the 1990s.[2]

Although NATO still does not perceive the manifold challenges of international energy security as its primary threat or future mission, it has increasingly recognized the need to cope with those geoeconomic and geopolitical implications for its security that has been defined much more comprehensively by including non-military security threats since the 1990s since its Riga summit of 2006. While the issue of energy security is still under discussion and controversy which specific roles NATO has to adopt, the Alliance has reached some consensus about its limited and complementary roles towards the "enduring energy challenge".[3] At its Bucharest Summit in April 2008, the Alliance Heads of States and Governments (HOSG) noted a report on "NATO's Role in Energy Security" that defined five key areas and principles of engagement on energy security in: (1) information and intelligence fusion and sharing; (2) projecting stability (through partnership and outreach programmes); (3) advancing international and regional cooperation; (4) supporting consequence management; and (5) supporting the protection of critical energy infrastructure (CEI).

On the background of the latest Russian-Ukrainian gas crisis in January 2009 that seriously affected the energy security of a number of Allies and Partner countries, the HOSG declared at NATO's Strasbourg-Kehl Summit of April 2009: "The Alliance will continue to consult on the most immediate risks in the field of energy security. ... The issues of a stable and reliable energy supply, diversification of routes, suppliers and energy sources, and the interconnectivity of energy networks, remain of critical importance. Today we have declared our continuing support for efforts aimed at promoting energy infrastructure security."[4]

Alliance's Strategic Concept, Approved by the Heads of State and Government participating in the Meeting of the North Atlantic Council in Washington D.C.', 24 April 1999 (http://www.nato.int/cps/en/natolive/official_texts_27433.htm; downloaded 5 January 2010).

[2] See also F. Umbach, 'Europe's Next Cold War. The European Union Needs a Plan to Secure Its Energy Security', in: Internationale Politik (Transatlantic Edition), Summer 2006, pp. 64-71.

[3] See also Andrew Monaghan, 'NATO and Energy Security after the Strasbourg-Kehl Summit'. Report, (Rome: NATO's Defence College, June 2009); idem, 'Energy Security: NATO's Limited, Complementary Role'. Research Paper No. 36 (ibid., May 2008); idem, 'Energy Security – What Role for NATO?', Research Paper No. 29 (ibid., October 2006); Nicolò Sartori, 'Can NATO Enhance Energy Cooperation in the Caspian Region?). NDC Forum Paper No. 5 (ibid., December 2008); Kevin Rosner, 'Twenty Years after the Fall of the Berlin Wall: NATO's Enduring Energy Challenge', Journal of Energy Security, 27 October 2009; Tom Lantos, Energy Security: A State Side View', NATO Review, Winter 2007 (http://www.nato.int/docu/review/2007/issue4/english/ interview2.html); Gay Luft vs. Christophe Paillard, 'Should NATO Play a Major Role in Energy Security', NATO Review, Spring 2007 (http://www.nato.int/docu/review/2007/issue1/ english/debate.html); Jamie Shea, 'Energy Security: NATO's Potential Role', ibid., Autumn 2006 (http://www.nato.int/docu/review/2006/issue3/english/special1.html), Johannes Varwick, 'NATO's Role in Energy Security', International Politics, Summer 2008, pp. 38-41; and Paul Gallis, 'NATO and Energy Security'. CRS Report for Congress, Washington D.C., 21 March 2006.

[4] See 'Strasbourg / Kehl Summit Declaration Issued by the Heads of State and Government participating in the meeting of the North Atlantic Council in Strasbourg / Kehl', 4 April 2009 (http://www.nato.int/cps/en/natolive/news_52837.htm; downloaded 5 January 2010).

Meanwhile, NATO has initiated numerous practical programmes of energy security both within the Alliance as well as with its Partner countries. They include exercises, workshops and research projects.

As mentioned above, NATO can, on request, support and advise on CEI protection (CEIP) and on its key risks. Since the mid-1990s and particularly since 2001, protection of critical infrastructures has been recognized as an important and rising national, as well as international, security risk that needs to be addressed by national states and international organizations. Recent history after 9/11 has demonstrated and highlighted that critical infrastructures have been increasingly damaged and that disruption of critical processes can have far-reaching political, social, and economic impacts. Critical infrastructures include installations and networks in the energy sector – especially the installations that produce electricity, oil and natural gas, storage and refineries, LNG terminals, and transport and distribution systems.[5] Serious damage can be caused by:

- natural events;
- technical failure and human error;
- intentional acts of terrorists or other criminal nature;
- armed conflicts, civil wars and wars with other countries; and
- cyberwarfare attacks as a new security risk phenomenon.

Although the worldwide energy industry has extensive experience with ensuring operational safety, managing natural catastrophes and prevention of damaging and disrupting energy flows, the increasing sophistication of global terrorism and the growing cyberwarfare capabilities of private hackers, organized crime and terrorist groups represent new challenges of a rapidly changing global security environments, in particular to those rather inexperienced European energy companies have started their energy business (exploration, production, exports, refinery processes, etc.) outside Europe (North and Central Africa, Middle East, Central Asia, etc.) just during the last decade. The traditional security measures of "guns, gates and guard" will still be needed, but they are insufficient to cope with the new risks and threats stemming from a new and rapidly changing security environment. This high-lights the security implications of our dependencies on the spread of information technologies in all areas of our daily life.

In this context, NATO also affirmed they would analyze developments in the electricity, gas, and oil sectors and would share best practices and experiences with partner nations. But like the EU, it recognizes that CEIP remains, first of all, a national and industrial responsibility. But it could also play an important preventive role for enhancing CEIP in the framework of its security concepts and efforts against international and home-grown terrorism or in the maritime environment. If invited, for instance, it can build on NATO's extensive experience of Alliance operations such as Operative Active Endeavour to deter and to prevent terrorist or piracy attacks or other disruptions to supply energy and other strategic resources

---

[5] See also European Commission, Green Paper on a European Programme for Critical Infrastructure Protection, COM(2005)9 576 final, Brussels, 17 November 2005.

and maintain security for key resource routes and strategic Choke Points of Sealanes of Communication (SLOCs).

A particular concern has been raised with regard to cyber attacks[6] on CEI as demonstrated by Russian electronic attacks on Estonia in April 2007 and on Georgian websites before and during the military conflict with Russia in July-August 2008. While Russia and China have been identified as being heavily engaged in cyberwarfare and cyberespionage, both countries are also at the forefront of nationalizing their cyberspace by creating new and more effective "great firewalls" due to their rising concerns of domestic political stability and the role the internet played in the political revolutions worldwide during recent years. In the West, the losses from cybercrime continue to outpace at a staggering rate the technical efforts as well as newly established laws and rules by legislators and in the courts. As Misha Glenny rightly pointed out: "The speed of technological change means that the traditional tools of state used to carve up the world in the nineteenth century, such as laws and treaties, are often inadequate, if not entirely irrelevant, when applied to this new domain."[7]

These new security threats prompted NATO's HOSG to declare at its Strasbourg-Kehl Summit of April 2009: "We remain committed to strengthening communication and information systems that are of critical importance to the Alliance against cyber attacks, as state and non-state actors may try to exploit the Alliance's and Allies' growing reliance on these systems".[8] In line with these concerns and new policies, the Alliance has established a NATO Cyber Defence Management Authority, improved its existing Computer Incident Response Capability and created the Cooperative Cyber Defence Centre of Excellence in Estonia. Furthermore, cyber defence has already become an integral part of NATO exercises.

This chapter focuses on the critical infrastructure security of the European gas supply systems. First, it analyses the changing international security environment for CEIP (i.e. towards cyber attacks) and their critical assets, in particular energy control centers, of gas supply systems under varying conditions – i.e. specific conditions that may impact on the effectiveness and security of critical assets, such as gas control centers. Second, the paper also summarizes and highlights some results of the EU's OCTAVIO- and INSPIRE-research projects. Third, it also tries to identify new emerging risks in the European gas supply system along the EU's evolving common Energy Security Strategy of March 2007 and subsequent documents, measures, and initiatives on energy security adopted as of spring of 2010. Finally, against this background, some conclusions and recommendations are made for the future role of NATO for protecting CEI.

---

[6] See also Johnny Ryan, "'iWar': A New Threat, its Convenience – and Our Increasing Vulnerability', NATO Review, Winter 2007 (http://www.nato.int/docu/review/2007/issue4/english/analysis2.html).

[7] Misha Glenny, 'States Embark on a Scramble for Cyberspace', Financial Times, 17 March 2010 (Internet edition).

[8] See again 'Strasbourg / Kehl Summit Declaration Issued by the Heads of State and Government participating in the meeting of the North Atlantic Council in Strasbourg / Kehl'.

The Rapidly Changing Risk and Security Environment: Implications for Critical Energy Infrastructure Protection (CEIP)

New forms of international terrorism are often characterized by loosely structured networks, connected by little more than common aims. They often act largely independent of each other and without a central command structure, but are able to operate quickly and flexibly without being detected. This is not just a problem of states and their governments. Business companies face the same global risks for their economic investment projects which change faster and more often than before. Furthermore, efforts directed against one radical group in a particular place are not as effective on an international level anymore, despite increasing cooperation among international authorities and intelligence services.

International experts have identified combined utilities-energy industries as very vulnerable by ranking high on the list of potential targets by terrorists[9] as recent developments have demonstrated:

- Terrorist attacks on oil and gas pipelines or crude thefts of oil have increased world-wide, albeit they hitherto had only local impact.[10] Those attacks on oil and other energy infrastructure are today at the heart of an economic jihad.[11] In October 2009, the nuclear scientist Adlene Hicheur (a Frenchman with a modest Algerian background), working at the prestigious CERN nuclear research laboratory near Geneva as well as in one in Oxfordshire, was arrested near Lyon after a 18-month surveillance operation with real fears that he, together with Al Qaeda's dangerous North African Units, planned a nuclear attack. But he was targeting more concretely a TOTAL oil refinery with an explosion that would have destroyed a city "the size of London", as the investigation found immediately afterwards.[12]

---

[9] See also F.Steinhäusler/P.Furthner/W.Heidegger/S.Rydell/L.aitseva, Security Risks to the Oil and Gas Industry: Terrorist Capabilities, Strategic Insights, Vol. VII, Issue 1, February 2008; M.Mihalka/D. Anderson, Is the Sky Falling? Energy Security and Transnational Terrorism, ibid., Vol. VII, Issue 3, July 2008, and Larry Ness, Terrorism & Public Utility Infrastructure Protection, in: Journal of Energy Security, October 2008, p. 1.

[10] See E. Blanche, Terror Attacks Threaten Gulf's Oil Routes', Jane's Intelligence Review (JIR), 6–11 December 2002.

[11] See N. Stracke, Economic Jihad: A Security Challenge for Global Energy Security, in: Gulf Research Center (Ed.), *Energy Security. Security & Terrorism Research Bulletin*, Issue No. 6, Dubai, August 2007, pp. 26-32; D.Moran/J.A. Russell, 'The Militarisation of Energy Security', Strategic Insights, Vol. 7, No. 1, February 2008 and F.Steinhäusler, 'Terrorists Threats to NATO Countries: A Bestiary', NATO Review, Spring 2007 (http://www.nato.int/docu/review/2007/ issue1/english/art4.html).

[12] See D. Overbye, 'French Investigate Scientist in Formal Terrorism Inquiry', The New York Times, 14 October 2009 (Internet-edition); B. Hall, 'Terrorist Suspect Linked to 'Big Bang' Facility, FT, 9 October 2009 (Internet-edition) and P. Allen/N. Ramdani, 'Big Bang Scientist Planned to Blow Up Oil Refinery in Terror Attack That Would Wipe Out City the Size of London', Daily Mail, 13 October 2009.

- Attacking even oil and LNG tankers on the sea is no longer a movie fiction, as terrorist attacks in November 2000 on the USS Cole warship, in October 2002 on the French-owned supertanker Limburg and most recently piracy attacks on an oil and gas tanker in Somalia's territorial waters and the waters of its Economic Exclusive Zone have highlighted. Thus security of transporting energy sources has become an ever more important security challenge to cope with.

- Global natural gas consumption is forecast to increase 70% from 2002 to 2025. In the EU, Liquefied Natural Gas (LNG) imports capabilities may double from 75 bcm to at least 150 bcm by 2020. The interregional LNG trade may even expand by 300% by 2030. By cooling to −260°F, natural gas can be transported by tankers in a liquid state. Although LNG shipping transports have been quite safe for over four decades, Al-Qaeda is believed to have identified LNG as a desirable target because of its explosive potential. It could result in a massive fire, melting steal at distances of 1,200 feet and second-degree burns on exposed skin a mile away and with thousands of deaths and upwards of 20,000 injuries near population centres.[13]

- Recently, security concerns about increasing cyberwarfare attacks for espionage or damaging and disrupting processes through malicious software programmes of critical infrastructure assets, and processes by individuals, crime and terrorist organisations, and governments have grown in both Western governments and industries. Those cyber attacks have risen to an unprecedented level of sophistication. As a result, the vulnerabilities of digital systems and networks have grown exponentially. However, public awareness has not kept up with those new threats and vulnerabilities in cyberspace, which have the potential to affect all sectors of private and public life, national and international businesses, and even defence policies of national states or multinational organizations such as EU and NATO. In the age-old struggle between attacker and defender, the attacker more than ever appears to have the advantage by being better armed, being able to freely choose the intensity of the attack as well as the target, being no longer constrained by any geographical distances and frontiers, as well as enjoying the stealth, anonymity, and inability to be identified. Those threats are challenging traditional assumptions and thinking of national as well as collective security and defence.

The emergence of botnets in particular – by implementing dormant viruses, unnoticed by Internet users, which the attacker can activate at any time (e.g. Trojan horses) and at any place in the world – allow criminal or terrorist attackers to launch massive hostile operations of data espionage, falsifying, destroying, or altering confidential data with extraordinary harmful effects to industry as well as critical national infrastructures. The newest botnet threat, Conficker, for instance,

---

[13] See Cindy Hurst, The Terrorist Threat to Liquefied Natural Gas: Fact or Fiction? and Eben Kaplan, Liquefied Natural Gas, A Potential Terrorist Target?, Council of Foreign Relations, Backgrounder, 27 February 2006 (http:www.cfr.org/publication/9810/liquefied_natural_gas.html).

is estimated to have infected 1.5 million computers. It has the ability to function autonomously by recruiting and commanding 5 million computers in 122 countries – without having yet any counter strategy to date – for coordinated simultaneous attacks on the economic system, critical national infrastructures, and the national defence structure of a country – all of them very interdependent of each other. Almost all industries and companies and even defence ministries are increasingly dependent on the use of the open Internet. Even protected infranets themselves are not immune to cyberattacks as Pentagon officials have admitted. The chairman of the Permanent Monitoring Panel on Information Security at the World Federation of Scientists, Henning Wegener, warned in 2009:

"A momentous, revolutionary, and probably irreversible shift is underway that concentrates computing, software management, and data storage in distant, undisclosed locations, without any transparency in their procedures and security measures, relegating traditional computers to mere portals of access, and depriving conventional fire walls of their function. The result is a huge integrated network structure with a universe of connectivities and vulnerabilities that defies quantification. It includes a myriad of important components open to attack."[14]

Hostile governments can hide behind "unholy alliances" with crime syndicates (with the blackmail goal of extorting money), terrorists or nationalist movements, and individuals without any risk of detection or identification, By blurring the border between cybercrime, cyberterrorism, and private or state-sponsored cyber war as a new form of "asymmetric warfare" in the twenty-first century, the threat of a "digital Pearl Harbor" is real and no longer science fiction.

- Russian hackers apparently penetrated a nuclear power plant near St. Petersburg in May 2008. While the website was taken offline, the intrusion did not affect the operation of the plant. But rumors of "radioactive emissions" from the plant were spread (deliberately?). For several hours, no effective communication existed between the plant and the nuclear corporation, Rosatom.[15]
- Massive denial-of-service attacks of viruses, worms, and other forms of malware on servers of government ministries, newspapers, banks, and other corporations, as well as on private web sites, and the countries' cell phones with the intention of crashing their sites are already an ever-increasing reality. Those attacks y lasted for 3 weeks in Estonia in April-May 2007, Lithuania in June-July 2008, Georgia in July-August 2008, in South Korea and the U.S. in July 2009 (attacking 12,000 computers in South Korea and 8,000 in other countries, originating from North Korea).

---

[14] Henning Wegener, 'Invisible Enemies. The New Magnitude of Cyberspace Threats: Internationale Politik (IP)-Global Edition 09/10 2009, pp. 50–55 (52). See also Johnny Ryan, '"iWar": A New Threat, its Convenience – and our Increasing Vulnerability'; Udo Helmbrecht, Haltet den Wurm! Wie Deutschland die sichere IT-Nutzung verbessern will, in: IP, September/Oktober 2009, pp. 58–63 and Jody Westby, Stabilität statt Cyberkrieg. Was für den Frieden im digitalen Raum notwendig wäre: ibid., pp. 64–69.
[15] See 'How Vulnerable are Energy Facilities to Cyber Attacks', Intelligence Report, Securing America's Future Energy (SAFE), Washington D.C., Vol. 3, Issue 1, 20 January 2010 (http://www.secureenergy.org), p. 2.

- Cyber attacks on the European CO2 Emission Rights Market at the beginning of 2010 have highlighted the potential cyber attacks can have on manipulating market prices and who will get the contract to produce energy to meet the projected demand. Those manipulations can also lead to energy supply shortages, disruptions, and power blackouts. The energy spot and other energy market operations (e.g. Amsterdam Power Exchange/ APX, the Paris Powernext and the European Energy eXchange/EEX in Germany are all fully operating via the Internet and, thus, are at risk of external computer manipulation. The World Economic Forum estimated in 2008 that there is a 10–20% probability of a major breakdown of critical information infrastructure (CII) in the next 10 years, with a potential global economic cost of approximately 250 billion US$.[16] Other experts have warned that a country being attacked electronically can lose electrical power for 3 months, leading to an economic loss of US$700 billion – "equivalent to 40 to 50 large hurricanes striking all at once. ... It is greater economic damage than any modern economy ever suffered ... It is greater than the Great Depression. It's greater than the damage we did with strategic bombing on Germany on World War II."[17]

The present U.S. government under President Barack Obama has announced a $17 billion programme for larger digital defensive efforts and appointed Howard A. Schmidt (a veteran computer specialist) as a new White House official to coordinate those efforts against cyberwarfare attacks. The rising electronic vulnerabilities and the insufficient digital defence capabilities in particular against the U.S. critical infrastructures, such as the country's power stations, the electric power grid, water supply systems, telecommunications, aviation systems (i.e. air traffic control systems), and financial markets have been recognized as "one of the most urgent national security problems facing the new administration", but the defence efforts are hampered by lacking a clear strategy and secrecy that hinders a greater national debate.[18]

Google's disclosure that Chinese hackers have stolen some of its intellectual property as well as of 30 other US high-tech companies (such as Adobe Acrobat, Cisco-Systems) have led to a paradigm shift in cyber security thinking in the US. These Chinese cyber-attacks have been perceived as being so aggressive and so pervasive that meanwhile many US companies demand political pressure from its own government against China instead of previous attempts to discourage it because of their concerns of undermining their own business interests in China.

---

[16] See Commission of the European Communities, Protecting Europe from Large-Scale Cyber-Attacks and Disruptions: Enhancing Preparedness, Security and Resilience.SEC(2009)399/ SEC(2009)400, Brussels, 30 March 2009, COM(2009) 149 final, p. 2.

[17] Thus Scott Borg, Chief Economist at US Cyber Consequences Unit, a private non-profit think tank, quoted in: J.N. Gordes/M.Myirea, 'A New Security Paradigm Is Needed to Protect Critical US Energy Infrastructure from Cyberwarfare', Foreign Policy Journal, 14 September 2009 (http: //www.foreignpolicyjournal.com/2009/09/14/a-new-security-paradigm-is-needed-to-protect-critical-us-energy-infrastructure-from-cyberwarfare/; downloaded 30 September 2009).

[18] David E. Sanger/John Markoff/Thom Shanker, U.S. Steps Up Effort on Digital Defenses, The New York Times, 28 April 2009.

Even more questions arose when Google had to ask the National Security Agency about the details of the cyber attacks 3 months after the first attacks. A comprehensive study by a Canadian university found in March 2009 that an automated cyberspying system run from servers based in China had accessed more than 1,300 computers in 103 countries, including many ministries, embassies, international organizations, NGOs, and news media.[19] The so-called GhostNet-systems not only searches computers for information and taps e-mails, but also turns them into giant listening devices. US experts believe that many US companies are not aware that they have already been attacked. These sophisticated computer attacks have been "devastatingly effective". Only a few organisations outside the defence and intelligence sectors can withstand them. The FBI traced more than 90,000 Chinese-originated attacks in 2009 on the Pentagon alone. Chinese hackers have also been seen as responsible for shutting down the British House of Commons computer system in 2006 and have attacked the networks of the British Foreign and other key departments in 2007.[20] Western intelligence sources estimate more than 500,000 hackers willing to engage in cyberwar.[21] But at the same time, China's government is increasingly concerned about the rise of cyberattacks on its own from Chinese computers as well as about physical and cyber attacks on its rapidly expanding oil and gas pipeline system.[22]

In addition to these new terrorist attacks, the vulnerability of the different sectoral infrastructures have also increased because they are much more linked with each other in some way due to the rapid spread of information technology. They can multiply in other locations, branches or sectors, with an impact that extends far beyond the original area of damage.[23] Due to those existing mutual dependencies of different types on different types of infrastructure systems and to interconnections as well as infrastructure services provided over large areas via physical (i.e. transnational gas pipelines from Russia to Europe), virtual or logical networks, growing in size and complexity, this can lead to a domino effect shutting down much of the government, economy, and society and can cause even regional, interregional, national or even global outages and failures. Thereby, many physical, virtual and logical dependencies are not apparent until a crisis occurs and a connection breaks down. Even smaller disruptions may cause dramatic consequences in complex systems (vulnerability paradox), whereas a high interdependence can lead to cascading shut-downs of critical (energy) infrastructures with more damage than any military offensive.[24]

---

[19] See also Malcolm Moore, 'China Globl Cyber-Espionage Network GhostNet Penetrates 103 Countries, Telegraph, 29 March 2009; K. Hille/J.Menn, 'Hackers in Frontline of China's Cyberwar', FT, 13 January 2010;

[20] See 'D. Gardham, 'Al Qaeda, China and Russia Pose Cyber War Threat to Britain, Warns Lord West, Telegraph, 25 June 2006.

[21] See also David Barboza, 'Hacking for Fun and Profit in China's Underworld', New York Times, 2 February 2010 and S. Nandan Andey, 'Red Guests. Hacktivism of Chinese Characteristics and the Google Inc. Cyber Atack Episode', Denkwuerdigkeiten, No. 63, April 2010.

[22] See X. Dingding/W.Zhihong, 'China Faces New Risk: Attacks on Pipelines'.

[23] See German Ministry of Interior, Protecting Critical Infrastructures, p. 9.

[24] See ibid., here p. 11.

Destroying a major refinery, damaging severely the natural gas delivery system, sinking oil and LNG tankers, conducting successfully a cyber attack on telecommunications and electricity delivery grids, sickening the operators and workers who keep the systems functioning and the attack on those systems can lead to cascading effects and even escalate them into catastrophic collapses – in particular by simultaneous, multi-pronged attacks intended to bring down the entire grid in a region or beyond with millions of people being affected. Recent examples are the following ones:

- Large-scale power outages in eight US federal states, including New York City, and Canada in 2003 largely due to problems with one electricity provider's transmission system incurred costs of up to US$10 billion. They affected as many as 50 million customers as well as a range of vital services and commerce: shut down of air and ground transportation systems, trapping people far from home, drinking water systems, stopped operations of sewage processing plants, disrupted manufacturing, and halted the functioning of emergency communications systems.
- The collapse of the entire EC debit card system in Switzerland in 2000 resulting from an error in one computing centre.
- A combination of heavy snowfall, icing and storms resulted in an electricity outage of up to 5 days, affecting more than 80.000 consumers in Germany, Belgium, and the Netherlands with a total damage of 20 million Euro.
- An electricity disruption in Emsland (Germany) on November 4, 2006 lead to cascading effects from north to the south of Germany, eleven neighbouring countries (Austria, Croatia, Hungary and others) and even Morocco. It affected 15 million people in Europe for up to 3 days.[25]
- According to some CIA sources, the series of power outages in Brazil in 2005, 2007, and 2009 (affecting 50 million people – ¼ of the entire population) were caused by cyber attacks on the Supervisory Control and Data Acquisition (SCADA) systems, though the Brazilian government has officially denied any cyber attacks occurred.[26]

In addition, natural disasters as the result of climate warming[27] are also rising and threaten existing and future critical infrastructures,[28] whereas changes in market activity (such as those caused by market liberalization and privatization of state-owned infrastructure operators as well as new regulations) have made private industry and government agencies increasingly dependent on external providers of goods and services. Those threats may further increase as the result of the impact

---

[25] See also Bundesnetzagentur für Elektrizität, Gas Telekommunikation, Post und Eisenbahnen, Bericht über die Systemstörung im deutschen und europäischen Verbundsystem am 4. November 2006, Bonn, February 2007.

[26] See 'How Vulnerable are Energy Facilities to Cyber Attacks', p. 2.

[27] Examples are the flooding of the River Elbe in 2002, the winter storm Kyrill in 2007, or the hurricanes Katrina and Rita during the summer of 2005 in the Gulf of Mexico. These hurricanes had an even more destabilizing effects on the global oil market than all terrorist attacks until today.

[28] See also C. Paskal, 'The Vulnerability of Energy Infrastructure to Environmental Change', Briefing Paper, Chatham House, London, July 2009.

of global warming as well as the growing concentration of exploration and production capabilities of the remaining conventional oil and gas resources in ever fewer, and at the same time politically unstable, countries and regions (i.e. the Greater Middle East) at a time when the world will need up to 40% more energy by 2030.

At the same time, the financial and personnel resources available to operators to protect their infrastructure systems are limited. Thus it is essential to use all available resources to efficiently and effectively assess risks and set priorities for adequate risk management. While it is impossible to protect a utility 100% from a physical or a cyber attack on a utility's facilities and infrastructures, those threats need be minimized as much as possible without compromising productivity and day-to-day operations. Nonetheless, a professional security and risk assessment needs to address physical and cyber security, SCADA and data acquisition distributed control systems (DCS), communications security, grid security, distribution security, generation security, and biological/chemical issues.[29]

Such an additional complementary approach is based on the assumption and recognition that the new and future technical and technological features of natural gas and electricity control centres will develop further by becoming even more important and complex for the EU's and NATO's political and physical crisis management as the result of those factors listed above. US expert Larry Ness has warned that Western societies and business companies need to overcome the traditional "let's-wait-until-something-happens-and-than-fix-it"-syndrome because "waiting versus acting will prove to be devastating".[30]

All these new security threats (i.e. linked with cyber threats) to critical infrastructures have produced a worldwide demand for new security technologies, services, and management capabilities. Thus the markets for civil defence products and services, including ICT and software producers, against terrorist, piracy, crime, and hackers are among the biggest growth markets in the world.

The EU's Programmes and Projects for Enhancing CEIP

"Europe's energy networks are the arteries on which we all depend for the energy to fuel our homes, businesses and leisure."

(European Commission, November 2008)[31]

The EU and its member states have discovered the need to secure critical infrastructure assets as part of their security strategies rather late. Moreover, the single member states have developed their own individual approaches, institutions, and programmes to cope with these new security challenges to protect critical infrastructures, including critical information infrastructures (CII), despite the perceived common risks, threats, vulnerabilities, and strategies for securing critical (information) infrastructures. A first step to address those common risks and vulnerabilities as well as to cope with the cross-border effects of damaged infrastructures

---

[29] See also L. Ness, 'Securing Utility and Energy Infrastructures, Hoboken/New Jersey 2006.

[30] Larry Ness, Terrorism & Public Utility Infrastructure Protection.

[31] European Commission, Green Paper. Towards a Secure, Sustainable and Competitive European Energy Network, SEC (2008)2869, COM(2008) 782 final, Brussels, 13 November 2008.

or its disrupted processes has been made by the establishment of the "European Network and Information Security Agency (ENISA)" in 2004 to enhance European coordination on information security.

A broader initiative has been made by the Commission of the European Communities at the end of 2005 by adopting a "Green Paper on a European Programme for Critical Infrastructure Protection".[32] In December 2006, the European Council adopted a "European Programme for Critical Infrastructure Protection (EPCIP)" that has defined principles, processes, and instruments for its implementation.[33] The EPCIP has been the nucleus for an EPCIP Action Plan, the Critical Infrastructure Warning Information Network (CIWIN), the use of CIP expert groups at EU level, CIP information sharing processes, a procedure for a common approach to the assessments of the needs to improve the protection of such infrastructures, and the identification and analysis of interdependencies between very different critical infrastructures. According to the EU's "Council Directive on the Identification and Designation of European Critical Infrastructures and the Assessment of the Need to Improve Their Protection" of 2008, "critical infrastructure means an asset, system or part thereof located in Member States which is essential for the maintenance of vital societal functions, health, safety, security, economic or social well-being of people, and the disruption of which would have a significant impact in a Member State as a result of the failure to maintain those functions."[34] In addition, European critical infrastructure has been defined as "critical infrastructure located in Member States the disruption or destruction of which would have a significant impact on at least two Member States. The significance of the impact shall be assessed in terms of cross-cutting criteria. This includes effects resulting from cross-sector dependencies on other types of infrastructure."[35]

Against this background, the Commission tendered a series of studies in the second half of 2007 under the EU's 7th Framework Programme for the Commission's General Directorate for Justice, Liberty and Security (JLS) that includes studies on specific sectoral infrastructures and assets. With regard to CEI, three recent research projects have been important and offer concrete results as well as technical solutions for enhancing CEIP and highly vulnerable energy control centers and their SCADA systems:

- The objective of the Octavio Project was described to develop a "Comprehensive Approach Definition to Improve the Security of Energy Control Centers based on Establishing Criteria and Methodology to Assess, Audit and Mitigate Risks for Electricity and Natural Gas Control Centers and their Interdependent ICT Infrastructures". The project focused on the structures, functionalities, and security of critical assets, particularly

[32] See ibid.
[33] See the EPCIP-website: http://ec.europa.eu/justice_home/funding/2004_2007/epcip/funding_epcip_en.htm.
[34] See Council Directive on the Identification and Designation of European Critical Infrastructures and the Assessment of the Need to Improve Their Protection", 2008/114/EC, Brussels, 8 December 2008.
[35] See ibid.

control centres, in electricity and gas supply systems. Furthermore, it has aimed to provide an accurate (risk) assessment regarding energy sector control centers (i.e. natural gas and electricity control centres) and their cyber structure requirements. More specifically, the project has also developed a comprehensive approach to improve the security of energy control centers based on establishing criteria and methodologies to assess, audit, and mitigate risks for the EU's electricity and natural gas control centers and their interdependent ICT infrastructures.

The Octavio Project consisted of three major parts: (1) the levels and uses of control centres in energy systems (electricity and natural gas), (2) the security of control centres in energy systems, (3) the security of different functionalities of control centres. The first Report described the structure and the control systems of electric and natural gas systems and their components as well as the threats and the potential impact of attacks on these systems. Special attention has been given to asset criticality, in particular to state-of-the-art control centres. The second Report focused on electronic and physical security (i.e. on the vulnerabilities and attack options) of IT systems controlling energy networks, in particular electric systems. The third Report concentrated on the vulnerabilities and interdependencies of gas supply systems and their components, i.e. the systems operator, maritime transport, regasification, plants transport (pipelines, compressor stations, storage CNG, metering stations, city gate stations) and distribution (DCS distribution, pipelines, customers, tanker, gas stations).

- The presently finalized INSPIRE-Project (in which CESS is involved as it was in Octavio) aims to mitigate the threats and to improve robustness as well as resiliency of Energy Control Centres and other Large Complex Critical Infrastructures (LCCI) by increasing safety and security of the LCCI's control systems. The project has developed technical solutions that increase the security and resilience of LCCIs by means of a self-reconfigurable architecture suitable for SCADA systems. It includes the main functional blocks of (1) monitoring by different means of probes, (2) diagnostic processes to identify damaged components and system reconfiguration actions and (3) the reconfiguration actions aiming to mitigate the effects of attacks as well as isolating the attacker.
- The Commission had issued a study on "Risk Governance of European Critical Infrastructures in the ICT and Energy Sector", whose final report was published in September 2009. The study focused on the manifold interfaces between the two sectors, based on the analysis of actors, regulatory, market and technical factors, decision processes, problem framing, development of risk scenarios and application to them of a Risk Governance Framework. Thereby, secure data is fundamental to correct operation of ICT systems both under normal conditions and even more critically under abnormal circumstances. Data corruption and/or lack of

access to required data can disrupt the balance of a system and implemented control measures. Hence recovery might be delayed or become impossible.[36]

Common and Different Characteristics Between Electricity and Gas Supply Sectors and the Functionality of Their Control Centers

In general, electricity grids and gas pipeline systems share a number of characteristics, e.g. the growing dependence on ICT. Besides the generation of electricity results in part from growing gas supplies and thus indirectly shares some of the risks: direct disruptions of gas supplies can translate into secondary effects in electricity grids. Yet the differences between the two kinds of energy systems are significant:

- Electricity is primarily generated within European countries and regions, and in several countries an import/export balance of electricity has developed.
- EU countries and regions are not directly exposed to disruptions of electricity supply by external producers except in case of major hostilities, although in the longer run importing electricity from the Southern Mediterranean and GCC countries is conceivable if large-scale industrialization of solar energy is achieved and effective solutions for long-distance transport have been realized.[37]
- For gas and oil pipelines the sources are mostly and increasingly outside EU territory.
- The distances to end-users are much longer for gas than is the case with electricity grids.
- The ratio of transmission and distribution displays rather different structures and vulnerabilities.
- The flexibilities for rerouting are incomparable for gas and electricity.
- The relative resilience of transport systems (nets or pipeline systems) and critical assets differs also pipelines are relatively easy to be repaired, control centers for pipelines much less so.
- Most importantly, electricity grids expand from local and regional systems to the EU at large and beyond. Gas and oil pipelines reach from increasingly distant producers which compete for demand security into EU territory.

The reports provide state-of-the art accounts of physical and electronic security of control centres. They concentrate on technical requirements of control centres should meet, i.e. for the functioning of control centres and their security vis-à-vis physical and cyber attacks. This is highly relevant for:

---

[36] See 'Study on Risk Governance of European Critical Infrastructures in the ICT and Energy Sector'. Final Report to European Commission/Directorate-General Justice, Freedom and Security, Brussels, 4 September 2009.

[37] See, for instance, the Desertec-project in the EUMENA Region (http://www.desertec.org). For the original study, see German Aerospace Center (DLR) et al., Trans-Mediterranean Interconnection for Concentrating Solar Power, Stuttgart, June 2006 (http://www.dlr.de/tt/trans-csp).

- the implementation of secure systems;
- a methodology for a standard assessment of safety and security of all EU gas and electricity control centers in order to get an overview for future security investments;
- decisions on modernizing aging systems, on investment choices and prioritizing in view of rapidly enlarging and extending systems;
- coping with the chances and risks of exponentially increasing dependence on ICT;
- preparing against natural and technical hazards, terrorist and criminal attacks; and
- for changing political and security conditions, e.g. the stability of producer countries, the uncertainties in transit zones, the heightened dangers in case of major crises and war-like conditions, growing asset criticality in case of progressing integration and interconnectors, increasing transport distances and the eventual transfer of authority with a potential for more centralized decision-making, the trade-offs between effectiveness versus security in large electricity grids and pipeline systems, etc.

The technical requirements for control centers are largely the same for up to 80% of the critical assets as displayed by pipeline demonstration centers of major companies (e.g. the Siemens Center in Zug, Switzerland and Fürth, Germany). The energy industry within the EU follows, by and large, the guidelines applied to US facilities.[38] But the extent of implementations and modernization, the limitations imposed by national postures, the divergent risks inherent in divergent suppliers, systems and transit zones, the uneven exposure to potential violence (be it by terrorists or in war-like situations), the competitiveness governing the European energy markets and the limitations on flexibility of adoptions to changing challenges inherent in gas pipeline systems pose additional challenges to energy industries as well as to national, EU, and international governmental authorities – be they producers, transit providers or suppliers.

Yet the functioning of SCADA systems is itself a condition that deserves analysis. In general, the security in process control systems is lagging 5–10 years behind the security of laptops or desktops. SCADA-systems have become a key element in the safe and secure operation of both installations and extended infrastructures. They control the operation of power plants as well as of networks. The operation of huge border crossing gas networks requires a network management and a control center hierarchy to ensure security of gas supplies:

- Main Control Centres (i.e. system and network control centres) responsible for generation coordination, load dispatching as well as monitoring and control of the storage sites and transmission network to provide reliable communication, to keep the integrity and security of the complete network and to guarantee the supply of the services;

---

[38] See 'Pipeline Security: An Overview of Federal Activities and Current Policy Issues', CSR Report for Congress, February 2004, here pp. 17 f.

- Regional Control Centres responsible for monitoring and control of the distribution network within a specific area;
- District Control Centres responsible for monitoring and control of the distribution network within a specific district.
- The efficiency of control centers by applying methods of data handling and processing is closely linked with the development and application of ICT. Their task is:
- Measurement and information gathering by sensors – including satellite based surveillance and control of pipeline systems, power plants, pump stations, storage sites, and networks;
- Acquisition: transmission of necessary information from the network to the Control Centre transmission of commands from Command Centres to "operational" components like substations;
- Processing, display and archiving of information from the network, generation of control information.

In contrast to the former auxiliary function for the control of operations of plants and networks, meanwhile control has been transferred to a centralized complex instrument with the central function in energy supply. Without this central function, any operation within the energy and gas supply chains ranging from production to distribution and supply would be impossible. The efficiency and reliability of those Control Centres, in particular the System or Central Command and Network Control Centres, is essential and the biggest threat in case of physical and electronic attacks, which could have extensive follow-on consequences on other critical infrastructures and lead to heavy losses at the stock exchange.

Acquisition and processing tasks are elements of a SCADA-system. With SCADA, control centres are able to identify and repair interferences, to take the necessary measures of repair centrally and to acquire data relevant for planning and further action. Originally, each power plant had its own Control Centre linked with others a part of a hierarchy of networks. The development of ICT enhanced the capabilities to combine not only the different tasks of command structure for the hierarchy of networks, but also for different media, such as electricity, gas, water, or district heating in a Central Command Centre. The latter have extended their capabilities by using Geographical Information Systems (GIS) to provide geo-referencing information of facilities, networks, vehicles, and geographical or political details. Modern SCADA systems use standard interfaces and standard components (of computers operating under UNIX or Windows). It has improved system interconnection and efficiency, but has also significantly increased the system vulnerability to outside electronic attacks.[39] Additionally, emerging network

---

[39] See also F. Umbach/U. Nerlich, 'European Energy Infrastructure Protection: Addressing the Cyberwarfare Threat', Journal of Energy Security (http://www.ensec.org), October 2009; F. Umbach, 'Critical Energy Infrastructure Protection in the Electricity and Gas Industries. Coping with Cyber Threats to Energy Control Centers', OSCE-CTN Newsletter, Special Bulletin: "Protecting Critical Energy Infrastructure from Terrorist Attacks", Vienna, January 2010, pp. 25–28 and F.Gaspard/ A. Hubrecht, 'Tackling Critical Energy Infrastructure Network Interdependencies', Journal of Energy Security, 23 March 2010.

technologies (i.e. Wireless Sensor Networks/WSNs) are being increasingly used in new generation SCADA systems.

Given the growing extension and complexity of energy systems (i.e. of gas systems), the requirements for the effectiveness and the security of control centers get more demanding and trade-offs between effective and secure solutions become more challenging: The increasing number of interconnectors between gas systems, the cost of ever larger numbers of sites along with the size of the system, the vast areas they cover, the inherent risks resulting from how administrative units and control centers are often connected, typically with needs for control engineers, ICS operators and IT security professionals to cooperate closely etc. render the requirements for effective and secure control centers even more critical.

A broad and systematic analysis is thus an important step. It follows similar analysis in the United States in the aftermath of 9/11. But the conditions for moving from highly decentralized to increasingly centralized energy systems differ from the US and the EU with regard to regional and state energy demands and decision-systems. In the US the prospects for extensive and homogeneous security solutions have allowed faster improvements than is feasible in a still enlarging EU that tends to integrate slowly and is increasingly dependent on oil and gas imports from outside the EU.

Security Requirements for Control Centers Under Varying Conditions

CONDITIONS FOR ASSET CRITICALITY IN GAS SUPPLY SYSTEMS

The criticality of assets, in particular of control centres, for the functioning of gas supply systems depends on both the degree to which technical security requirements are met and on the conditions under which they are expected to function. Technical security requirements as described in the Octavio project are necessary, but their criticality depends also on a variety of additional conditions:

- assumed general security conditions of gas pipeline systems;
- the size, length, and expected growth of pipeline systems;
- design parameters;
- the given security status;
- geographical conditions;
- conditions of social-political stability;
- economic conditions;
- strategic conditions; and
- cost and investment choices

The gas sector's key infrastructures are compressor stations, control centres, pressure control systems, and export stations. It needs to be mentioned that completely destroyed compressor stations would take more than 2 years to rebuild and replace, though some mitigation capabilities can be introduced by enabling reverse flow in pipelines. But currently very few pipelines are enabled to do it.

GENERAL SECURITY CONDITIONS AND THE FUNCTIONALITY
OF CONTROL CENTERS

Gas pipeline systems consist of three types of pipes depending on their purpose:
(1) gathering pipelines: typically short and with small diameters, (2) transportation
pipelines: mostly very long and large diameters, with several compressor stations
moving gas between cities, countries, regions, and continents, (3) distribution
pipelines: typically combining several interconnected pipelines with smaller
diameters.

The critical components of gas pipeline systems are: initial injection or inlet
stations; compressor stations with locations depending on terrain topography, and
the operational conditions of the network (e.g. for the North Stream pipeline, one
compressor in the Baltic Sea is foreseen); intermediate stations; block valve
stations as first lines of protection for sectionization; regulator stations at the
downhill side to release pressure once needed; final delivery station.

A critically important additional component to support remote operations is
the combination of (1) field instrumentation along the pipeline to measure flow,
pressure, temperature, etc.; (2) local Remote Terminal Units (RTUs) for the real
time transmission of data to a central database (through cellular phones, satellite
channels, microwave links, etc.); (3) a control center or Main Control Room
receiving all data from RTUs. Most pipeline systems use a SCADA system at their
control centre to monitor the hydraulic conditions of the line and to send
operational commands. In some cases Advanced Pipeline Applications are added
to provide extended functionality for leak detection, leak location, etc.

Depending on the type of attack, all elements of pipeline systems can be targeted.
Attacks on control centres (in addition to compressor stations) are, however, among
the most attractive targets for sabotage, terrorists, multiple attacks, pressure games,
etc. As has been pointed out, however, this needs to be supplemented by analysis
of how security measures for control centers can be sustained under varying
conditions.

Despite its ever growing importance and existing vulnerabilities of energy
control centres, the present trend signals limiting spending on them. These findings
do display different views on security and asset criticality in gas pipelines.

Some characteristics relevant for pipeline security in general are:

- Gas pipeline systems are considered to be the safest and most economical
  mode of transportation in terms of accidents and safer than most critical
  infrastructures.
- They are ranked among the most critical systems with regard to possible
  disruptive effects on industries and societies, in particular cascading and
  secondary effects.
- The dependence of pipeline systems and especially their critical assets on
  IT has exponentially increased.[40]
- Large pipeline systems are impossible to be completely protected.

---

[40] The number of IT incidents increased between 2000 and 2007 from 31% to 70% externally
caused, but decreased from 31% to 20% accidentally, and from 38% to 10% internal incidents caused.

- They are the easiest to attack depending on the element. Pipelines, in particular remote pipelines, are easy to attack, but attacks can be detected quickly and they are easy to be repaired.
- They are typically underground (1–1.5 m deep), but in populated areas they are required to be marked to avoid damage through excavation damage (the largest share of unintended incidents along with corrosion, explosions, fire, car accidents, etc.), but such indications can also be useful for attackers.
- Pipelines with smaller diameters are easier to attack because the steal walls tend to be thinner and they are more exposed to external inter-ference, but the impact of attacks would also be more limited because such pipes are likely to be used only in the distribution part of the systems.
- Critical assets like control centres are more difficult to be attacked and they are more difficult to be repaired, in particular if key components needed for replacements (valves, compressors) are not instantly available.
- Given the dependence of control centres on electricity supplies, increasing IT vulnerability can be critical. But the US systems are assumed to be most attractive targets for terrorists, yet given the degree of protection, attacks would need to be extremely sophisticated. Thus far, however, no successful attacks on control centers in the US have been detected or made public.
- As with other complex systems, gas supply systems face typical trade-off-problems between effectiveness and security. This is particularly critical in case of needed investments under conditions of limited resources.

THE SIZE OF GAS SUPPLY SYSTEMS AS A SECURITY CONDITION

The European gas supply system is overwhelmingly based on pipelines. In contrast to the EU's oil supply security (based on flexible shipping imports), the much more inflexible pipeline gas supply system creates much more dependencies, risks, and vulnerabilities – particularly during crisis situations as Europe experienced with the Russian-Ukrainian gas conflicts in 2006 and 2009 when the gas flows had been cut. In general, the size, length and expected growth of pipeline systems will impact on both the need for control assets and the security requirements of control centers and other critical components:

- Except for LNG transport, a global gas supply system does not exist. But enabled through IT developments and driven by both increasing demand and supply as well as increasing competitiveness within the gas market, gas supply systems grow steadily in terms of indentified resources, length of transport lines, transit zones, diversity of geophysical conditions, and distribution of critical assets across ever wider regional differences.

- The growing size of gas supply system, the length of pipes, the diversity of regional conditions, the increasing exposure to both accidental and intentional hazards, the vast amount of critical information from far away locations, the vulnerability of systems for controlling the flow of gas as well as the security of the system acquirements the need to integrate warning signals from the given system with higher-level crisis information, the fact that awareness is the single most important aspect of preparedness all up to increasingly demanding security requirements for gas pipeline systems.
- Increasing size, length, and complexity of pipeline systems is one of the most critical factors. There is no direct link between the overall size, i.e. the total number of kilometers of gas pipeline systems and the increase in security requirements (e.g. between 2002 and 2005 the totals in kilometers globally increased by more than 30%). While the overall size indicates global trends, it is particularly important to recognize the trends in major gas markets like the EU, the US, the Persian Gulf, or East as well as South Asia.

DESIGN PARAMETERS

Security requirements for gas pipelines increasingly determine the design parameters: the wall thickness, the diameters, the depth cover, the locations and distances of sectioning valves, the distribution of critical elements across the whole pipeline, the redundancy of control centers, the routing and remoteness and site-specific surroundings like terrain and population density, and, of course, the specific safety precautions like SCADA systems.

GIVEN SECURITY STATUS

During the last few years, there is not a single transnational gas pipeline that has not been interrupted in the wider Middle East. There are important producer countries that are considered at risk with a high security threat to assets, e.g. Algeria, Bolivia, Caucasus, Indonesia, Iran, Iraq, Libya, Nigeria, Russia, Venezuela, etc. Most of them have experienced numerous attacks in the past for different reasons (sabotage, war-like situations, insurgency, terrorism, etc.). On the other hand, the United States arguably is the most attractive target area for terrorists, but so far US energy facilities have only been hurt outside the US. The reasons for this are manifold ranging from asset protection to access to the US. Not all countries concerned are willing or even able to afford similar degrees of security. But there are lessons to be learned for others, like the EU.

The vast extent of existing gas pipeline systems requires and allows prioritization of threats, targets, and protective measures. Intentions and access are not sufficient to determine attack motivations, but rather the high-leverage damage done to preferred opponents and the expected secondary effects.

The effectiveness from the pipeline operator's perspective, i.e. the usefulness of information from state authorities (for the US from the DHS), depends largely on whether the collective data provided by pipeline operators provide a common operational picture. This, in turn, results in part from awareness and communication capacity of pipeline operators, i.e. above all the functioning of SCADA systems.

For specific pipelines the status of the systems will also depend on:

- the degree to which security requirements have been fulfilled or neglected;
- failures due to corrosion, ground movement, material or construction defects or external interference;
- the record of leaks, ruptures, ignitions, etc;
- the conditions for sufficient awareness (helicopters, smart pipes, tele-control systems, etc.);
- the likely impact of disasters on industrial and/ or social environments.

GEOGRAPHICAL CONDITIONS

The vast extent of today's gas pipeline systems implies that a variety of rather different geographical conditions are considered for the security of a given pipeline pertain in terms of climate, geological conditions, remoteness, crossing waters, depth of sea-based pipes, likelihood of earthquakes and other natural disasters, density of populations, etc. Scenarios for security risks of gas pipelines will in part be shaped by those conditions.

CONDITIONS OF SOCIAL-POLITICAL STABILITY

The security of gas pipelines is critically dependent on the social and political stability of the majority of oil-producing countries and of countries in transit zones. US-Canadian or Norwegian-EU are currently positive exceptions. Domestic turmoil, hostility to receiving countries, lacking governability (as in Iraq), sanction regimes (as in Iran or formerly Libya), insurgency, external pressures exacerbating domestic tensions, etc., can impede the functioning of gas pipeline systems – often with intended disruptive effects in receiving countries.[41]

ECONOMIC CONDITIONS

Increasing dependence of industries and societies on gas supply generates several conditions that impact on the security of gas supply systems:

---

[41] A good example is the planned, but highly risky 4,000 km Trans – Sahara Gas Pipeline (TSGP) that will send gas from the Niger Delta in Nigeria, though Niger, the Sahara Desert, and to Algeria's export terminals – see also R.Fabiani, 'Is the Trans-Sahaa Gas Pipeline a Viable Project? The Impact of Terrorism Risk', Terrorism Monitor (ed. by the Jamestown Foundation), Vol. 7, Issue 25, 13 August 2009. See also F. Steinhaeusler/P.Furthner/A.de la Cruz, 'Applying Advanced Technology for Threat Assessment: A Case Study of the BTC Pipeline', Journal of Energy Security, 27 August 2009.

- Gas supplies can be interrupted as a result of disruptive circumstances in producing countries and/or transit countries or of policies with specific geopolitical objectives. This increases the need for redundancy, reserves, flexible rerouting, and effective control systems that will be critical for producing and receiving countries.
- Such interruptions can be intentional on the part of producing countries. Preparations for such contingencies require an even higher degree of energy crisis management capacity on the part of receiving countries and companies, e.g. through interconnectors. Effective and secure control systems with sufficient capacity for flexibility within a system that generally allows little flexibility is critical.
- Conversely, economic conditions in receiving countries that reduce the financial capacity and/or willingness to invest in the extension and modernization of pipeline systems and their security can jeopardize the security of pipelines and exacerbate security risks for all parties concerned. The current financial crisis indicates some elements of this type of crisis. On the part of producing countries and pipeline operators this tends to reinforce difficult choices between effectiveness and security, most likely at the expense of security.
- Producing countries (with implications for transit zones) may face problems with demand security as a result of increasing competitiveness of gas suppliers and temporary or even structural reductions in demands which will have similar consequences like lowered requirements for and investments into security of pipeline security.
- In an otherwise stable environment it is, however, conceivable that pipeline security is considered a decisive advantage over exporting competitors, in particular if this is conducted to enhanced cooperativeness between producing, transit and consuming countries as well as companies and their shared interests in enhancing pipeline security. The criticality of assets is likely to be central in any such development.

STRATEGIC CONDITIONS

Strategic conditions range from accident-prone circumstances without open violence via criminal access (e.g. vandalism, theft), hostile local attacks, to uncertain threat environments, war-like situations, enforced regime changes with hostile consequences, and non-governmental circumstances. This can pertain to producing countries, transit zones, receiving countries, and combinations thereof.

- Criminal activities against gas pipelines are most likely in countries with generally low governability like Nigeria where theft and vandalism (in addition to incompetence, e.g. of gasoline scavengers) have killed several thousand people.

- Accidents are most frequent, though pipeline accidents represent a small part of transportation accidents. They are relatively easy to control (e.g. unintentional release of gas), and only in exceptional circumstances have strategic consequences. However, in the absence of state-of-the-art protection large-scale explosions can happen. Given the density of pipelines in the US, accidents do happen, but with low casualties and limited damage. Serious accidental disasters like the pipeline explosion near Brussels in 2004 or at the end of June 2009 in Viareggio (Italy) are rare exceptions – albeit they may happen more often in the future.
- Hostile local attacks do happen in instable producing and transit regions, e.g. to weakened governments to enable overthrows. They can occur as seizures, combined physical and cyber attacks ignitions, etc. by insurgents, terrorists, and third party agents.
- Threats of strategic terrorism have occurred in the past against neuralgic points in Saudi Arabia, Algeria, Nigeria, Pakistan, etc., at targets that are critical for supplies to Spain, Italy, France, Portugal, and other European states. Suicide strikes are the most obvious attack mode. While protection of critical elements of pipeline systems will make a difference, in cases of strategic terrorism trained forces (police, paramilitary) are indispensible. In Saudi Arabia a special force of 35,000 is being established.
- Attacks in war-like conditions like the massive ignitions in Iraq following the Gulf Wars can also affect the gas supply. This would be targeted on criticality assets rather than pipelines. Moreover, especially in crisis situations physical protection will be in place. But combinations of electronic and physical attacks as part of more wide-ranging conflicts could be severely reduced in such contingencies. The value of such increased security measures was clearly demonstrated in the failed terrorist attack on the world's largest oil refinery (with a capacity of up to 6.8 mb/d) in Abqaip (Saudi Arabia) on February 24, 2006.[42]
- More extreme circumstances are denials of supply and/or transit, over-throws of governments and regimes, or imposed regimes on gas exports.

While protection of critical assets is always relevant, it is directly relevant in accident-prone situations, criminal activities against pipeline systems and local and direct attacks, i.e. in lesser cases. In higher-intensity cases it is a means of crisis management along with other instruments.

COSTS AND INVESTMENT CHOICES

There is a need and there are opportunities to prioritize protective measures and to ensure that investments do not go to the wrong places and aim too much or too little. There exist needs for major investments in modernization (e.g. in Ukraine, Russia, etc.) and for additional and new pipelines, and some of the major

---

[42] See 'Bombers Attempt Attack on Saudi Oil Facility', International Herald Tribune, 24 February 2006.

developing new pipeline systems would serve fiercely competing interests under conditions of an increasingly uncertain demand security.[43]

SECURITY CONDITIONS IN PERSPECTIVE

Asset security in pipeline systems is an important requirement in many cases much more so than protection of the pipes themselves. This is a prerequisite for effective mitigation against accidents and incidents caused by criminals. The same is true for local hostile attacks along with other means, like the speed of response and the means to cope with aggressors. Protection against strategic terrorism requires a broader spectrum of protective means and measures, but effective control centers and other critical assets remain an indispensible mean of crisis management. In major contingencies the continued functioning of gas pipeline supplies will depend on a wide variety of circumstances. Agreed definitions of criticality of pipeline assets still need to be refined, whereas definitions are needed to define security requirements for assets in pipeline systems in relation to conditions that apply to a given contingency. The Octavio Project has laid some useful foundations on which to base more comprehensive sets of security requirements for gas pipelines and their critical assets, on top control centres.

The EU's Evolving Energy Strategy: Implications for Asset Criticality in European Gas Pipeline Systems

TOWARDS AN EU COMMON ENERGY SECURITY STRATEGY

An accurate risk assessment of the projects needs to be based on a "comprehensive security" understanding as the European Commission and the European Council developed in December 2003 in its "European Security Strategy" (the main document of its Common Foreign and Security Policies/CFSP) and in 2006 for its common energy security policies. Therefore, it was useful for the Octavio-Project and others beyond it not just to focus on an analysis on the present technical features of the control centres and to improve its technical resilience requirements. In an additional complementary approach and analysis to those agreed technical aspects for "identifying survivability and resilience requirements for every function-ality in natural gas control centres", it is important to include a strategic perspective of forthcoming political and technical aspects of newly built gas and electricity

---

[43] See F. Umbach, 'Motor oder Bremsklotz? Deutschland und die Energie(außen) politik der Europäischen Union' (Motor or Brake Block? Germany and the Energy (Foreign) Policy of the EU), in: WeltTrends, May-June 2009, pp. 45-55; idem, 'Future Transport, Pipelines and Energy Security." PPT-Presentation at the Public Hearing on "Security of Energy Supplies"; organised by the Committee on Industry, Research and Energy of the European Parliament, European Parliament, Brussels, 2 April 2009 (http://www.europarl.europa.eu/activities/committees/hearings.do;jsessionid=3CB1098FA347D9B83AEA65B827F808F0.node1?page=2&body=&language=DE) and idem, 'Diversifizierung statt Protektorat. Energiepartnerschaft zwischen Russland und der EU (Diversification instead of a Protectorate. Energy Partnership between Russia and the EU'), in: Die Politische Meinung, September 2008, pp. 25-30.

control centres by taking into account the EU's strategic developments of the agreed common energy policies (i.e. gas policies) of 2006 such as:

- The wide-ranging March 2007 decisions of the EU's Energy Action Plan (EAP) and the 20-20-20 Programme and their implications;
- The 2nd Strategic Energy Review of November 2008);
- The implications of the new Russian-Ukrainian gas conflict of December 2008–January 2009; and
- The European Council's decision of March 20, 2009 on financing energy infrastructures for the very first time, in particular gas and electricity interconnectors for enhancing the EU's crisis capabilities with a total budget of 4 billion Euro.

In this context, the following factors of the EU's agreed energy objectives between its 27 member states (since March 2007) need to be taken into account:

- The creation of liberalized common energy markets for the individual energy resources (i.e. oil, gas, and electricity) and to build up the physical infrastructure for the EU's crisis management mechanisms required not just more gas storage facilities, but also new gas pipelines and other interconnectors
- The lack of an adequate infrastructure, as highlighted during the latest Russian-Ukrainian gas conflict in January 2009, has been identified as a major physical weakness that hinders the EU to have an effective political crisis mechanism as agreed and institutionalised after its March decisions of 2007.
- Those interconnectors include not just new transnational pipelines and electricity grids between individual EU member states (such as between Germany and Poland or between Baltic states and Poland as well as Sweden), but also new gas and electricity control centres for transnational supplies, connecting the 27 EU countries and their national energy markets to subregional and regional markets.
- Furthermore, the agreed common energy policies for creating unified energy markets and building new transnational interconnectors for transnational gas and electricity supply within the EU-27 will also determine many technical and technological dimensions of those future control gas and electricity control centres.

The building of those new gas and other interconnectors has manifold implications for the EU's critical infrastructure security and its vulnerabilities:

- As a result of the creation of those liberalized common energy markets and transnational physical interconnectors, the supply of particular gas and electricity becomes, to a certain extent, even more centralized and transnational in the common EU energy markets for its 27 member states – alongside, simultaneously, of more decentralized supply structures as the result of the expansion of renewable sources (wind, biomass, solar for electricity production). The mix of centralized and decentralized structures enhances and decreases vulnerabilities and security efforts.

- Given the strategic importance of those gas and electricity control centres for the EU's future energy supply security and effective crisis mechanisms, the infrastructure safety and security of those gas and electricity inter-connectors, including transnational gas and electricity control centres, will acquire both a higher strategic value as well as new dimensions of vulnerability and dependencies of transnational infrastructures. That, how-ever, is not just true for the EU, its member states, governments, and industries. International terrorists, too, may reflect the political and public attention to the present Russian-Ukrainian gas conflict as well as the past oil and gas crisis in Europe of 2006. Hence those transnational gas and electricity control centres could attract much more attention as a potential terrorist target of European vulnerability with much more devastating economic, social and psychological effects on governments and public opinion in EU member states than national gas and electricity control centres. Thus the growing interconnectedness between EU member states offers better perspectives for future supply crisis and its management. But in case of terrorist and cyber warfare attacks, it simultaneously increases the vulnerability and dependency on this critical infrastructure with potential cascading effects on various EU member states.

In this context, the project needs to take into account the rapidly changing nature of transnational crimes terrorists and cyber threats on critical energy infrastructure since 2001.

THE EU'S AGREED COMMON ENERGY POLICIES SINCE 2007: IMPLICATIONS FOR CEIP

Until 2006, with the privatisation of the gas sector and new emerging companies, the EU had neither a common energy policy nor a single actor, which had assumed overall responsibility for the security of gas supply, mostly transported by politically and technically inflexible pipeline systems during supply crisis. The EU's long-term strategy for energy supply security needs to cope with uninterrupted physical availability of energy products on the market, at a price which is affordable for all private and industrial consumers. At the same, the EU needs to balance its future energy supply policies with growing environmental concerns, which has become an even more important objective in the light of the Kyoto Protocol.

In November 2000, the European Commission warned in its first 'Green Paper' that in the next 20–30 years up to 70% of the Union's energy demand (presently 50%) will have to be imported. With regard to oil, EU's dependence could reach up to 90%, for oil, 70% for gas, and 100% for coal. In 2006, the EU-27's total primary energy supply was generated by oil (37%), gas (24%), solid fuels (18%), nuclear energy (14%), and renewables (7%). The future new capacity will still be predominantly generated by fossil resources with the rising percentage of gas, while the number of oil and solid-fuel power stations will continue to decline.[44]

The expansion of natural gas as an environmental clean energy source is widely considered as the most problematic factor in the next two decades for the EU member states' energy policies. Today, Europe is already the largest natural gas import market and will continue to be the world's champion of gas importers until 2030. But today, almost half of the EU's gas consumption is being imported from only three countries: Russia (23%), Norway (14%), and Algeria (10%). The new EU members and former allies of the Soviet Union, are, in particular, still very much dependent on gas imports from Russia.

Meanwhile, the EU's great hopes of a real strategic energy partnership with Moscow appear to be rather a long-term vision. Furthermore, it has also become increasingly uncertain whether Moscow will be able to increase its gas exports beyond 180–200 bcm after 2020 due to an emerging domestic gas crisis by 2010.[45]

Just 1 year after the Russian-Ukrainian gas conflict in January 2006, the European Council under the German Presidency agreed in March 2007 on the worlds most ambitious integrated climate and energy policy with an 'Energy Action Plan' (EAP) for the years 2007–2009. With the world's most comprehensive action plan on climate protection and energy supply, the EU-27 were able to agree on 17 individual measures and three 20% targets at the March Summit of 2007 for increasing energy efficiency, expanding renewable energy sources to one fifth of the energy mix, and reducing carbon emissions until 2020 compared to 1990 (if other industrialized countries such as the USA, India, and China commit themselves to similar policies, the EU would be willing to reduce emissions by 30%).

In November 2008, the new European Commission's "2nd Strategic Energy Review" and its new "EU Energy Security and Solidarity Action Plan" have identified major weaknesses and problems which need to be overcome on the way to a real common energy (foreign) policy and by enhancing the energy supply security of its 27 member states. It has proposed five key areas for joint cooperation and projects in the forthcoming years:

- Infrastructure needs and the diversification of energy supplies;
- external energy relations;
- oil and gas stocks and crisis response mechanisms;

---

[44] See F. Umbach, 'Global Energy Security and the Implications for the EU', Energy Policy, Vol. 38, Issue 3, March 2010, pp. 1229-1240.

[45] See also F. Umbach/A.Riley, "Out of Gas. Looming Russian Gas Deficits Demand the Readjustment of European Energy Policy", in: Internationale Politik-Global Edition, pp. 83–90; F. Umbach, "Zielkonflikte der europäischen Energiesicherheit. Dilemmata zwischen Russland und Zentralasien (Conflicts of Objectives of European Energy Security: Dilemmas between Russia and Central Asia). DGAPanalyse No. 3, Berlin, November 2007, pp. 16–18; and idem., "Memorandum: The European Union and Russia – Perspectives of the Common "Strategic Energy Partnership". Personal Analysis for the Sub-Committee on Foreign Affairs, Defence and Development Policy, House of Lords/Great Britain, in: House of Lords/European Union Committee (Ed.), The European Union and Russia. Report with Evidence, 14th Report of Session 2007-2008. HL Paper 98, The Stationary Office: Norwich-London 2008, S. 185–188.

- energy efficiency; and
- making the best use of the EU's indigenous energy resources.[46]

The EU's 2nd Strategic Energy Review of last November, for the very first time, has offered new concrete figures for the EU's forecasted energy consumption and imports, including on gas, by taking into account the EU's March decisions of 2007 and its "20-20-20" strategy. If the EU is able to implement these strategies, it may drastically decrease its gas import demand by 2020 at the current levels or even lower (around 300 bcm) in contrast to previous forecasts (490 bcm). This new forecast has not just implications for present energy investment programmes, including gas pipelines, LNG-terminals, and diversification plans. It has also impacts on the EU's gas supply security as well as for critical infrastructure vulnerability and protection needs because it reduces the numbers of critical infrastructures, but increases the strategic value of the remaining ones, especially during crisis and conflicts.

But all the planned new pipelines and LNG projects also require new or more sophisticated gas control centres in order to guarantee a secure transnational gas supply in crisis across the borders of EU member states in order to enhance common energy supply security within the EU-27. In context of the EU's future infrastructure security, it is important to note the following structural facts of the changing energy infrastructure:

- Europe's energy networks and infrastructure to transport electricity, gas, oil and other fuels from producers (often from beyond the EU) are aging and are part of large, centralized production.
- The lack of suitable network links is a barrier of investment in renewable energy and decentralized generation. It makes the freely moving of energy more difficult and some EU regions more vulnerable to supply disruption.

In this light and in light of the Russian-Georgian war, the European Council called on the Commission in October 2008 to "reinforce and complete critical infrastructures". The strategic aim is focused on promoting the interconnection and interoperability of national networks, as well as access to such networks in order to strengthen political solidarity and security of supply in a "truly European energy network", which also includes improving coherence between different national network plans of its member states. Within the EU internal market, regional (cross-border and multi-country) networks are important for security of supply and solidarity and are seen as a first step towards a fully interconnected internal energy market.

According to the European Commission's "Priority Interconnection Plan" of 2006[47] and the "2nd Strategic Energy Review and its Energy Security and Solidarity Action Plan", six strategic priority infrastructure projects have been identified:

---

[46] See European Commission, An EU Energy Security and Solidarity Action Plan. Second Strategic Energy Review. Communication from the Commission to the European Parliament, the Council, the European Economic and Social Committee of the Regions, Brussels: November 2008.
[47] European Commission, Priority Interconnection Plan, COM (2006)846.

- A Baltic Interconnection Plan for connecting the remaining isolated energy markets of Baltic countries as a priority;
- A new Southern Gas Corridor with the possibility for importing gas from the Caspian region and the Middle East (i.e. via the planned Nabucco-Pipeline);
- Expansion of LNG projects, including at the Adriatic Coast and by building sufficient gas storage capacity;
- A Mediterranean Energy Ring for connecting not only new fossil fuels supplies but also for renewables;
- North-South Gas and Electricity Interconnections within Central and South-East Europe (building notably on the NETS-initiative) as part of the Energy Community Gas Ring; and
- A Blueprint for a North Sea Offshore Grid to interconnect national electricity grids and plug-in the numerous planned offshore wind projects as one of the building blocks of a future European supergrid.[48]

On January 28, 2009, directly after the end of the latest gas conflict between Russia and Ukraine, the European Commission felt confirmed in his warnings and programmes for enhancing the EU's energy supply security and in particular gas supply by proposing concrete energy infrastructure projects with a total sum of 3.5 billion Euro for 2009 and 2010 as part of the EU's overall economic stimulus programme in coping with the global economic-financial crisis. 1.75 billion Euro had been foreseen for supporting to build new gas and electricity interconnectors in order to enhance energy supply security and the EU's energy crisis management capabilities.[49]

Whereas the resilience and technological capabilities of gas and electricity control centres will further been improved and enhanced with the new infrastructure initiatives in order to guarantee a transnational and cross-border supply of gas and electricity between EU member states, technical failures or terrorist attacks on those transnational gas and electricity control centres may have European-wide instead of hitherto just national effects on supply security.

Summary and Perspectives

Whereas the financial and personnel resources available to operators protect their infrastructure systems are limited, it is essential for both the energy industry and governments to use all available resources efficiently and effectively by assessing risks and setting priorities for adequate risk management. While it is impossible to protect a utility 100% from a physical or a cyber attack on a utility's facilities and infrastructure, those threats need be minimized as much as possible without compromising their productivity and day-to-day operations. Nonetheless, a professional

---

[48] See European Commission, Green Paper towards a Secure, Sustainable and Competitive European Energy Network, p. 13.

[49] See European Council, Brussels European Council. Presidency Conclusions, Brussels, 19–20 March 2009.

security and risk assessment needs to address physical and cyber security, super-visory control, data acquisition (SCADA), and distributed control systems (DCS), communications security, grid security, distribution security, generation security, and biological/chemical issues in new integrated security concepts such as the TAAS Industrial Corporate Security Awareness Programme (ICSAP).[50] But the criticality of the energy-ICT interface depends on the scale of the energy supply disruption and the length of time it takes to return to normal. With well protected infrastructure programmes and well trained and equipped security forces (e.g. in Saudi Arabia), the oil and gas industry and their governments are able to foil or mitigate terror attacks on critical oil, gas, and other energy infrastructures.

Looking ahead with regard to the EU's emerging new safety and security challenges of its critical energy infrastructure, the lack of political solidarity on the way to create liberalized common energy markets inside of the EU is not just the result of the traditional preference on national energy policies and its sensitivity to pass more sovereignty and authority to the EU, but also the lack of infrastructure between the individual EU-27 member states (i.e. between the "old" and "new" EU member states). An important prerequisite to overcome the historical legacies of physical infrastructure and the traditional policies, the EU has agreed to create numerous new interconnectors for both trans-border electricity and gas supplies. Due to the Russian-Ukrainian gas conflict of December 2008-January 2009, the EU's future gas supply security is widely seen as the Achilles heel of European energy security. An important part of this new infrastructure is to improve the individual national energy supply and a common crisis management system, particularly during a supply crisis, with control centres for gas and electricity supply.

In this respect, any future risk assessment needs to include the wider political-strategic policies and intentions of NATO; the EU and its member states for analysing the concrete risks and future vulnerabilities of the existing and newly build critical energy infrastructure. In this context, the March 2007, November 2008, and March 2009 decisions of the EU's energy policies and newly built energy infrastructure are of utmost importance. Any analysis of a comprehensive risk assessment of these gas and electricity control centres would benefit by including these dimensions and new policies in a strategic perspective of the EU's future energy infrastructure security. If the EU's agreed energy policies and projects are implemented, they will greatly enhance a common energy security of the EU, a common crisis management system, and a common energy market as well as common energy foreign policies. On this way, transnational gas and control centres are becoming an integral part and a highly important and critical infrastructure as a physical pre-condition of an already institutionalised political crisis mechanism of the EU's common energy policies.

The recent Russian-Georgian war in August 2008 and the European Commission's "2nd Strategic Energy Review" with its new "EU Energy Security and Solidarity Action Plan" of November 2008 have highlighted the improved, but

---

[50] See, for instance, P. Furthner, F. Steinhäusler, 'Integrated Security Concept for the Oil and Gas Industry, Strategic Insights', Vol. VII, Issue 1, February 2008.

still insufficient political solidarity between NATO and the EU's member states in speaking with one voice towards external energy partners. But by building new transnational gas and electricity interconnections inside the EU-27 as part of enhancing energy supply security and the creation of common energy markets, those processes will ultimately lead to more common energy (foreign) policies in the future and strengthen the EPCIP efforts. In this regard, the future safety and security of gas control centres and any discussions of critical gas infrastructure need to take into account:

- The new transnational dimensions of interconnecting gas supplies and national gas markets within the EU's internal market.
- The implications of terrorist and cyber attacks on these new or modernised control centres with their much higher strategic value, which could have much more wide-ranging cascading effects on transnational gas supplies.
- The overall dependence of European gas control centres on external gas infrastructures outside the EU (i.e. Russian or other foreign gas pipelines, gas control centres, etc.) – particularly in the light of the EU's further growing dependence on gas and other energy imports from outside Europe, and thereby often from much more unstable political regions.

Thus safety and security issues of gas control centres and other gas and energy infrastructures should become part of the various energy dialogues as an integral part of NATO's energy security dialogues within the Alliance as well as its partner countries. It should also encourage a new security paradigm to include security into design and operational criteria of new key technologies such as Smart Grids as well as a new security culture by exchanging and sharing information, technologies and best practices. But in general, CEIP is first and foremost the task and responsibility of companies and their member states themselves. But above all, there is clearly a need for an enhanced transnational cooperation within the Alliance as well as with its partner countries and other international organizations such as the EU, the OSCE, and others.[51] Those talks and joint analyses by NATO and its partner countries should be easier in the future as the U.S. and Russia have started secret talks and negotiations to fight cybercrime. But in general, NATO's "niche" role in energy security and CEIP will remain limited and complementary, but also needs to be more focused to and deepened with other actors, in particular the EU.

---

[51] For an overview of international organizations dealing with energy security and CEIP see F.Umbach, 'The Role of International Structures in Protecting Non-Nuclear Critical Energy Infrastructure from Terrorist Attacks', Public-Private Expert Workshop of the OSCE, Office of the Secretary General/Action against Terrorism Unit, supported by the U.S.A., Vienna, 11–12 February 2010.

OPERATIONAL ANALYSIS SUPPORT TO ENERGY SECURITY IN SOUTH EAST EUROPE: A BULGARIAN ACADEMIC COMMUNITY APPROACH

ZLATOGOR MINCHEV*[1],
VELIZAR SHALAMANOV[2]
Institute for Parallel Processing-Bulgarian Academy of Sciences, C4I Department

Abstract. This paper presents the authors' understanding of the problem of energy security in South East Europe (SEE) and the Wider Black Sea Area (WBSA) as a result of the region's geo-strategy; the development of the relevant capabilities in Bulgarian academic community linked with administration, industry, and non-governmental organizations in the area of analytical support to energy security, as well as the concept of a Center of Excellence for Military Support to Civilian Authorities as an instrument for addressing: energy security, cyber defense, and other new challenges to security through the concept of an integrated security sector; and finally – the integration of national capabilities in the international network for operational analysis and foresight/strategic planning.

Introduction

The new challenges related to the twenty-first century energy supplies have opened a vast field for searching of new alternative energy sources and technologies, for the global resilient society such as: nuclear power, biofuels, fluorophore-nitroxide compounds (FNO), wind power, solar power, (sea) wave power, etc.

Naturally, this is strongly related to the world's fossil non-retrievable fuel reserves diminishing and the common understanding of their usage influence to CO_2 emissions augmentation. This phenomenon is closely related to the global

---

* To whom correspondence should be addressed: Dr. Zlatogor Minchev Institute of Parallel Processing, Bulgarian Academy of Sciences C4I Department, e-mail:zlatogor@abv.bg

[1] Dr. Zlatogor Minchev is the Director of JTSAC, C4I Department at the Institute of Parallel Processing, Bulgarian Academy of Sciences (BAS) and currently is an Executive Head of C4I Department.

[2] Dr. Velizar Shalamanov moved in July 2009 to NATO C3 Agency as Director Sponsor Account "NATO and Nations" after being head of C4I Department at the Institute of Parallel Processing, BAS for 5 years and in the period 1998–2001 – Deputy Minister of Defense – Bulgaria on Defense Policy, Planning and Integration.

A. Gheorghe and L. Muresan (eds.), *Energy Security: International and Local Issues, Theoretical Perspectives, and Critical Energy Infrastructures*,
DOI 10.1007/978-94-007-0719-1_15, © Springer Science+Business Media B.V. 2011

warming effect and climate changes in the Earth. Above all, it should be clear that the world's demand/supply ratio, regarding natural oil and gas usage, is getting strongly unstable variable by means of a constant increase in the global world's developed economies needs, the growth of free market prices, and general but complex supply limitations.

In this situation, both the North Atlantic Alliance and EU are prioritizing the energy security as an important element of their future security/defense and economic plans.

Today the New NATO Strategic Concept will be based on the development of a comprehensive approach with relevant technological support. The transatlantic policy within the next 20 years will be closely related to EU/NATO dialogue on security and defense topics and priorities that exists in both their agendas.

According to the Alliance's Comprehensive Approach for an Integrated Security (that encompasses both the EU and the UN) the areas of Consultation, Command & Control (C3) will support NATO and Nations. These C3 areas are gathered around new challenges such as: energy security, climate change, piracy, and cyber defense – all problem areas that are adding new dimensions for operational analysis and technology support to the already traditional areas of common defense situated around Article 5, crisis response/emergency management, fighting terrorism, and maintaining the partnership and enlargement process for NATO.

The new EU agenda (ESRIA[3]) is also considering these problems in the horizon of the next 10–15 years, when defense and security boundaries will be less distinct and the security will encompass defense with respect to society's social security and the global context for a "non-isolated world".

Here it should be noted that nowadays the transatlantic role of the Alliance is getting more and more similar to the one of UN and the EU will have to be responsible and to develop its own capabilities, according to ESRIA, in five clusters: (1) security cycle – preventing, protecting, preparing, responding, and recovering; (2) countering of different means of attack; (3) securing critical assets; (4) securing identity, access, and movement of people and goods; (5) cross-cutting enablers.

Where is energy security situated within ESRIA and NATO's New Strategic Concept?

Generally the energy security in a changing world[4] is a challenge in itself because of the complexity, speed of change, and uncertainty, i.e. a multicriteria problem which could be addressed by an effective use of operational analysis (operations research) in a multidisciplinary environment for foresight and strategic planning.

An extremely important dimension of this approach is to be able to effectively govern the process starting from foresight, decision making, and planning as main aspects of consultations down to the technical solution development and its

---

[3] European Security Research and Innovation Forum, Final Report, December, 2009.
[4] Energy Security in a Changing World, Romanian Journal of International and Regional Studies, Volume IV, Bucharest, 2–3/2008, EURISC Foundation – Romanian Institute of International studies, IRSI, 286 pp.

integration in existing command and control systems. A solution requires a really comprehensive C3 support to the security comprehensiveness.

But, energy security is also directly related to crisis management, critical infrastructure protection, security of economics and indirectly to the rest of ESRIA, which makes the task of energy security even more uncertain and multi-aspect.

So, even in the relatively short time horizon of 10 years the global energy security could be appointed as: C3 comprehensive, uncertain, complex and thus difficult to tackle, control, and predict.

This paper will present the authors' understanding of the problem of energy security in South East Europe (SEE) and the Wider Black Sea Area (WBSA) as a result of the region's geostrategy; the development of the relevant capabilities in Bulgarian academic community linked with administration, industry, and non-governmental organizations in the area of analytical support to energy security, as well as the concept of a Center of Excellence for Military Support to Civilian Authorities as an instrument for addressing: energy security, cyber defense, and other new challenges to security through the concept of an integrated security sector; and finally – the integration of national capabilities in the international network for operational analysis and foresight/strategic planning.

Bulgaria in SEE/WBSA Context

The right and reliable security and economic environment as well as the adequate strategic context is an important topic to be studied in order to address the proper way of providing energy security and critical infrastructure protection. A model of influences and dependences relations of the Bulgarian regional and political context is presented in Figure 1.

As it is clear from Figure 1, Bulgaria has a unique position – both geographic and geostrategic in South Eastern Europe (to extend from Western Balkans to Caucasus with Black Sea in the middle). It is on the crossroad between EU and Turkey/Russia, between NATO and Russia, between "functioning core" and "gap of nonfunctioning good governance" and many other dichotomies. The key influences are related to EU, NATO, US, Russia and Turkey, but there are also many key bilateral relations between Bulgaria and NATO or EU countries as well as relations between these factors of influence. The country is surrounded by active conflict zones in Europe, Asia, and Africa in the perimeter range of 1,000–2,000 km.

This environment and its dynamics is an important factor in planning for energy security and critical infrastructure protection. As a result of such an environment and dynamics of external (flows of: money, people, resources, and security guarantees) and internal factors (majority in the Parliament of the Euro-Atlantic or Euro-Asian forces) Bulgaria has had a very interesting trajectory of transition during the last 20 years (see Figure 2). It is a unique situation in comparison to any other new member of NATO or EU for the last 20 years and very similar as model with the countries that are now in PfP in SEE and may be for other NATO partner countries outside of Europe. This internal dynamics of country "climate" is the key factor in shaping decisions on energy security and infrastructure protection.

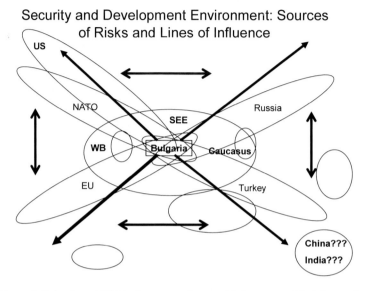

Figure 1. Bulgaria and SEE in the center of powers' influences and their interactions.

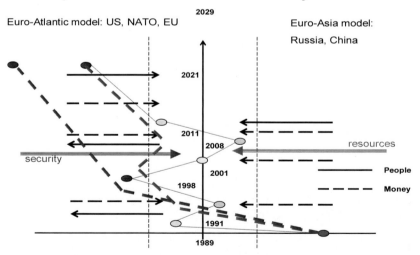

Figure 2. Trajectory of the new NATO/EU members (*left line*) and oscillations of Bulgarian transition (*right line*).

The key point for the future is that economy will be more and more integrated in the Euro-Atlantic community and political and security sector, as well as energy/infrastructure sector, have to follow or may have to, after a certain time, lead this transition in order to be effective and to keep the integrity of the country and even the integrity of people at high political level.

So, the challenge is to achieve a high level for integration[5] – horizontal (between different infrastructures and security services) and vertical – from policy and governance to management and command and control. The key instrument for such integration is the combination of operational analysis (OA) for concept development and decision making support together with Computer Assisted eXercises (CAX) for experimentation/training and deeper/larger understanding of situations, concepts, and plan validation.

The main goal of the established in 2006 Joint Training Simulation and Analysis Center – Civil Security (JTSAC-CS)[6] (at the Institute for Parallel Processing – Bulgarian Academy of Sciences[7]) is related exactly to this challenge – to provide tools and a test environment for OA and CAX in support of the change management process of the security sector, including the improvement of energy security and critical infrastructure protection.

Bulgaria being in the middle of Adriatic – Black Sea – Caspian Sea Bridge and being a member of NATO and EU with close cooperation with US (including joint military facilities) could support regional cooperation trough OA and CAX in JTSAC-CS in the transition period for the region.

Considering the disclosed context of the Bulgarian position in SEE and WBSA the role of OA/CAX implementation is inevitable by means of possible scenario developments, simulations and analyses that are related to Critical Infrastructure Protection (CIP) and Emergency/Crisis Management (EM, CM). This is based both on world known best practices for future situational analysis from one hand and from another – this shows great flexibility in the methodology. Generally, the nature of CAX is multidimensional and in the most extreme, best case, allows a combination of live, virtual, and constructive simulations of the environment (CIP) model. In the next paragraph a description of JTSAC-CS capabilities will be given.

Joint Training Simulation and Analysis Center – Civil Security Capabilities

The Joint Training Simulation and Analysis Center – Civil Security (JTSAC-CS) is an academic R&D center founded in 2006 with the support of NATO, EU, the Bulgarian Government, and the Bulgarian Academy of Science (BAS) as a part of the Institute for Parallel Processing (IPP) – BAS.

---

[5] Niemeyer, K., Shalamanov, V., Tagarev, T., Tsachev, T., and Rademaker, M. NATO Operations Research Support to Force and Operations Planning in the New Security Environment - SfP 981149 Final Report, Sofia, Artgraf, 2008.

[6] Shalamanov, V. Computer Assisted Exercise Environment for Terrorist Attack Consequence Management, In Transforming Training and Experimentation through Modelling and Simulation - Meeting Proceedings RTO-MP-MSG-045, Rome, Italy, October 4-7, 22-1 - 22-18, 2006.

[7] http://www.bas.bg/clpp/en/indexen.htm

310

Z. MINCHEV AND V. SHALAMANOV

The Center specializes in integrated Live, Virtual and Constructive (LVC) simulations for Computer Assisted eXercises (CAX) and operational analysis (OA). Generally, JTSAC studies the application of OA, CAX, and information technologies in the new security challenges of the twenty-first century.

The mission of JTSAC is to be a key partner of the integrated security sector institutions planning and analysis processes with the help of OA & CAX. The vision focuses on ability with high-professional teams to support the complete life cycle of OA & CAX. The strategy is based on knowledge and technology integration for better education and training.

The main objective of JTSAC – CS is to provide scientific and educational support to the Integrated Security Sector on the bases of OA and CAX, conducted jointly by subject matter experts, scientists from BAS, and leading national and international high-tech companies from the security sector.

The basic JTSAC – CS capabilities are integrated around the Basic low-cost Environment for Simulation & Training – BEST. This environment has been developed since 2005 within a series of projects and tested with the EU TACOM SEE 2006, Struma 2008, and in 2010 will be part of Phoenix 2010 exercises.

BEST is integrating CAX simulation via CAX-ENVironment (CAX-ENV) and six additional modules depicted in Figure 3:

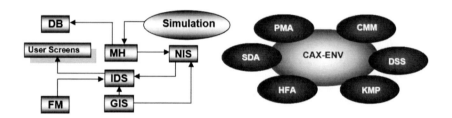

Figure 3. CAX-ENV and other BEST modules.

CAX-ENV is an element of BEST that encompasses a network system for: Message Handling and Instant Messaging chat (MHS); Integrated Display System (IDS) for displaying different fused information about simulated events': geographical, seismological, and meteorological information (via Geographical Information Systems,), exchanged messages log via a web information system integrated into a network information system (NIS) that allows remote Field Modules integration for mobile C2 Center construction, including WAN, LAN, and satellite TCP and VoIP communications assurance and video surveillance (including night vision cameras). Finally the completed simulation is archived in to a database (DB) for After Action Review (AAR) and Post Mission Analysis (PMA).

A detailed implementation of BEST methodology[8] is shown in Figure 4:

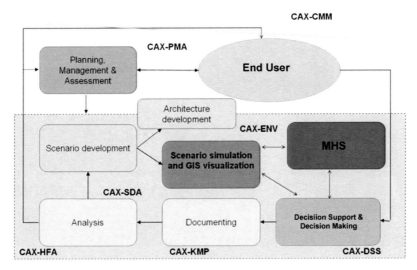

Figure 4. JTSAC-CS BEST methodology.

Figure 4 illustrates the processes implemented in BEST elements[9]: The Change Management Model[10] (CMM) is giving the context of the security sector transformation in the sense of security sector integration and validation of the process via the active legislation testing through CAX. In this sense, CMM also provides the link with the end-user of CAX. The Project Management and Assessment (PMA) implements tools and methods for economical evaluation planning and control on the bases of COTS such as: MS Project®, QPR Balanced Score Card®, and ad-hoc developed software solutions. The Scenario Development and Assessment (SDA) implements a four-step process: structural analysis, system analysis (both developed within ad-hoc developed software I-SCIP[11]), dynamic risk forecasting (developed with the COTS Powersim Studio®), and agent-based simulation (developed with NC3A software for agent based simulation – GAMMA®). Following developed scenarios requirements, a CAX

---

[8] Niemeyer, K., Shalamanov, V., Tagarev, T., Tsachev, T., and Rademaker, M. OR Support to Force and Operations Planning in the New Security Environment - NATO SfP 981149, Final Report, April, 2008.

[9] Shalamanov, V., Penchev, G., and Nikolova, I. The Role of Center of Operational Analysis in Integration of Science, Industry, Government Capacity to Support Integrated Security Sector in Bulgaria, International Conference for Security and Defence Industrial R&D, NATO SfP 982063 "Management of Security Related R&D in Support of Defence Industrial Transformation", May 14-15, Sofia, 76-95, 2009.

[10] Shalamanov, V., Minchev, Z., et al. Security Studies in Bulgaria 1999-2008, G.C. Marshall, Bulgarian Academy of Sciences, Demetra Publishing House Ltd., October, 2008.

[11] Minchev, Z. Intelligent Scenario Development for CAX, NATO ARW Scientific Support for Decision Making in the Security Sector, Velingrad, Bulgaria, IOS Press, 16-24, 2006.

ENVironment (ENV) architecture was designed via System Architect, OpNet (for communications), ARIS and NAF, and DoDAF principles.[12]

The Decision Support Systems (DSS) package provides a set of distribution tasks solvers for emergency delivery of resources (water, food, medicines, blankets, clothes, etc.), people evacuation, rescuing and network (electrical, water or road) distribution problems. The Knowledge Management Package (KMP) provides an integrated space for archiving results in electronic form, from ongoing or already completed CAXs, available in a WWW environment (see: http://www.caxbg.com/). Finally, the Human Factor Analysis (HFA) gives possibilities via surveys, a battery of psychological tests (including: alertness, attention, stress, fatigue, etc.) and neurofeedback tracking for evaluation in a qualitative manner of the real involvement of the trained participants in CAX and for improvement of their results/performance, i.e. an ability to learn and improve their knowledge and reactions for hypothetical, plausible scenarios based on hypothetical/future situations.

JTSAC Capabilities Application

Figure 5 presents the Battle Lab Architecture for C2 in crisis/emergency management (CM/EM) applicable to energy security scenarios. This architecture was developed under NATO RTO MSG-049 "M&S Support for Emergency Response Planning and Training" in preparation of CAX Phoenix 2010 and is part of the EU FP7 MACRToolset Project.

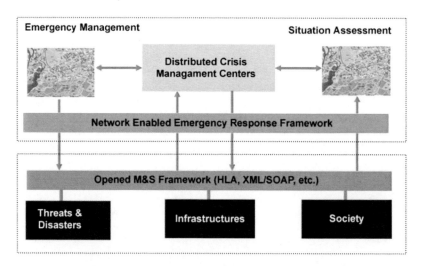

Figure 5. Battle Lab Architecture for C2 in crisis/emergency management.

---

[12] Stoykov, M., Shalamanov, V., Kirov, G., Stoyanov, V., Ivanov, I., Tsankov, A., Integrated System for Emergency Management (Architectural Methodology), Change Management Series, Softrade, 2006 (in Bulgarian).

Using JTSAC as a base for the development of CM/EM C2 Battle lab is essential for validation of the products envisioned in the MACROToolset Project and at the same time will be used to consolidate cooperation between Bulgarian Academy of Sciences, industry, ministries, and NGOs in the area but also as an element to be connected with CFBLNet (Combined Federation of Battle Laboratories).[13]

Referring to exercises as an instrument it must be recognized that EU TACOM SEE-2006[14] was the first full scale exercise to involve all the national institutions in Bulgaria (including media and NGOs) with responsibilities to security (more than 10) together with EU, NATO, UN OCHA, and seven SEE countries with observers from 17 EU countries. After several national exercises led by MoD, MoI or Ministry of Health, JTSAC is currently supporting Phoenix 2010 with the architecture following the one prepared for MACROTools Project (see Figure 5).

This research led to development of a framework for integration of new technologies in a creative environment for CAX that is used for education and training (E&T) and research and development (R&D) for the integrated security sector (considered a complex adaptive system) change management as presented Figure 6:

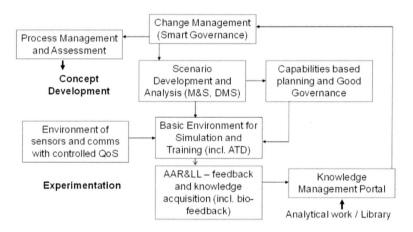

Figure 6. Integration of new technologies in a creative environment E&T/R&D for the integrated security sector change management.

The authors of this chapter are currently working on the development of a framework for the Regional Center for Energy Security and Network of Knowledge Centers on Energy Security for SEE and Black Sea Region, based on their

---

[13] Combined Federation of Battle Laboratories (CFBLNet) – www.CFBL.info with NATO and European PoP: CFBLinfo@nc3a.nato.int is secure, established, multi-national, and on the cutting edge of research, development, trials, and assessment in the Net Centric Warfare.

[14] EU TACOM SEE 2006, Technical Report, Institute for Parallel Processing, BAS, Sofia, December, 2006.

experience in building JTSAC and Center for Security and Defence Management[15] established within the key support of the NATO SfP 981149 Project.

Finally, it must be noted that the institution building for the last 10 years in the area of comprehensive approach to security in Bulgaria, which is related to the Bulgarian Academy of Sciences (BAS) was organized in accordance with the following key event list:

- 1999 – Situation Center in MoD/Government, Framework Agreement with BAS;
- 2002 – Center for National Security and Defense Research (CNSDR) in BAS;
- 2003 – Science Coordinating Council to Standing Government Committee on Civil Protection (in BAS);
- 2004 – C4I Department development in IPP, BAS;
- 2005 – Center of Operational Analysis in BAS (related to the NATO SfP 981149 Project);
- 2006 – Joint Training Simulation and Analysis Center in BAS (Civil Security);
- 2009 – Center for Security and Defence Management in BAS;

The next steps, the authors presume, will be focused on:

- 2010 – Allied Command for Transformation EPOW, SPP Program exercises;
- 2011 – 2013 – Center of Excellence on Military Support to Civilian Authorities (CoE MSCA).

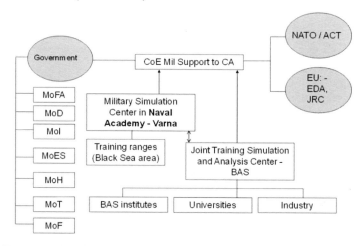

Figure 7. Model of center of excellence on military support to civilian authorities.

---

[15] http://www.caxbg.com/is/

The role of the Bulgarian Academy of Sciences as an "Advisor" to the Nation since 1869 on Science and Technologies, including security related issues, could further support MoFA, MoI and MoD for a larger regional role in the context of the NATO/EU role in SEE/WBSA with joint Bulgarian – US projects added to the process.

This will logically result in development of a Center of Excellence (CoE) on Military Support to Civilian Authorities accredited with Allied Command for Transformation, providing planning and testing of civilian capabilities, supported by military capabilities in maritime security. A model of the general architecture of the CoE MSCA is presented in Figure 7.

Conclusion

This paper presents a brief overview of the Bulgarian position in SEE/WBSA that was disclosed both in the geographic and strategic context. As a part of this, a methodological framework based on OA and CAX for plausible futures validation is presented via the academic JTSAC-CS. The disclosed framework addresses energy security and CIP in the region with CAX integration and validation, allows for a good and explanatory view point from a scientific perspective of the problem at hand.

All of the above developments are sustainable only in the larger regional and NATO/EU context of further development of the integration processes in SEE and WBSA.

NC3A is a key operational analysis and C3 support, at large, agency in NATO. After a very successful first Regional Chief Information Officers' Conference for SEE and accepting the proposed by the regional nations' role of technical advisor to both the SEDM process and SEEBRIG, there is a possibility for improvement of the regional capabilities in the area of OA and CAX and development of a regional center in support of security studies and training. An established framework MoU with MoD of Bulgaria, negotiated the current Cooperation Agreement with the Bulgarian Academy of Sciences and recently signed MoU between MoD and the Bulgarian Academy of Sciences are providing sound groundwork for trilateral cooperation – MoD-NC3A-BAS.

IIASA in Austria is a leading applied system analysis institute in support of global problem analysis, so regional cooperation in the area of OA and CAX to support energy security and environmental issues could follow this model.

EDA is the EU instrument in the area of capabilities planning for defense and security, promoting multinational projects, so could be a good EU instrument to further foster regional cooperation in addition to the above developments.

JRC of the EC is a main asset for research in the civil security area for EU so it could be key player in the FP7 format to involve regional nations in EU large research effort in the area of energy security.

US – GlobalEESE (Global Energy and Environment Strategic Ecosystem)[16] is another forum to exploit the developments presented in the paper for the strengthening of the energy security for the SEE and Europe as a whole.

Last, but not least, a series of ARW/ASI supported by NATO Science Committee, especially HSD Panel on security matters in SEE and WBSA for the last several years, logically set up a base for this more focused discussion on energy security as one of the key challenges for the region. The next logical step is institutionalizing of the process through the use of JTSAC.

---

[16] The Energy and Environment Strategic Foresight laboratory – US Department of Energy: Final Report from the event 27–28 April 2009, August, 2009, http://globaleese.org/

ENERGY SECURITY FOR INDUSTRIAL AND MILITARY INSTALLATIONS: EMERGENT CONDITIONS THAT INFLUENCE THE STRATEGIC SELECTION OF TECHNOLOGIES

JAMES H. LAMBERT*, CHRISTOPHER W. KARVETSKI
*Center for Risk Management of Engineering Systems,
University of Virginia, Charlottesville, VA, USA*

RENAE D. DITMER
Stratcon, LLC

TAREK ABDALLAH, MELANIE D. JOHNSON,
IGOR LINKOV
*Engineer Research and Development Center (ERDC),
US Army Corps of Engineers*

Abstract.   We describe recent efforts integrating scenario analysis with multiple criteria decision analysis in support of strategic planning for the energy security of industrial and military installations. Energy security is an increasingly important issue for industrial and military installations. Disruptions of the grid and other outages for key buildings, facilities, and entire installations jeopardize critical activities and missions. Cost and supply volatilities of traditional energy sources and backup technologies increase the need for innovation in meeting energy demands. Part of such demands should be met with renewable energy sources. Each of the hundreds of installations of a large industrial or military organization presents a unique challenge in the attainment of energy security goals. This Chapter describes a framework to highlight what science, engineering, and other conditions most influence the planning of strategic investments in innovation for energy security. The framework aims to avoid surprises that could result from a failure to account systematically for the emergent conditions that affect industrial and military installations, including emergent conditions of regulation, technologies, economics, geopolitics, environment, and other topics. Science, engineering, and other investigative resources can be focused on the future conditions that most matter to the selection of technologies and their operations plans.

* To whom correspondence should be addressed: James H. Lambert, Center for Risk Management of Engineering Systems, University of Virginia, Charlottesville, VA, USA, e-mail:lambert@virginia.edu

A. Gheorghe and L. Muresan (eds.), *Energy Security: International and Local Issues,*
Theoretical Perspectives, and Critical Energy Infrastructures,
DOI 10.1007/978-94-007-0719-1_16, © Springer Science+Business Media B.V. 2011

Introduction

Recent efforts of the US Army to address energy security are illustrative and can be generalized and transferred to other large industrial and military organizations.

> "In an age of terrorism, combustible and explosive fuels and weapons-grade nuclear materials create security risks. World market forces and regional geopolitical instabilities broadly threaten energy supplies. Infrastructure vulnerabilities pose further risks of disruption to Army installations." The U.S. Army Energy and Water Campaign Plan for Installations, 2007

Energy security is the capacity to avoid adverse impact of energy disruptions caused by natural, accidental, or intentional events affecting energy and utility supply and distributions systems (United States Army, 2007). For the United States Army, energy security is reflected in the level of assurance that the critical missions of installations and operational units can be accomplished in the face of disruptions to electricity or fuel supplies (United States Army, 2009). The criticality of mission, the fluctuating cost of energy, integration of multiple fuel sources, aging buildings and equipment, the diversity of energy efficiency strategies and technologies, the use of renewable resources, and cultural considerations underwrite the complexity involved in energy security planning. Related goals can be formalized into energy security goals, objectives, and metrics.

The US Army spent $4.1 billion for fuel and energy during 2008 (United States Army, 2009). They face difficulties in priority setting and decision making for achieving energy security, much like any large organization. Each of the installations has different mission objectives and the energy security priorities and strategic planning could be unique across the installations. The imperative need to consider rising energy costs along with other emergent conditions of the energy environment further complicates the evaluation of alternative investment strategies. In this Chapter, we define an emergent condition as a future event or trend, likely or unlikely, that could impact decision making for energy security.

Combinations of emergent conditions create different scenarios that can be used to evaluate and select robust portfolios of actions and investments. These different combinations of emergent conditions affect how an energy manager will pursue the different energy security objectives. For example, the consideration of different regulatory policies will increase the value of different renewable energy sources. Vulnerability to intentional or accidental destruction to centralized energy sources may warrant an increased investment in short or long-term storage systems or generators. Further assumptions of deteriorating geopolitics may lessen the desirability of diesel generators. There is a need for preliminary engineering to efficiently filter the scenarios to a manageable subset for decision-making and prioritization of investment.

Multiple criteria decision tools are available for installation energy managers to deal with the multiple goals, objectives, and metrics and evaluate the diverse portfolios of energy technologies and sources. When using these tools, it is imperative to integrate scenarios of emergent conditions into the analysis. Such conditions may be factual and/or may be the advocacy positions of various stakeholders in a large-scale infrastructure system. The present pages describes i) the past and current efforts for installation energy security, ii) the assumptions of emergent conditions of the energy environment related to energy security of Army installations; iii) the integration of emergent conditions into a formal decision model; iv) indicates the conclusions and recommendations for further knowledge.

A Framework for Understanding Energy Security of Installations

While energy usage and costs along with other priorities were certainly a concern prior to 2005, *The US Army Energy Strategy for Installations* (United States Army, 2005) recognized the growing dependence of the United States in general as well as the Army on imported fossil fuels. Environmental and climate factors were also expected to play a larger role in future energy policy. Five initiatives were established for installations going forward in energy planning until 2030: (i) eliminate energy waste in existing facilities, (ii) increase energy efficiency in renovation of new construction, (iii) reduce dependence on fossil fuels, (iv) conserve water resources, and (v) improve energy security. The first two initiatives imply increased monitoring of consumption and decreased use of peak energy as part of a plan for controlling the variability of energy costs. Utilizing and improving renewable energy technologies for geothermal, solar, biomass, wind, and other energies will curtail the sensitivity of mission vulnerability to sudden shocks to the global energy market and geopolitical instabilities. Renewable energies will also reduce emissions. State, regional, and local regulations related to renewable energies and other sources and technologies affect the choice of energy supplies. Availability of renewable energies also factors into the selection of the energy portfolio. For example, Figure 1 shows a group of installations that consider solar energy as an available and cost-effective source of energy. Likewise, Figure 2 and Figure 3, respectively, show a group of installations that consider wind and biomass as suitable additions to the energy portfolio of an installation. While alternative sources may reduce cost and supply volatility, utility and infrastructure systems may still be vulnerable to destructive forces or deterioration. A reliability and vulnerability assessment procedure along with the ability to island an installation can ensure continuity of energy supplies.

The *Army Energy and Water Campaign Plan for Installations* (United States Army, 2007) was established to implement the visions of *The US Army Energy Strategy for Installations*. Short, mid, and long-term tools, technologies, and projects provide a framework for establishing this vision. Army facility energy consumption in 2007 consisted of approximately one-third electrical energy. Natural gas and fuel oil accounted for the remainder of the remained of consumption. Increasing coal, nuclear, and renewable energy lessen the use of imported energy fuels.

Figure 1. Installations that have sufficient solar resource availability [Source: United States Army, 2009].

Figure 2. Installations that have sufficient wind resource availability [Source: United States Army, 2009].

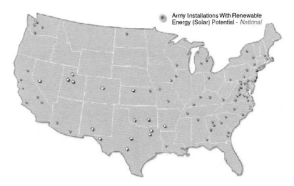

Figure 3. Installations that have sufficient biomass resource availability [Source: United States Army, 2009].

While other energy management programs focus on minimizing cost, consumption, and emissions, programs aimed at increasing energy security are more difficult to quantify and monitor. These energy security programs are focused on uninterrupted and reliable service. Survey methodology and standards for utility systems aid installations in measuring energy security. Facilities can then be grouped by mission criticality (United States Army, 2007).

The recent *Army Energy Security Implementation Strategy* (United States Army, 2009) presented future vision, mission, and goals for determining energy security objectives (ESOs) and corresponding metrics for measuring energy security progress. This energy security framework is displayed in Figure 4. The expressed energy mission is "make energy a consideration in all Army activities in an effort to reduce demand, increase efficiency, seek alternative sources, and create a culture of energy accountability, while sustaining or enhancing operational capabilities" (United States Army, 2009).

The ESOs support the energy security goals (ESGs). The ESOs are measured by corresponding metrics. Five ESGs (United States Army, 2009) expressed are: ESG1. Reduced energy consumption, ESG 2. Increased energy efficiency across platforms and facilities, ESG 3. Increased use of renewable/alternative energy, ESG 4. Assured access to sufficient energy supplies, and ESG 5. Reduced adverse impacts on the environment.

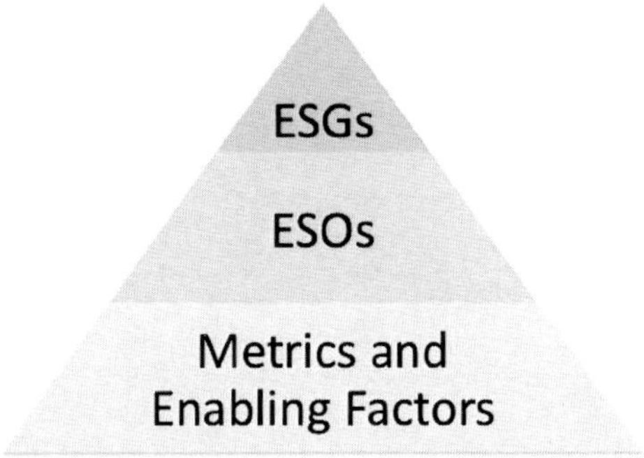

Figure 4. Hierarchy of energy security framework components. Metrics and enabling factors measure and promote energy security objectives (ESOs), which support the broad energy security goals (ESGs).

Enabling factors encourage energy awareness, innovation, and accountability. For example, one installation objective for installations includes *Improvement of Energy Performance of Current Infrastructure*. This objective supports the goals of *Reduced Energy Consumption* and *Increased Energy Efficiency across Platforms and Facilities*. A *completed metering program by 2012* serves as a performance target as well as an enabling factor. The percent completion of the metering program serves as a metric. Figure 5 illustrates this example.

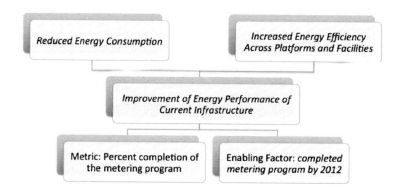

Figure 5. Example of energy security framework, where metrics and enabling factors support two ESGs.

Uncertainties and Emergent Conditions of the Energy Environment

The local, regional, national, and international energy environment for Army installations and other large facilities is increasingly affected by geopolitical and economic as well as other events and trends (Constantini et al., 2007; Nakićenović, 2000; Mintzer et al., 2003). Army installation managers are concerned with understanding what energy security investment alternatives or portfolios of investment are robust in their performances across emergent conditions. If a robust investment is not available, an integrated scenario and decision analysis should illuminate which scenarios (defined as portfolios of emergent conditions) warrant additional modeling and resources to inform the selection of a preferred investment alternative.

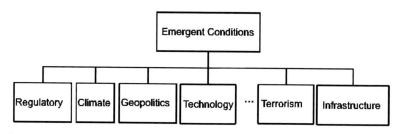

Figure 6. Various perspectives for identifying emergent conditions for evaluating in energy security of installations.

The many diverse categories of emergent conditions reflect the complexity of the energy environment. Methodologies from the field of risk analysis can help in identifying and understanding obvious and non-obvious sources of risk (Haimes, 2009). Figure 6 provides a partial representation of the different system viewpoints that can aid the identification of relevant emergent conditions. For example, analyzing energy security through a climate perspective will highlight potentially disruptive weather events such as ice and wind storms that can disrupt distribution from the public energy grid.

Technology-related emergent conditions include potential developments in new nuclear energy technologies, coal technologies, or promising renewable energy technologies. Political, socio-cultural, and regulatory emergent conditions include new energy guidelines, emerging regulatory and industry interests, or quickly evolving national or international energy policies (World Energy Council, 2007).

Societal emergent conditions include changes in societal viewpoints on energy production and use. The degree of cultural acceptance of changing energy guidelines is of notable importance for Army installations (United States Army, 2009). International emergent conditions include wildly vacillating energy costs or availability due to changes in global demand and shifts in the geopolitical landscape (United Nations University – Millenium Project, 2008).

Emergent conditions at the installation-level that can impact mission execution derive from these regional, national, and international events as well as from local disruption of energy services, commercial energy grid failures, destruction of energy systems, weather and climate changes and events, fuel and material supply chain matters, workforce availability, organizational and institutional issues, deterioration of other infrastructures, and changing security requirements.

Both the Army at large and individual installations have adopted a broad range of strategies and a varying array of technologies for achieving energy efficiency. The additional need to achieve energy security calls for a methodical and integrated approach of systems analysis. To this point, the quantifiable metrics for success in achieving energy objectives have been based primarily on energy cost savings. With energy security now a critical factor in achieving installation energy goals, incorporating emergent conditions to this analysis becomes a necessary rather than optional feature.

The emergent conditions are identified individually, either through the different viewpoints of Figure 6 or as perspectives of one of the many stakeholders involved in the decision process (Karvetski et al., 2010a). These emergent conditions can however occur jointly. For example, the confluence of a damaging storm during a period of fossil fuel shortage could be a particularly disruptive event for a critical mission. Therefore the consequences of the emergent conditions should be considered jointly as a *scenario* assumption. Table 1 displays several emergent conditions on the left that could compose different scenario assumptions. A "+" indicates that the emergent condition of the row is one assumption of the column scenario. For example, scenario s_1 consists of one descriptive emergent condition while scenario s_2 consists of four.

Future scenarios of the external environment as well as policy scenarios are common practice in the evaluation of energy alternatives (Kowalski et al., 2009). Different ideal energy technologies are identified for the given scenarios. For example, Tonn et al. (2009) address United States energy portfolios with a scenario perspective-based methodology to increase energy independence and energy security. They incorporate economic, environmental, political, and other perspectives to create and evaluate different energy portfolios for a 20-year horizon. They suggest that desirable energy portfolios eliminate energy imports and include variety in domestic energy resources.

TABLE 1. Key scenarios are defined by combining diverse emergent conditions of the energy environment

Emergent conditions	Scenarios				
	S_1	S_2	S_3	S_4	S_5
Large carbon emissions tax					
Large government subsidies for renewable energy				+	
Reemergence of nuclear technology					
Abandonment of nuclear technology					
Newly established Renewable Portfolio Standards					
Short-term national/regional energy blackout					
Long-term national/regional energy blackout					
Increased volatility in oil and gas prices and supply			+		
Oil and gas remain available and cost-effective	+				
Deterioration in geopolitics and war/peace/terrorism					+
Few changes in geopolitics and war/peace/terrorism					
Improvement in geopolitics and war/peace/terrorism					
Attack on national power grid					
Low growth in energy technology					
Moderate growth in energy technology					
High growth in energy technology		+			
Low environmental-movement impacts					
Moderate environmental-movement impacts					
High environmental-movement impacts				+	
Low national economic growth					
Moderate national economic growth					
High national economic growth		+			
Early realization of climate change					
National switch to solar energy					
Increase in National/International demand for energy security			+		
Stimulated demand for distributed energy					
Increase in demand for domestic energy sources			+		
Accelerated commercialization of renewable energy		+			
Aggressive public investment in R&D in hydrogen and fuel cell technologies		+			
Prolonged drought/Inclement weather					
Improved battery technology					
Switch to SmartGrid Technologies					
Changing demand for food-based agriculture					

Across management fields, scenario analysis is typically a qualitative exercise (Wright et al., 2009; Goodwin and Wright, 2001). A defining trait of scenario analysis, which distinguishes scenario analysis from forecasting or predicting, is the absence of probabilities in characterizing possible future states of nature. Scenario analysis emphasizes creativity and communication and formulates unlikely but plausible and consistent, albeit imprecisely specified, stories of how the future may unfold. These stories seek to characterize system vulnerabilities and opportunities and aid in current and adaptive decision making. In the next section, we discuss the pairing of scenario analysis with multiple criteria decision analysis.

Scenario and MCDA Methodology for Analyzing Energy Security of Installations

It is important to consider both the immediate, local emergent conditions along with national and international emergent conditions in the identification and valuation of energy security alternatives. More robust investments strategies can be formulated by grouping a diverse set of future scenarios of emergent conditions by their relative impact on decision making (Karvetski et al., 2010a; Karvetski et al., 2010b; Groves and Lempert, 2007; Lempert et al., 2006; Mason, 1998). Multiple criteria decision analysis (MCDA) can provide guidance to organizations making resource allocation decisions under uncertainty (Kleinmuntz, 2007; Keeney, 1992). The tools of decision analysis can handle a set of multiple, often competing energy security objectives (Montibeller and Franco, 2010).

Multiple criteria decision analysis tools have been effective in selecting energy alternatives in multiple case studies with a diverse set of objectives (Chatzimouratidis and Pilavachi, 2009; Kowalski, 2009; Loken 2007; Papadopoulos and Karagiannidis, 2006). For example, Nigim et al. (2004) advocate the use of multicriteria decision making to rank renewable energy initiatives at a community level. While conventional electrical energy is cheaper, the implementation of different regional renewable energy sources and activities at the community level has many positive consequences and externalities. The analytic hierarchy process is one possible MCDA model.

Practitioners of multiple criteria decision analysis should incorporate a variety of assumptions of emergent conditions that are essential to the understanding of and, ultimately, the comparative evaluation of alternative investments in energy security. Representing uncertain future scenarios describing the joint occurrence of emergent into a formal decision model is an important challenge for applying traditional decision analysis tools for prioritizing environmental and energy alternatives.

For example, quantifiable performance criteria or metrics for energy investments can be linked to ESGs. Figure 7 displays an example of seven criteria on how they can be used to evaluate alternative investments and monitor progress towards the energy security goals (United States Army, 2009). While the success of all ESGs is highly desirable, the relative importance of performance criteria or metrics associated with the ESGs is likely to vary with the assumptions of emergent conditions.

Performance criteria/metrics

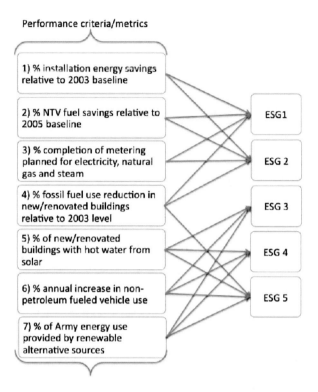

Figure 7. Relationship of metrics to some goals of energy security of installations.

Our recent efforts to integrate scenario-based analysis with multi-criteria decision analysis (MCDA) for energy security include Karvetski et al. (2010c) and Lambert et al. (2010). Our efforts were motivated in part by Wright et al. (2009) and Goodwin and Wright (2001), who suggest a multi-criteria decision making framework as a quantitative approach for performing scenario analysis, using a multiattribute value function. Montibeller et al. (2006) implement the Goodwin and Wright approach in practice for two case studies unrelated to energy security. Ram et al. (2010) extend the multiattribute value function approach to account for twelve scenarios, also unrelated to energy security. In the pairing of MCDA with scenario analysis for energy security, elicitation of value and other judgments needs to be efficient, as the number of possible scenario assumptions is large.

While there are other tools to consider the effects of diverse scenarios, we briefly outline one approach, focusing on the desirable information output. In this approach, like many multiple criteria methods, the performance criteria associated with metrics are linked to different ESGs and are used to differentiate the benefits of drawbacks of different alternative investments. Selecting an investment requires tradeoffs and preferences. A way to classify the scenarios is to considering how the scenarios affect the elicited preferences and tradeoffs among these performance criteria. These scenario-derived changes in preference are represented in Table 2. For example, if scenario s_1 = {Oil and gas remain available and cost-effective} were to play out, the importance of a performance criteria associated with reducing foreign oil would not be as important (Criterion 4 could, for example, denote the performance criterion 4 from Figure 7). This translates to a *major decrease* in the relevance of this criterion. Formal decision analysis methods quantify this change in preference and a new function is derived to quantify the value of each investment for the scenario. Different investment alternatives may be prioritized higher under this scenario (Figure 8).

The technical details of the approach (using the analytic hierarchy process as an example) can be found in Karvetski et al. (2010c) and Lambert et al. (2010). An alternative application of the approach applied to infrastructure development using multi-attribute value theory can be found in Karvetski et al. (2010a). Using the new value function for each scenario, the alternatives are ranked for each scenario. Ideally, there are a few candidate investments that are robust across the scenarios. This may not always be the case. Some scenarios threaten the performance of different technologies while increasing the value of other alternatives.

TABLE 2. Changes in the relevance of the performance criteria as they are influenced by different scenario assumptions of the energy environment

Criteria	Scenarios of the energy environment				
	s_1	s_2	s_3	s_4	s_5
Criterion 1					
Criterion 2					
Criterion 3					
Criterion 4	Major decrease		Minor increase		
Criterion 5					
Criterion 6		Minor increase		Major increase	Minor increase
Criterion 7		Major increase		Major increase	Major increase

Figure 9 displays an example output of such an approach. In this example there are over twenty diverse sources and technologies being prioritized individually (alternatives can also be prioritized jointly as portfolios). These are listed across the top of the figure. The ranking of the alternatives is represented on the left axis (1 being the best). The diamond represents some baseline scenario and additional scenario assumptions of emergent conditions can increase or decrease the ranking of the investments. The two rectangles of the approach direct attention to possible insight from the approach. The left rectangle highlights an investment that is highly prioritized and is robust with respect to the other alternatives across different scenario assumptions. The right rectangle highlights an alternative that is not as highly prioritized and has a relative ranking that is highly correlated or influenced with the scenario assumptions.

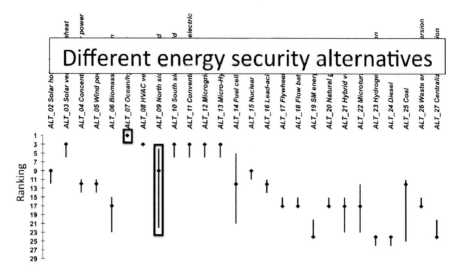

Figure 8. Illustration of the ranking of energy-security investment alternatives across diverse scenarios. The heights of the vertical bars indicate large sensitivity. The diamond icon indicates a baseline or no-scenario result while the top of the bar indicates the highest ranking across scenarios, the bottom of the bar indicates the lowest ranking across scenarios.

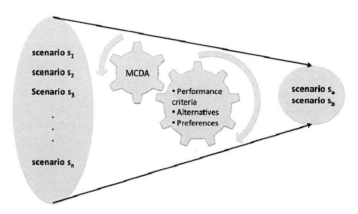

Figure 9. The reduction of the available scenarios to a few scenarios that are most influential to the selection among alternatives for energy security.

With a large number of scenarios, if a robust investment cannot be found, it is often needed to reduce the number of scenarios to a more manageable set. Figure 10 illustrates this approach. On the left of the figure, the machinery of MCDA reduces n scenarios down to two scenarios that are deemed the most influential on the performance of select alternative investments (n >> 2). Influential scenarios could be those that represent vulnerabilities to a most preferred investment. Once these scenarios are identified, they can be further studied with other tools of modeling and simulation and decision analysis.

Table 3 displays an example of the final output of the using scenario assumptions with MCDA (This case using portfolios of investment alternatives generically labeled X_n). In this output, top-performing portfolios are highlighted, but equal effort is dedicated to identifying influential scenarios. These scenarios can be further classified as vulnerabilities for certain portfolios as well as opportunities. These influential scenarios should direct future science and engineering efforts towards final decision making.

TABLE 3. A synopsis of outputs that should aid in future engineering and other efforts for making strategic energy security investments. Portfolios are generically labeled Xn

What scenarios are most influential or disruptive?	Scenario s_1, disrupts portfolio X_{03} from being the top prioritized portfolio.
What portfolios perform best?	X_{03} performs best under all but one considered scenario, s_1. Portfolio X_{02} ranked best under s_1.
What portfolios have upside potential to any of the additionally considered scenarios, $s_1,\ldots,s_5$?	X_{03} has upside potential to scenarios $s_2,\ldots,s_5$ and X_{05} has large upside potential to scenarios s_2 and s_4.
What portfolios have large downside potential to any of the additionally considered scenarios $s_1,\ldots,s_5$?	X_{01} has downside potential to scenarios s_2 and s_4 and X_{02} has large downside potential to the scenarios $s_2,\ldots,s_5$.

Conclusions

Strategic decision making in energy security is challenging for large organizations with hundreds or thousands of constituent installations. Regional differences, mission aims, and other sources of diversity across the industrial or military organization make it challenging to distinguish what factors should be influential to the selection of energy-security technologies for particular installations.

This paper has illustrated our recent efforts to define and highlight the assembly of emergent conditions into scenarios that most matter to strategic choices about energy security for large industrial and military installations. Our approach has been to combine tools of multiple criteria decision analysis with scenario-informed analysis. We are able to identify and combine and analyze a diverse nature of emergent conditions including energy/environmental regulation, climate change, technology innovation, economics, geopolitics, environment, demographic, and other conditions. Our effort assures that modeling and simulation and other scientific/engineering investigative resources are directed to where they are most needed for the informed comparison and selection of technologies.

Considering the large variety of emergent conditions will increase the modeling burden, but neglecting certain emergent conditions could decrease the ability of selecting robust solutions. This Chapter has addressed a practical need to identify and process the diverse assumptions of emergent conditions that should influence strategic decisions about energy security.

Acknowledgments

Preparation of this chapter was supported in part by American Recovery and Reinvestment Act funding to the University of Virginia through the US Army Corps of Engineers, and by funding provided to US Army Corps of Engineers at the Engineer Research and Development Center by the Installation Technology Transfer Program managed by the Assistance Chief of Staff for Installation Management.

References

Chatzimouratidis, A.I., Pilavachi, P.A., 2009. Technological, economic and sustainability evaluation of power plants using the Analytic Hierarchy Process. Energy Policy 37, 778-787.

Constantini, V., Gracceva, F., Markandya, A., Vicini, G., 2007. Security of energy supply: Comparing scenarios from a European perspective. Energy Policy 35. 210-226.

Goodwin, P., Wright, G., 2001. Enhancing strategy evaluation in scenario planning: a role for decision analysis. Journal of Management Studies 38(1), 1-16.

Groves, D.G., Lempert, R.J., 2007. A new analytic method for finding policy-relevant scenarios. *Global Environmental Change* 17, 73-85.

Haimes, Y.Y., 2009. Risk Modeling, Assessment, and Management, 3rd edition. Wiley and Sons, Inc. New

Karvetski, C.W., Lambert, J.H., Linkov, I., 2010a. Emergent Conditions and Multiple Criteria Analysis in Infrastructure Prioritization for Developing Countries. Journal of Multicriteria Decision Analysis, In Press.

Karvetski, C.W., Lambert, J.H., Keisler, J.M., Linkov, I., 2010b. Integration of Decision Analysis and Scenario Planning for Coastal Engineering and Climate Change. To appear in *Systems, Man, and Cybernetics Part A*.

Karvetski, C.W., Lambert, J.H., Linkov, I., 2010c. Scenario and Multiple Criteria Decision Analysis for Energy and Environmental Security of Military and Industrial Installations. Submitted to *Integrated Environmental Assessment and Management*.

Keeney, R.L., 1992. Value-Focused Thinking, a Path to Creative Decisionmaking. Harvard University Press, Cambridge.

Kleinmuntz, D.N., 2007. Resource Allocation Decisions. In: Edwards, W., Miles, R.F., von Winterfeldt, D. (Eds.), Advances in Decision Analysis. Cambridge University Press.

Kowalski, K., Stagle, S., Madlener, R., Omann, 2009. Sustainable energy futures: Methodological challenges in combing scenarios and participatory multi-criteria analysis. European Journal of Operational Research 197, 1063-1074.

Lambert, J.H., Karvetski, C.W., Linkov, I., Abdallah, T., 2010. Energy Security of Military and Industrial Facilities: A Scenario-Based Multiple Criteria Decision Analysis to Identify Threats and Opportunities. Proceedings of the Tenth International Conference on Probabilistic Safety Assessment and Management (IAPSAM), Seattle.

Lempert, R.J., Groves, D.G., Popper, S.W., Bankes, S.C., 2006. A General, Analytic Method for Generating Robust Strategies and Narrative Scenarios. *Management Science* 52(4), 514-528.

Loken, E., 2007. Use of multicriteria decision analysis methods for energy planning problems. Renewable and Sustainable Energy Reviews 11, 1584–1595.

Mason, D.H., 1998. Scenario planning: mapping the paths to the desired future. In: Fahey, L., Randall, R. (Eds.), Learning from the Future. John Wiley & Sons, New York. 109–121.

Mintzer, I., Leonard, J., Schwartz, P., 2003. US Energy Scenarios for the 21st Century. Pew Center on Global Climate Change. <http://www.pewclimate.org/docUploads/EnergyScenarios.pdf> (Accessed May 20, 2009).

Montibeller, G., Gummer H., Tumidei, D., 2006. Combining Scenario Planning and Multi-Criteria Decision Analysis in Practice. Journal of Multi-Criteria Decision Analysis 14, 5-20.

Montibeller, G., Franco, A., (forthcoming) 2010. Multi-criteria decision analysis for strategic decision making. In: Zopounidis, C. and Pardalos, P.M. (eds.), Handbook of Multicriteria Analysis, Springer.

Nakićenović, N., 2000. Energy Scenarios. Chapter 9 in United Nations Development Programme. United Nations Department of Economic and Social Affairs. World Energy Council. World Energy Assessment. New York 2000.

Nigim, K., Munier, N., Green, J., 2004. Pre-feasibility MCDM tools to aid communities in prioritizing local viable energy sources. Renewable Energy 29, 1775-1791.

Papadopoulos, A., Karagiannidis, A., 2006. Application of the multi-criteria method Electre III for the optimisation of decentralised energy systems. Omega 36, 766-776.

Ram, C., Montibeller, G., Morten, A., 2009. Extending the use of scenario planning and MCDA: An application to food security in Trinidad and Tobago. In progress.

Tonn, B., Healy, K.C., Gibson, A., Ashish, A., Cody, P., Beres, D., Lulla, S., Maxur, J., Ritter, A.J., 2009. Power from Perspective: Potential future United States energy portfolios. Energy Policy 37, 1432-1443.

United Nations University-Millennium Project, 2008. Global Energy Scenarios. <http://www.millennium-project.org/millennium/scenarios/energy-scenarios.html> (Accessed May 20, 2009).

United States Army, 2009. Army Energy Security Implementation Strategy. < http://www.asaie.army.mil/Public/Partnerships/doc/AESIS_13JAN09_Approved%204-03-09.pdf> (Accessed April 26, 2010).

United States Army, 2007. The U.S. Army Energy and Water Campaign Plan for Installations. http://army-energy.hqda.pentagon.mil/docs/AEWCampaignPlan.pdf (Accessed July 2, 2009).

United States Army, 2005. The U.S. Army Energy Strategy for Installations. <http://army-energy.hqda.pentagon.mil/docs/strategy.pdf.> (Accessed on: July 24, 2009).

World Energy Council, 2007. Deciding the Future: Energy Policy Scenarios to 2050. <http://www.worldenergy.org/documents/scenarios_study_online.pdf> (Accessed September 10, 2009).

Wright, G., Cairns, G., Goodwin, P., 2009. Teaching scenario planning: Lessons from practice in academe. European Journal of Operational Research 194, 323-335.

THE ENERGY INCIDENT DATA BASE

ROBERT K. MULLEN*
Independent Consultant

Abstract. The Energy Incident Data Base (EIDB) is concerned solely with the activities of sub-national actors directed against government and private sector energy enterprises worldwide. It is unclassified and its contents are drawn from a wide array of sources developed over more than 35 years of monitoring and analyzing these activities

The Energy Incident Data Base

The Energy Incident Data Base (EIDB) is concerned solely with the activities of subnational[1] actors directed against government and private sector energy enterprises worldwide. It is unclassified and its contents are drawn from a wide array of sources developed over more than 35 years of monitoring and analyzing these activities.

An energy enterprise target is considered to be any entity associated in any way with the exploration, development, production, refining, generation, storage, transmission, distribution, or transportation of energy in any form, by any means, and for any purpose. This includes any physical or fungible asset that is owned, leased, rented, or contracted for by an energy enterprise; the employees, contract employees, executives, consultants, private contractors, or other persons; and the infrastructure elements that support energy enterprises that are subjected to attack, attempted or planned attack, or threatened by subnational activities. The infrastructure supporting energy enterprises includes, but is not limited to, railways, ports, suppliers and service firms and assets, financial, telecommunications, architectural and engineering firms and assets, transportation firms and facilities, security firms, government institutions and agencies (central, provincial, and local), and others.

---

* To whom correspondence should be addressed: Robert K. Mullen, Independent Consultant, e-mail: bezoar@earthlink.net

[1] Subnational is defined as any individual or group not known to be working in an official capacity for a recognized government. Due to plausible denial, state sponsored subnational entities may or may not be included in this definition, depending on circumstances. (Not included in this definition, for example, are persons engaged in espionage who have been publicly exposed as working for a national government, while those engaged in espionage for a subnational entity are included.)

A. Gheorghe and L. Muresan (eds.), *Energy Security: International and Local Issues,*
Theoretical Perspectives, and Critical Energy Infrastructures,
DOI 10.1007/978-94-007-0719-1_17, © Springer Science+Business Media B.V. 2011

333

As of 1 February 2009, the EIDB contained over 29,700 records: These records hold an estimated 200,000 incidents in over 70 Mb of data.[2]

The EIDB consists of five linked and searchable subsidiary data bases consisting of the following:

- Adversary identities,
- The means used by adversaries to perpetrate and commit acts,
- The types of incidents,
- The locations of the incidents, and
- The types of targets affected by the incidents.

Adversaries are identified by specific names of individuals or groups (e.g., Animal/Earth Liberation Front, People's Revolutionary Army (Mexico), Anser al-Sunnah, etc.); by generic terms (e.g., terrorist, militia, insurgent, saboteur, employee, contract employee, former employee, tribal, pirate, etc.); or as unknown-unspecified when accounts or events do not provide sufficient information to identify an adversary or adversaries.

The means used to perpetrate, commit, attempt to commit, plan, or threaten an act describe what the adversary used or attempted, planned or threatened to use, such as explosives, firearms, chemicals, radiologicals, tools, land or water vehicles, forged/counterfeit/fraudulent documents, computers, etc.

The types of act include bombing, arson, homicide, abduction, sabotage, fraud, theft, embezzlement, blockade, violent demonstration, hacking and other computer crime, etc: i.e., what the adversary did or attempted, planned, or threatened to do.

The location is by country and the next lowest political subdivision (state, province, department, canton, island, etc.); and the named place of the incident. Locations that occur on the high seas or in territorial waters are designated – on the high seas – by the name of the body of water: i.e., ocean, sea, strait, etc., and by their latitude and longitude (when available). In territorial waters locations are designated by the name of the bight, bay, harbor, port, river, channel, roadstead or road, anchorage, etc.[3]

The type of target involved in an incident includes any physical or monetary asset, person, company, or other entity that is associated in any manner, and in any capacity, with any energy enterprise, worldwide.

---

[2] In the 29,000+ records in the Data Base are 1,557 separate specific and generic adversaries who used some 137 different means to perpetrate 347 types of incidents in 247 countries (taking into consideration name changes and fissions) and the high seas and territorial waters against some 661 types of targets in the energy sector.

[3] Targets on the high seas and in territorial waters include energy carriers as maritime tankers, lighters, barges, gas carriers, coal carriers, and certain bulk carriers (such as oil/bulk/ore carriers, or OBOs), floating production storage and offloading (FPSO) vessels; nuclear carriers (not necessarily Lloyd's classification) of new, spent, and reprocessed power reactor fuel and radioactive waste from power reactors; oil field service, supply, crew and workboat vessels; and offshore oil or gas drilling, production, and gathering platforms and offshore terminals. Subnational offences against them include piracy, hijacking, boarding, occupation, theft of cargo, kidnapping for ransom, and killings committed with edged and primitive weapons, smallarms and crew served machine guns, rockets, grenades, bombs, and explosives laden boats.

The date of an incident, or of the account(s) of an incident when no incident date is supplied, or the approximate date of the incident when given as such in the account(s) or is deduced from associated information, is entered into each record.

The data in each subsidiary data base are coded for search purposes, and may be queried in each, singly or in combination, by specific date or date range, or unrestricted by date.

An accession number is generated automatically for each new record. Its only function is to serve as a unique identifier for each record in the EIDB. The EIDB records can be searched by accession numbers.

A narrative discussion is part of each incident record in the EIDB. It includes, when available, date of incident, geographic specifics concerning the location or locations of an incident or incidents; the physical, social, and monetary consequences in local and U.S. currency (at the time of the incident), and times to repair and restore service from an incident, where appropriate, collateral damage; the number of adversaries involved in incident execution or attempted execution; the level of technological expertise demonstrated by the adversary; detentions, arrests or later dispositions concerning the adversary or adversaries; references to other records in the EIDB, where pertinent; and other matters of interest.

The record narratives, including their source citations, are searchable by key words, word phrases, numeric characters, or alpha and numeric characters in combination, for specific dates or date ranges, or unrestricted by dates.

As new data become available for previously recorded incidents, the appropriate EIDB records are updated to include the new data or to revise old data. Over time, this can occur on multiple occasions for a single record.

Hard copies of the accounts used in preparation, inputting into the EIDB, and updating EIDB records are preserved, as are selected background accounts on tactical, strategic, geopolitical, and economic matters related to adversary activities, their resources and other characteristics, and the social and political environments in which the adversaries operate. Descriptions of the structures, operations, capacities, and other data on energy systems impacted by adversary activities are also preserved, as are descriptions of structural and operational changes in these systems as they occur over time.

Country maps, and maps of major political subdivisions and major cities within them, are kept for reference purposes. Maps of electric power systems that identify generation facilities, transmission lines, major substations and switch-yards; and maps of natural gas, crude oil, and petroleum product pipelines, with pipeline specifications and identification of compressor and pumping stations and other features also form part of the reference library supporting the EIDB. These and other reference materials are continuously acquired for use in the incident verification process; to flesh out accounts used in the preparation of incident records for the EIDB; and for other purposes, such as trend and pattern analyses associated with assessing subnational activities.[4]

---

[4] Further supporting the EIDB and reports derived from it is a 300+ volume reference library on: adversary capabilities and characteristics; particular and generic energy system structural and functional information; arms, ammunition, ballistics, and explosives; terrorism and insurgencies, as well as ethnic, indigenous, environmental, and political militants whom impact the energy sector;

Reports on specific topics (trend analyses, threat and vulnerability assessments; consequence analyses; and adversary group size statistics, tactics and technology assessments, etc.) are uniquely possible with the EIDB and have been provided to various U.S. Federal, state, and local official organizations and commissions; industrial security organizations, selected elements of U.S. and foreign energy industry organizations; and used in testimony before a U.S. Senate Subcommittee.

---

organized crime, criminals and criminal gangs, cults and sects that impact this sector; nuclear fuel cycle matters; maritime ship and ship owner's registries and directories; and an extensive private map library of virtually every country in the world and many of their administrative elements. The map library also includes energy system (pipeline and electric power transmission line) maps. These latter often serve to confirm reported incident locations, and to provide more detail than that available in original incident accounts.

AN ENERGY SECURITY STRATEGY FOR ROMANIA: PROMOTING ENERGY EFFICIENCY AND RENEWABLE ENERGY SOURCES

ADRIAN GHEORGHE*
Old Dominion University, Norfolk, VA, USA

LIVIU MURESAN, SERGIU CELAC*, SEPTIMIU CACEU,
CLAUDIU DEGERATU, LEONELA LENES,
CRISTIAN KANOVITS, RADU BORES
EURISC Foundation, Bucharest, Romania

CHRIS PREBENSEN, TRYGVE REFVEM
Norwegian Atlantic Treaty Association, Oslo, Norway

Abstract. The national interests reflect the most important, stable, and institutionalized perception of the values that the Romanian nation aims to preserve, promote, protect, and defend by legitimate means, the values on which the Romanian nation builds its future, through which it guarantees its existence and identity, and for which it is integrated in the European and Euro-Atlantic community and takes part in the globalization process.

The International Crisis and Its Impact on Energy Policies

FINANCIAL CRISIS AND EU'S ENERGY SECURITY

The onset of the global financial crisis in late 2007 set in motion a series of events that caused rising political, economic, and social tensions around the world. The obvious failure of the prevailing financial and economic system further translated into a political and institutional crisis, which was described as a global leadership gap. Structural instability is likely to persist until a new equilibrium emerges in the form of global power.

_____

* To whom correspondence should be addressed: Adrian Gheorghe, Old Dominion University, Norfolk, VA, USA, e-mail: adriangheorghe9145@gmail.com

* To whom correspondence should also be addressed: Sergiu Celac, EURISC Foundation, Bucharest, Romania, e-mail: Sergiu.celac@ncsd.ro

A. Gheorghe and L. Muresan (eds.), *Energy Security: International and Local Issues,*
Theoretical Perspectives, and Critical Energy Infrastructures,
DOI 10.1007/978-94-007-0719-1_18, © Springer Science+Business Media B.V. 2011

The continued dependence of the world economy on traditional sources of energy and the challenges of climate change are making energy security a strategic priority for most nations. The current economic crisis may have slowed economic growth – and therefore energy consumption – but the projected long-term rise in demand is likely to put additional pressure on energy markets.

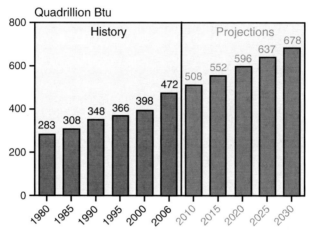

Sources: **History:** Energy Information Administration (EIA), *International Energy Annual 2006* (June-December 2008), web site www.eia.doe.gov/iea. **Projections:** ELA, World Energy Projections Plus (2009).

Figure 1. World marketed energy consumption 1980–2030.

In historical retrospect, the competition for access to increasingly scarce resources frequently caused regional crises, conflicts, and military interventions. The effort to achieve energy security has transcended national boundaries and became a matter of geopolitical concern. Recent events, such as the Russian-Ukrainian spats (in 2006, 2007, and 2009) over the transit of gas to Europe, revealed the continued impact of strategic considerations in which major energy producers and consumers determined their choice of energy sources and routes.

The global crisis brought about some additional difficulties for the accomplishment of the European Union's strategic designs. In the absence of a common EU policy on imported energy supplies, the burden of decision-making concerning the diversification of sources still rests with the member states. It is a fact that many EU countries, especially the new members from Central and Eastern Europe, are massively dependent on a single supplier of natural gas – The Russian Federation. The dramatic consequences of the interruption of the gas flow at times of peak demand in wintertime led to serious consideration of how to tap the considerable gas resources of the Caspian Basin and Central Asia as alternative sources of supply.

The forecasts depict a bleak picture. While the energy consumption of the developed world is bound to rise, an even higher growth of demand is expected in dynamic emerging economies such as China and India. This, combined with the inability to develop reliable and affordable alternatives to oil and natural gas, has encouraged the belief that the power to control the access to the international energy resources is shifting from energy consumers to energy producers.[1]

According to the scenarios developed by the European Commission, Directorate of General Energy, the gas import requirements in the EU, Switzerland, and the Balkan countries will increase from 221 billion cubic meters (bcm) in 2010 to 427 bcm in 2030 in the best-case scenario and will reach 652 bcm in the worst-case scenario. This would require, in the next few decades, increasing imports from the traditional gas suppliers of Europe (Norway, Russia, and North Africa) as well as the development of additional sources of supply. In the Second Strategic Energy Review "An EU Energy and Solidarity Action Plan" published in 2008, a calculation was presented implying a reduction of EUs import requirements in 2020 to 284–337 bcm – assuming implementation of the 20-20-20-plan.

The total amount of gas potentially available to Europe was assessed to be about 450 bcm in 2010, 640 bcm by 2020 and 715 bcm by 2030, compared to 304 bcm imported in 2005. Pipelines are expected to remain the dominant means for gas transport in the next few decades. However, investments of €126 billion are required for gas infrastructure to cope with the expected demand. This figure includes pipelines and storage facilities as well as liquefaction and gasification terminals.[2] Estimates published by the European Commission showed that the total investment needed for the upgrading of electric power transmission infrastructure would amount to at least €300 million to construct four alternating current lines between Turkey and the EU, €2,000 million to build four sub-marine high-voltage electric links directly connecting North Africa with Southern Europe, and about €200 million for a sub-marine HDVC link between Cyprus and Turkey. In terms of funding all these projects, as stated by the then EU Energy Commissioner Andris Piebalgs, "although the EU could provide loans and guarantees, it should not consider providing capital financing. The EU can facilitate getting loans in such a difficult situation of the credit crunch, but it would not go beyond that, because then it turns into a public private partnership instead of a consortium's project"[3]. The almost exclusive reliance on private financial resources for the implementation of vital projects may pose some problems in the coming decades.

The inherent difficulties related to efforts to ensure EU energy security are compounded by the impact of the global economical crisis. The international agenda is being adjusted according to the new requirements. 2009 started with a series of events (European Council, G20 Summit, NATO Summit, US-EU Energy Council) during which the topic of energy security held a prominent place in the discussions, with an emphasis on seeking alternative sources and transit routes for

---

[1] Scaroni, Paolo "Europeans Must Face the Threat to Energy Supplies", in *Financial Times,* January 18, 2006, p. 15.

[2] Energy corridors. EU and Neighboring Countries, European Commission, Directorate General for Research. Sustainable Energy Systems.

[3] http://bbjonline.hu/index.php?col=1004&id=46633

energy supply. The credit crunch added an element of uncertainty with regard to some major European energy projects. Besides the important political decisions that are still pending at EU level, enormous investments are required, to be provided mostly by consortia of private multinational companies, which need adequate guarantees against what they regard as an enhanced factor of risk.

THE EU RESPONSE TO THE CHALLENGES OF THE FINANCIAL CRISIS

In early 2009, the regional Summit in Budapest gave the go-ahead to the Nabucco[4] project. The final Declaration, which all participants approved, stated that they would strive to create a transparent and cost-based delivery system along the entire length of the gas pipeline, encourage direct foreign investments in supply and transit countries, and mobilize the concerned parties to promote efficient energy cooperation. The Prime Minister of Hungary at the time, Ferenc Gyurcsány, mentioned that the EU would support the project with €2 billion, including €200–300 million in advance financing in the form of a capital injection. The EU support could be supplemented with loans from the European Investment Bank (EIB) to the tune of 25% of the pipeline construction cost and from the European Bank for Reconstruction and Development (EBRD) to cover most of the investment (in order to participate in this project, EBRD requires an inter-government agreement for the pipeline).[5] After a long series of negotiations, the €3.98 billion package of measures was finally agreed to in May 2009.

The European Council held on 19–20 March 2009 emphasized that "the global economic and financial crisis is one of the most important challenges EU ever faced". The Council further agreed, as part of the European Recovery Planning, to support projects in the field of energy. It specifically endorsed the setting up of a crisis management mechanism to deal with disruption of supplies. Energy security was described as a key priority; it had to be enhanced by improving energy efficiency, diversifying energy suppliers, sources and supply routes and promoting EU's interests *vis-à-vis* third countries. The European Council endorsed the broad initiatives set out in the Second Strategic Energy Review (SER) and

---

[4] The Nabucco project envisions the construction of a gas pipeline with a capacity of 25–30 bcm/year to connect the Caspian Basin and Central Asia to the EU market. The project was initiated by specialised companies from five nations: BOTAS from Turkey, Bulgargaz from Bulgaria, Transgaz from Romania, MOL from Hungary, and OMV Erdgas from Austria. The five companies signed, on 11 October 2002 in Vienna, a Cooperation Agreement for the foundation of a corporation charged with the preparation of a feasibility study for a completely new pipeline designed to transport natural gas from the production areas in the Caspian, Central Asia and Middle East through the territories of the five countries. The starting points would be at the borders of Turkey with Georgia and Iran, and the terminal point at the Baumgarten junction in Austria (an important staging post for the transit of Russian gas towards Western Europe). The transport pipeline could also be linked to other sources of gas available in the area (Syria, Iraq, Egypt). The European Commission also recognized the importance of the project by including it among the priorities of the TEN Programme (Trans European Networks). This involves the financing by the European Commission of 50% of the costs for the preparation of the feasibility study.
[5] The Hungarian Ministry of Foreign Affairs, at http://www.mfa.gov.hu/kum/en/bal/actualities/spokesman_statements/Nabucco_Declaration_090127.htm

further refined in the Council (TTE) conclusions of 19 February 2009.[6] The Presidency Conclusions stressed that the European Commission, acting in co-operation with Member States, should present a detailed set of actions required to implement the priority objectives identified in the SER by completing the list of other agreed crucial energy projects. Moreover, the European Commission was asked to prepare, in 2010, a new EU Energy Security and Infrastructure Instrument. A revision of the Energy Efficiency Action Plan is also expected.

Although the G20 Summit in London held on 2 April 2009 had no specific resolutions concerning energy security, it is interesting to note that the Joint Statement made by President Dmitri Medvedev of the Russian Federation and President Barack Obama of the United States of America during the conference stated their common pledge to promote cooperation in implementing Global Energy Security Principles, as adopted at the G-8 Summit in Sankt Petersburg in 2006, including improved energy efficiency and the development of clean energy technologies.

While positive signals have been given at the level of political declarations reflecting the prominent place of energy security on the international agenda, the financial implications of energy-related investments need to be taken into consideration pragmatically in the context of the global crisis. At the heart of the financing issue is the relationship between uncertainty, cost of investments and profitability. Europe has a number of projects that remain sustainable by themselves, while others are more difficult to accomplish and may need further support.

For example, the Nord Stream, a major offshore gas pipeline across the Baltic Sea, aims to directly connect the Russian Federation to Germany. While EON, Wintershall, and Gasunie are now official partners, the Nord Stream AG project operator is 51% owned by the Russian company Gazprom. In September 2008, the Russian Prime Minister Vladimir Putin announced that five EU member states would receive natural gas through the pipeline: Germany, France, Great Britain, Holland, and Denmark.[7] At the time when the decisions were made, there were no signed contracts with the importers, even though the first section of the pipeline with a capacity of 27.5 bcm/year was planned to become operational in 2010. Members of the European Parliament asked for a new inquiry regarding the project's impact upon the environment. Poland, Lithuania, and Estonia, which felt excluded from the benefits of the pipeline and were concerned about their energy security, also criticized the project on environmental grounds. At the present stage, however, since the project has the support of important promoters, it is not likely to face major difficulties in order to be implemented. The pipeline is now under construction and its capacity is likely to be doubled by the construction of a second parallel pipeline. This will contribute to an improved security of gas supply for EU and reduce the dependence on Ukraine transit. Gazprom is planning another major new gas pipeline –"Southstream" – together with ENI and other interested gas buyers in southern Europe.

---

[6] Council of the EU, The Presidency Conclusions, http://www.consilium.europa.eu/uedocs/cms_data/docs/pressdata/en/ec/106809.pdf

[7] Financial online magazine, http://www.wall-street.ro/articol/International/48609/Putin-Numarul-participantilor-la-proiectul-Nord-Stream-ar-putea-creste.html

The Galsi pipeline from Algeria to Italy via Sardinia is a joint initiative of Sonatrach, Enel, Edison, and several other partners. Unlike the Nord Stream, the Galsi project does not enjoy yet the support of one or two major gas importers, which could provide some guarantees for its future activity.

On 4 March 2010, the European Commission assigned €2.3 billion for investments in gas pipelines and power interconnectors aimed at upgrading the energy routes from Russia and to encourage the development of new ones from North Africa and the Caspian. The Nabucco pipeline project from the Caspian region will receive a contribution of €200 million towards its total cost of €7.9 billion, while the ITGI link between Italy and Greece gets €100 million. A further €175 million was allotted to a gas link between Spain and France that could have a considerable impact. Spanish energy companies have been pressing for that connection, which would allow Algerian gas to flow into mainland Europe soon after the 8 bcm Medgaz pipeline starts pumping in the summer of 2010.

In all, the European Commission would allocate €1.39 billion for 31 gas pipeline projects and €910 million for 12 electricity interconnection projects."I see this as a turning point in Europe's energy strategy," noted the current EU Energy Commissioner Guenther Oettinger. "Europe's energy and climate objectives require large and risky infrastructure investments. In the present economic situation, the risk is that businesses would have preferred to postpone such investments." The plan to subsidize hundreds of kilometers of gas pipelines and power cables was part of a €4 billion package aimed at helping the bloc member states assist each other during energy crises. €1 billion was already announced in December 2009 for pioneering technology to trap and bury carbon emissions from power stations and €565 million was earmarked for offshore wind farms.

TYPES OF RISKS IN PROMOTING INVESTMENTS

The barriers to investment in capital-intensive energy infrastructure projects can be related to their exposure to three different types of risk, namely market risk (uncertainty about price and volume), regulatory risk (impact of market rules and regulations) and political risk (uncertainty relating to international relations and involvement of several actors along transit routes. The perception of such risks influences the expectation of profitability and therefore the decision to invest.[8]

To promote investment in energy corridors the action of policy makers should in particular focus on reducing market risks, regulatory risks and political risks. The completion of the EU Internal Market is bound to help wholesalers and corridor developers hedge market risks: it would involve broader access to national markets and pipeline capacity by developing interconnections and homogenizing regulations. It would also remove the barriers to entry related to the excessive market power of incumbents and to the lack of transparency on infrastructure capacity usage and allocation. The creation of a liquid spot market and secondary financial markets should be encouraged in order to allow proper hedging of volume and price risks,

---

[8] *Energy Corridors. EU and Neighboring Countries*, European Commission, Directorate General for Research. Sustainable Energy Systems, http://ec.europa.eu/research/energy/pdf/energy_corridors_en.pdf

and to render long-term contracts more compatible with a competitive and integrated European market.

Concerning regulatory risk mitigation there is a need to clearly define the limits of the European internal gas market and improve coordination of projects across the borders in order to address the different regulation regimes, interests, risks, and expected revenue of an investment project for the countries involved.

In addition to the uncertainty related to European markets and regulation, the international dimension of gas trade increases its exposure to political risks, in particular institutional instability in producing and transit countries, conflicts between countries, and political fallout of the evolving international economic situation. Financial support to priority corridors will be needed also for political reasons. In case the investments that are considered important for the security of supply cannot be completed exclusively on the basis of commercial market considerations (e.g. lack of throughput guarantees), they should be supported by institutional loans (EIB, EBRD) or sovereign guarantees. Specialized international agencies can also play an important role. In this respect the EU would be well advised to support the further development of activities under the Energy Charter Treaty and other transnational institutions.

As recent events have proved, there is a need for enhanced dialog to improve international stability and build up confidence between importing, producing, and transit countries. The development of strategic partnerships between the EU and the major supplier and transit countries is also important.

NEW CHALLENGES TO ENERGY SECURITY

Romania, much like the other countries of the Central and East-European region, faces the very real prospect that investment in energy supply will not be sufficient to meet future demand.

To close the gap between supply and demand, the countries of the region will have to rethink their approach to energy supply. That (according to a World Bank report[9]) translates into doing five things: (1) build the capacity for reliable electricity and primary energy supply, attract the huge investment the region needs to achieve this—$1.8 trillion in primary energy and $1.5 trillion in electricity by 2030—by creating better market conditions and more reasonable tariff regimes; (2) deepen regional cooperation on energy development; (3) reduce the enormous waste on the production side, especially that associated with flared and vented gas; (4) undertake major energy efficiency measures at both the supply and the consumption ends; (5) address potential environmental concerns and minimize the carbon footprint of the new capacity to be added.

Investing in energy efficiency achieves three of these goals, simultaneously and at minimum cost: lower greenhouse gas emissions, better energy security, and more sustainable economic growth. Energy efficiency is thus a triple-win for governments, end users, market participants (public and private), and society in

---

[9] http://siteresources.worldbank.org/ECAEXT/Resources/258598-1268240913359/Full_report.pdf

general. An additional $1 invested in more efficient electrical equipment and appliances could avoid more than $2 in supply-side investment. Energy efficiency should therefore be considered as an energy resource, on a par with—and even preferred over—supply-side resources.

The total projected energy sector investment requirements for the region over the next 20–25 years are huge. They amount to about $3.3 trillion (in 2008 dollars), some 3% of accumulated regional GDP during that period.

Although the public sector will have to finance a portion of these investments, it will not have the capacity to meet full investment needs. The countries in the region will therefore need to call on the financial depth and technical know-how of private sector investors and energy companies. Although the current financial crisis is a serious impediment to private sector investment in any activities or countries seen as high risk, as the financial crisis passes the prospects for such investment will improve. However, in order to attract these investors, countries will need to create an enabling environment that provides secure ownership rights, is subject to the rule of law, fosters transparency, and enables reasonable risk mitigation. In addition, individual sectors will have to be viewed as financially and commercially viable. This will be particularly critical in those sectors, such as electricity and heating that are largely dependent on their domestic markets.

Realistically, however, in spite of strict deadlines to comply with environmental requirements, Romania will continue to encounter certain difficulties in reshaping its energy security approach due, among other things, to the high prices and global limitation in equipment production. Investments in local energy production capacities are also affected by expectations of a drop in consumption and the volatility of the price of oil, which influence the prices of other raw materials and products, and implicitly make business plans and access to financing increasingly difficult. Energy companies will be more careful with the money they allocate to project development because of the international crisis, considering that investments in this field take a very long time to recoup.

The global economic crisis seriously affected the energy sector, causing investments to plunge in a harsher financing environment. Particularly the oil and gas sector has been faced with a steady stream of announcements of cutbacks in capital spending and project completion delays.

In contrast with other fields in which the international, institutional, legislative, or investment context has been a favorable one allowing for rapid development during the past years, the energy security sector and climate change mitigation are now facing difficulties that were not encountered decades ago. The impact of the financial crisis has been more strongly felt in this area. The attempts to reset the worldwide priorities indicate an emerging awareness about these issues, but we hardly see, for the time being, a strong determination by the major global actors to seek viable solutions to the combined challenges of energy security and climate change.

Energy Security: A Key Element of Romanian National Security

GENERAL PROVISIONS OF THE NATIONAL SECURITY STRATEGY (NSS) OF ROMANIA

As a tool attaching weight and practical value to the requirements related to the protection and safe functioning of all the compartments of a national system, the National Security Strategy is an integrating concept; it becomes operational through an aggregate of plans, measures, and actions aimed at effectively preventing and countering the risks and threats that may jeopardize the national values as well as the values that lend identity and unity to the European construction. The goal of the Strategy is both to deal with the dangers brought about by the international environment and to guarantee the state of domestic security, personal safety of individuals, and security of the communities; the Strategy also addresses energy and food security, transportation and infrastructure security, and cultural and environmental security.

As fundamental policy documents for Romania, the successive versions of the National Security Strategy have provided an important framework over the past 20 years not only for national defense but also for generating specific policies in the vital areas of economic and social development. Immediately after the regime change of December 1989, all strategic documents included economic interests and objectives among their top priorities.

The national interests reflect the most important, stable, and institutionalized perception of the values that the Romanian nation aims to preserve, promote, protect, and defend by legitimate means, the values on which the Romanian nation builds its future, through which it guarantees its existence and identity, and for which it is integrated in the European and Euro-Atlantic community and takes part in the globalization process.

The Romanian National Security Strategy now, in force, acknowledges that, in the modern world, security and prosperity are inseparable terms of the same equation. Romania's progress, prosperity, and national security can only be the result of a complex process designed to promote and guarantee national and community interests. The range of those interests basically covers: the rights and responsibilities arising from Romania's membership of the European Union and the North-Atlantic Alliance; maintaining the integrity, unity, sovereignty, independence, and indivisibility of the Romanian state; developing a competitive, dynamic, and highly-performing market economy; radical modernization of the educational system and effectively turning to good account the human, scientific, and technological potential; increasing the well-being, living standards, and health conditions of the citizens; asserting and protecting national identity, culture, and spiritual life in the context of active participation in building up the new European identity.

The entire spectrum of risks and threats identified by Romanian National Security Strategy is relevant for the purpose of this paper. The assessment of the current security situation starts from the premise of Romania's recently acquired status as a member of NATO and EU. The analysis highlights a range of new, asymmetric threats of military or non-military nature, including information or intelligence threats stemming chiefly from the international and to a certain extent domestic environment, which may have a serious impact on the security of the Romanian citizens, the Romanian state or the organizations of which Romania is part. The danger posed by such negative developments may be substantially increased if they combine, considering especially that the dividing lines between global and regional threats and between foreign and domestic ones tend to become ever more blurred.

The main risks and threats to Romania's national security and to its values and interests as a member state of the EU and the Euro-Atlantic community are identified as follows:

International Terrorism: The September 11, 2001 terrorist attack was a turning point for international security. The subsequent similar strikes in Europe and other areas of the world proved that terrorism is a real and present threat. When structured in cross-border networks, international terrorism poses a most serious threat to human lives and freedom, democracy, civilization, way of life, and the other fundamental basic values that make the foundations of the demo-cratic community of Euro-Atlantic states. The impact of terrorist acts affects world trade and economic development, including energy security, distorts the business environment in certain regions and generates political instability. Among the main targets of terrorist attacks are infrastructure facilities, transportation routes, mari-time shipping lanes, ports, and terminals.

Cyber Threats and Their Impact on Energy Security: Bruce Averill and Eric A.M. Luiijf approached the connection between energy security and cyber defense in a more visionary view. The authors wrote about recent "highly sophisticated and targeted" cyber attacks on Google and Yahoo, which brought the issue of cyber security to the top of the agenda. Internet companies, however, are not the only ones that are vulnerable to such attacks. Over the past decade, a series of events highlighted the vulnerability of electric grids and other elements of energy infrastructures to both cyber disruptions (due, *inter alia*, to malware) and outside attacks using cyber methods. Development of "Smart" electric grids is a very promising technology, and its vulnerability to Cyber attacks need to be addressed.

Both the general public and official regulators are becoming increasingly aware of the vulnerability of energy facilities and infrastructure to cyber disrup-tions and attacks. The study stated that "a concerted and cooperative effort by academia, manufacturers, industry leaders, and policymakers is required to secure our energy systems (oil, gas, power, and district heating) against such disturb-ances".[10] In the absence of firm and positive actions by the energy industry and national regulatory systems, the international community may prove to be too slow to respond to rapidly evolving threats. Above all, the top management of energy companies needs to be aware of the risk and take appropriate action. Rather than being neglected or downplayed, incidents need to be reported, investigated, and acted upon in a cooperative international fashion just as near misses and crashes are in the aviation industry.

---

[10] Bruce Averill and Eric A.M. Luiijf, "Canvassing the Cyber Security Landscape: Why Energy Companies Need to Pay Attention", *Journal of Energy Security,* 18 May 2010.

One potentially useful approach would be to bring together CIOs (and possibly CEOs) from energy companies, government agencies, and network security providers to discuss in detail the actual incidents behind closed doors. The key is to convince participants that revealing weaknesses that they have overcome is an effective way to learn about potential threats before they happen. The goal would be to drive the creation of a cross-sector national plan to improve cyber security in each participating company or agency, thus strengthening the entire sector.

Proliferation of Weapons of Mass Destruction: Nuclear, chemical, biological, and radiological weapons constitute another very serious threat in view of their destructive capacity. The main concerns are the control of fissionable material, nuclear reactor safety, and access to dual use materials and technologies.

Regional Conflicts: Despite the positive developments that have made Europe a safer and more prosperous place, the strategic area where Romania is located is still fraught with local conflicts, impacting heavily on regional and European peace and security. Some inter-ethnic or religious conflicts were the consequence of the more or less violent dissolution of multinational states, but they also have strong political undertones. They still pose a serious threat to regional stability even though, due to important efforts by the international community, some of them have been contained so far. Such conflicts and other sources of tension, disputes, separatist trends, territorial differences, or local crises in the proximity of Romania generate uncertainty, waste resources, and perpetuate poverty. They also feed other forms of violence and criminality and may favor the spread of terrorism. As a borderland state of the EU and NATO, Romania has three problematic areas in its vicinity: the Western Balkans, Caucasus, and Eastern Europe (Moldova). All these areas are of strategic and economic significance for all existing and projected European southern energy supply corridors. The Strategy states: "The Black Sea region is Europe's most saturated and, probably, one of the densest parts in the world in terms of separatist conflicts, tense situations and disputes. The separatist conflicts in the eastern province of the Republic of Moldova (Transnistria), in the western and northern parts of Georgia (Abkhazia and South Ossetia), in the south of Azerbaijan (Nagorno-Karabakh), in the south of the Russian Federation (Chechnya and other republics or autonomous regions in the North Caucasus), other smaller and less intense separatist movements as well as tensions related to disputes over territory or borders pose serious threats to the security of the region and run the danger that violent confrontations may reignite".[11]

Cross-Border Organized Crime: International criminality is a developing global threat that can affect state policies and the performance of democratic institutions. It is both an expression of the proliferation of negative phenomena that swell in the wake of globalization and a direct consequence of the mismanagement of the profound political, economic, and social changes that occurred in Central, Eastern, and South-Eastern Europe after the dissolution of the communist regimes.

---

[11] ***, The National Security Strategy of Romania, 2007, www.mapn.ro, p. 33.

The Quality of Governance: It is a well established fact that deficient govern-ance undermines the citizens' confidence in democracy and public institutions and can pose a major threat to the security of states. In most cases it is a consequence of democratic deficit and institutional corruption, which translate into political patronage and cronyism, ineffectiveness of public administration, authoritarian tendencies and a lack of transparency and public accountability. In Romania's area of strategic interest, ineffective governance often endangers the normal exercise of fundamental human rights and affected the fulfilment of some international obligations – as well as the obligations pertaining to safeguarding national identity – bringing about the risk of humanitarian crises with cross-border impact.

Geophysical Threats and Natural Disasters: National security can also be jeopardized by the effects of climate change, earthquakes, or extreme weather events such as cold spells or heat waves, inundations, environmental degradation, air, water, or soil pollution, some of them caused by dangerous, harmful or irres-ponsible human activities. One can add to these natural or man-made catastrophes, the gradual but irreversible depletion of vital resources, and the increasing possibility of pandemics. For some time there has been a lively debate in Romania on the environmental impact of the envisaged regional energy and transport infrastructure projects affecting the Danube Delta and the Black Sea. "The eco-nomic dimension will play a key role in the process of Black Sea cooperation. Romania will promote a pragmatic foreign and economic cooperation policy using the tools provided by the bilateral and multilateral formats in order to take part in the economic projects that have a substantial impact on consolidating democracy, security and stability. The efforts will be mostly aimed at developing energy and transport corridors that can link, in economic and commercial terms, the Black Sea region to the Euro-Atlantic community and allow for a better use of the potential offered by the sea and river ports in the region, while also protecting the environment and restoring the multiple functionality of the Danube-Black Sea eco-system".[12]

The risks and threats to national security can be further aggravated by other vulnerabilities such as excessive dependence on imported vital resources that are hard to access; persistent negative demographic trends and massive outward migration; insufficient provision of social security; chronic poverty and widening discrepancies in terms of disposable income; the slow emergence, fragmentation, and poor involvement of the middle class in the country's socio-economic affairs; the frailty of civic spirit and solidarity; the poor condition and low effectiveness of the healthcare system; organizational flaws and lack of proper resources for the educational system to meet the requirements of the labor market; inadequate crisis management structures and resources; a weak commitment of the civil society to the debate on national security issues. The problem of dependence on imported vital resources was first mentioned in the 2007 version of the National Security Strategy: before that it was regarded as a matter of access. This notion was further developed as a subchapter of Romania's Energy Security Strategy 2007.[13]

---

[12] ***, The National Security Strategy of Romania, 2007, www.mapn.ro, p. 36
[13] ***, Romania's Energy Security Strategy 2007, www.minind.ro

In order to cope with these risks and threats, the National Security Strategy of Romania set the following lines of action:

- Participate actively in building international security;
- Promote Romania's new European and Euro-Atlantic identity;
- Achieve regional security in the context of a new political paradigm;
- Take the role of a dynamic vector of security in the Black Sea area;
- Approach comprehensively and properly the homeland security issues;
- Consider good governance as an essential tool in the process of building national security;
- Build a competitive and highly-performing economy as a solid pillar of national security;
- Transform the central agencies having national security responsibilities;
- Develop and ensure active protection of strategic infrastructure.

Romania's status as a member of the European and Euro-Atlantic communities has brought about new responsibilities and requires enhanced political, organizational, and financial efforts on the part of public authorities and security institutions. NATO membership has given Romania not only certainties for its security, real defence guarantees, and access to strategic decision-making but also corresponding obligations.

Even before its accession to NATO, Romania was an active promoter of energy security in the South-East European and the Black Sea regions, including regional cooperation in military and civilian areas and promotion of Euro-Atlantic values and security culture. Such concerns were duly reflected in basic NATO documents. The Riga NATO Summit Declaration supported "a coordinated, international effort to assess risks to energy infrastructures and to promote energy infrastructure security." It further directed the Council in Permanent Session to consult on the most immediate risks in the field of energy security in order to define those areas where NATO may add value to safeguard the security interests of the Allies and, upon request, assist national and international efforts.[14]

Romania contributed actively to the process of shaping the new NATO Strategic Concept by supporting in particular the inclusion of three new dimensions: energy security, cyber-defence, defense against ballistic missiles. The link that the new Strategic Concept is seeking to establish between the dimensions of regional security and energy security was aptly emphasized by Bogdan Aurescu, State Secretary at the Romanian Ministry of Foreign Affairs: "First, the reference to Strategic South indicates a complex region which encompasses the Black Sea and the Caucasus. Second, this region has a considerable energy potential, since it brings together energy producer, transit and consumer countries. And third, energy security is a common challenge to all of us, and NATO should address this issue within its own means and capabilities. [...] This is Romania's perspective on

---

[14] ***, Riga Summit Declaration, 29 November 2006, http://www.nato.int/cps/en/natolive/official_texts_37920.htm?selectedLocale=en

NATO's role in energy security and the New Strategic Concept".[15] It is to be expected that Romania, alongside other new NATO members, will be among the main supporters of new missions for the Alliance in the field of the energy security.

Taking into consideration the importance of the economic dimension of the EU enlargement, Romania has consistently advocated the need to give special attention to the Black Sea and Caspian regions. Lying at the crossroads of three areas of paramount importance – Europe, the Middle East, and Central Asia – the Black Sea region is a main transit area for energy resources and, at the same time, an important source of asymmetric risks and hotbeds of conflict. Far from being considered a mere buffer or peripheral area, the Black Sea region is a connector of strategic importance, linking the Euro-Atlantic community (as a security provider and energy consumer) to the Middle East area, the Caspian, and Central Asia (as energy providers and security consumers). From the energy point of view, the Black Sea region is not only the nexus of the main transit routes but may also become a significant producer in the following decade.

It is a priority for Romania to harmonize and render more efficient the institutional processes of regional cooperation for development and to establish a new framework of dialogue involving both state actors and interested democratic organisations. For this purpose Romania actively promoted the idea of a Black Sea Forum for Dialogue and Cooperation, an initiative chiefly aimed at championing democracy and economic development, energy security, confidence building, regional stability, peace, and security.[16]

In addition to its efforts at a regional level, Romania also initiated concrete steps in order to highlight the importance of energy security in the context of policy debates in the framework of the EU and its institutions. Romania has supported vital European projects aimed at diversifying energy sources and routes and reducing one-source dependency such as Nabucco pipeline for gas and the Pan-European Oil Pipeline from the Black Sea port of Constanta to the Adriatic. Besides its advocacy in Brussels in favor of the Caspian region, Romania added substance through concrete projects: the comprehensive strategic partnership with Azerbaijan which includes energy sector cooperation and the trilateral memorandum of understanding with Azerbaijan and Georgia (AGRI) in the field of LNG. In response to EU priority objectives concerning the interconnection of national electric power infrastructures, Romania has developed its own national program for regional connectivity.

The National Security Strategy of Romania also emphasizes the importance of attracting strategic investments for sustained economic development that generates jobs and produces resources for social protection. Directly related to this, the Strategy stresses the close relationship of energy security to operational adjustment and optimization of the structure of primary energy resource consumption and improved energy efficiency: "For this reason, it will be a priority to take

---

[15] Aurescu, Bogdan, *NATO's Role in Energy Security and the New Strategic Concept*, speech at the EAPC seminar on Energy Security and Critical Infrastructure Protection, Tbilisi, 5 May .2010, http://www.mae.ro/index.php?unde=doc&id=15414&idlnk=&cat=4&lang=en

[16] ***, The National Security Strategy of Romania, 2007, www.mapn.ro, pp. 32–34.

action towards reducing dependency on the supplies from unstable regions or states that use energy as a lever for political pressure; speeding up the programmes to produce nuclear energy; increasing the production of hydro-energy and energy produced from coal using modern, clean technologies. An essential role is also assigned to energy produced from renewable or alternative sources and enhanced energy efficiency at the level of industrial consumers and households. The measures in this field will be adjusted to the European Union's strategies regarding energy security".[17]

At the same level of importance the Strategy highlights the special significance of enhancing the protection of critical infrastructures: "Action will be taken in order to modernize and adjust the legislation and institutions in accordance with the real dynamics of the process, the demands imposed by the new risks and threats and the requirements resulting from cooperation with NATO, the European Union and strategic partners [...] The main fields of interest are transport infrastructure, especially the construction of new highways, modernization of railway infrastructure and traffic management; energy and communications infrastructure; hydrographical basin management and protection against floods; the upgrading of the military bases system. The main directions for the adjustment and modernization of infrastructure will be: to rehabilitate, modernize and develop the transport infrastructure in order to improve the quality of the services, render passenger and goods transport more efficient and adjust it to the European transport system; to develop, secure and build alternative routes for new energy supply networks in agreement with the European Union's strategic projects and Romania's interests in the Black Sea area; to promote new technologies and develop the integrated IT system in the field of security in conformity with the European standards; to adapt the national system of aeronautical and port security to the new types of threats; to modernize the specific military infrastructure and also to prepare and upgrade the capabilities offered within NATO, the European Union and the strategic partnerships; to identify high risk areas, draw up maps of probable risks and consolidate the infrastructure for the defence of the population against natural calamities, disasters, and environmental accidents; to diminish the risk of nuclear incidents or accidents and to provide the security of nuclear infrastructure".[18]

It results that, in political and strategic terms, the National Security Strategy of Romania takes a comprehensive view of energy security at the national and European level in a way that is synchronized with the major trends and driving forces. The Strategy covers a broad range of goals and priorities providing a coherent vision from the perspective and within the context of the time when it was adopted in 2007. Given the complexity and the dynamics in the field of energy security, it is expected that the new Energy Strategy of Romania, which is currently being prepared, would be updated and restructured in accordance with the new trends and foreseeable developments in the field.

---

[17] ***, The National Security Strategy of Romania, 2007, www.mapn.ro, p. 46
[18] ***, The National Security Strategy of Romania, 2007, www.mapn.ro, p. 52

THE ROMANIAN ENERGY STRATEGY 2007

In 2007, the Romanian Government endorsed the Energy Strategy 2007–2020[19] covering all current and projected activities of public and private entities in that sector. The Ministry of Economy and Finance, together with the Ministry for Environment and Sustainable Development, took responsibility for drafting the first national Energy Strategy after Romania's accession to the EU. Although the document was conceived only 3 years ago with a long-term view to the year 2020, the current Romanian authorities (as of June 2010) announced that a new, improved Strategy had become necessary. According to the incumbent Romanian Minister of Economy, Adriean Videanu, "Romania, has an energy strategy for 2007–2020, which was developed immediately after Romania's accession to the EU. That strategy took into consideration the realities prevailing at that moment, such as an economic growth of even 7%. Nobody estimated that the financial crisis will have such an important impact. Nowadays, the realities are different, and the prognosis of rising energy consumption, which was the basis of the existing strategy, was not confirmed. Under these circumstances it becomes utterly important to review the existing energy strategy of Romania for a longer period, until 2025, a document which will cover the new realities. Probably before the end of the year (2010) we will have a draft of the new strategy".[20]

The main reason why it was felt that a fundamental revision was required was due to the fact that the existing Energy Strategy had been designed in accordance with the data that were available in 2007. Since then major changes have occurred and new challenges have emerged in the European energy scene. Starting with the Russian-Ukrainian gas crisis of 2006 and considering its effect on the European security of supply, all European actors proceeded with a thorough review of their priorities in terms of energy security. The energy dialog acquired a deeper political connotation for all the actors involved in this process. Even before the onset of the financial and economic crisis, certain differences emerged between some member states of the European Union that had established their own strategic partnerships with the main energy suppliers (chiefly Russia) and other EU members which did not have such special arrangements. The current crisis revealed serious vulnerabilities in the energy sector both within the EU and in its relations with other international actors.

In the case of Romania, which experienced the effects of the global crisis with a certain delay and is still in recession at the time of writing (June 2010), the need to reconsider some of the basic premises and resulting conclusions of the Energy Strategy prepared in 2007 is, therefore, warranted and opportune.

The main provisions of the 2007 Strategy covered medium and long-term objectives, described the international context at that time and the state of play in the Romanian energy sector, including projections for an adequate energy mix, and formulated conclusions and policy recommendations. The principal thrust of

---

[19] ***, Hotarare privind aprobarea strategiei energetice a Romaniei pentru perioada 2007-2020, Monitorul official al Romaniei, partea I, Nr. 781/19.XI.2007.

[20] ***, "Romania pregateste o noua strategie energetica", 1 februarie 2010, *Focus Energetic*, http://focus-energetic.ro/?p=334

the strategic design was to make sure that the Romanian energy sector should be able to meet the energy needs of the country in the short, medium and long term, at affordable prices and in compliance with the requirements of a modern market economy.

Strategic Objectives

The Energy Strategy of Romania for 2007–2020 spelled out strategic objectives in three main areas: Energy Security, Sustainable Development, and Competitiveness.
    In the area of energy security:

- Enhancing energy security by ensuring the supply of required energy resources and reducing dependence on imports;
- Diversification of sources of supply and transport routes;
- Increasing the compatibility of national power, gas and oil networks;
- Protection of critical infrastructures.

In the area of sustainable development:

- Increasing energy efficiency;
- Promoting energy production from renewable sources;
- Promoting highly efficient cogeneration capacities;
- Supporting research and development activities;
- Reducing the negative impact of the energy sector on the environment;
- Rational and efficient use of primary energy resources.

In the area of competitiveness:

- Development of an open energy market for electricity, natural gas, oil, uranium, green certificates, CO_2 certificates, and energy services.
- Ensuring the free transit of energy and free, permanent, and non-discriminatory access of market participants to energy resources, transport and distribution networks, and international interconnections.
- Continuation of the restructuring and privatization of power generation, district heating, and natural gas facilities.
- Continuation of the restructuring process in the coal sector with an aim to increasing profitability and getting access to capital markets.

International Context

Relying on the data available in 2007, the Energy Strategy described the prevailing global trends in energy production and consumption and highlighted the main objectives of the agreed EU energy policy.

(a) Main Energy Sector Trends: Evolutions and Challenges

By 2007, the estimated total global demand for energy in 2030 was about 50% higher than in 2003. The increase in oil demand was deemed to be almost 46%. Forecasts indicated an economic growth that implied an increased consumption of energy resources. The International Energy Agency reported that the next decade would witness a dynamic increase of renewable energy sources and natural gas in the overall structure of primary energy consumption. Approximately a quarter of the demand for primary energy resources would continue to be covered by coal.

The Strategy also took into consideration the geopolitical factors at play in the Middle East. During the first decade of the twenty-first century, the increase of oil prices was reflected in a corresponding hike of the prices for natural gas. Another factor that influenced the price for oil products was a shortage of refining facilities. Adequate technologies were needed to process primary energy resources that had varying levels of quality and were located in different geographic areas. In addition to these factors there was a tendency to build up stocks for emergency situations. With the advent of the economic crisis, the tendency to resort to resources protectionism also intensified.

Another geopolitical aspect that influenced energy policies was the inclination of importer states to be increasingly involved in upstream activities overseas. For instance, this applied to Austria in Central Asia, British Petroleum in the Caucasus, France and Germany in the Russian Federation, etc. At the same time, major producers tended to become more actively involved in downstream operations, including infrastructure and distribution systems. This applied to the acquisitions of Gazprom in a number of European states, Socar (Azerbaijan) in Romania, Kazmunaigaz (Kazakhstan) in other parts of Europe.

The implications of the Ukrainian gas crisis highlighted the increasing role of the transit states as a link between producers and consumers that shapes the configuration of the global energy network. Especially in Europe, the transit states (mostly former communist countries) are also consumers with a high single-source dependency rate.

(b) The Relevance of the EU Energy Policy for Romania

The Romanian Energy Strategy for 2007–2020 noted that the European Union was the second largest energy consumer in the world. The level of import dependency was expected to rise from 57% in 2007 to 84% in 2030 for natural gas and from 82% in 2007 to 93% in 2030 for oil. Security of supply was therefore regarded as an important factor for the sustainable development of the European Union. The relationship between energy policies and climate change was also emphasized.

Although the Romanian Energy Strategy of 2007 contained several references to the EU directives relative to energy and climate change, few of the actual provisions of that document were correlated with the EU objectives and the resulting obligations of Romania as a member state.

(c) Renewable Energy as a Strategic Objective

In 1997, the European Commission proposed that the EU should aim to reach a 12% share of renewable energy by 2010. The directives that were adopted for the electricity and transport sectors set specific national sectoral targets. In 2006, the share of renewables went up to 7% of the EU gross energy consumption. However, interim progress reports indicated that the EU was unlikely to reach either the electricity or transport target for 2010. More recently, the EU agreed on a revised target for renewables of 20% by 2020. The national targets are given below[21]:

In March 2007, the EU heads of state and government endorsed the first EU energy action plan and called on the European Commission to prepare a new action plan for the post-2010 period.

Member state	Share of renewables in 2005 (%)	Share required by 2020 (%)
Austria	23.30	34
Belgium	2.20	13
Bulgaria	9.40	16
Cyprus	2.90	13
Czech Republic	6.10	13
Denmark	17	30
Estonia	18	25
Finland	28.50	38
France	10.30	23
Germany	5.80	18
Greece	6.90	18
Hungary	4.30	13
Ireland	3.10	16
Italy	5.20	17
Latvia	32.60	40
Lithuania	15	23
Luxembourg	0.90	11
Malta	0	10
The Netherlands	2.40	14
Poland	7.20	15
Portugal	20.50	31
Romania	17.80	24
Slovak Republic	6.70	14

---

[21] ***, European Commission, *Renewable energy targets*, http://ec.europa.eu/energy/renewables/targets_en.htm

Originally intended to take effect this year, it has already been delayed and the EU executive is now expected to adopt the plan at the beginning of 2011.[22]

Some offshoots of the initial action plan have included far-reaching energy liberalization proposals, the climate and energy package, and the Strategic Energy Technology Plan (SET Plan).

The 'Europe 2020' strategy proposal, presented by the Commission in March 2010, incorporated the 2020 climate goals in its flagship initiative to promote a resource-efficient Europe. The EU energy ministers conducted, on 31 May 2010, an early examination of the forthcoming EU Energy Strategy for 2011–2020 and agreed that it should be ready for endorsement by an EU Summit in March 2011. The Energy Strategy had been originally intended for the 2010–2014 timeframe. The EU Executive decided to extend the horizon to 2020 in order to align it with the 'Europe 2020' strategy and to provide a stable framework for long-term energy investments,

It is envisaged that the implementation of the third internal energy market package and the Strategic Energy Technology (SET) Plan should be the key priorities of the new EU Energy Strategy. Some of the objectives that are likely to be included in the new EU Energy Strategy are: reducing greenhouse gas emissions by 20% compared to 1990 levels; increasing the share of renewables in final energy consumption to 20%; and moving towards a 20% increase in energy efficiency. The EU is committed to move to a 30% reduction of emissions by 2020, provided that other developed countries commit themselves to comparable reductions and that developing countries contribute adequately according to their responsibilities and respective capabilities.

It is to be expected that, in early 2011, the European Union may make major decisions on energy policy including ways to support the shift towards an efficient low-carbon economy and greater security of supply.

Other Perspectives on Energy Security of Romania

Directions Regarding Romania's Energy Security, a study[23] published in 2008 by the European Institute of Romania (EIR) noted: "As far as Romania is concerned, the country is for the first time faced with a situation in which both its geographic location and its geopolitical position provide opportunities that should be fully capitalized upon. The geopolitical and geo-economic context in the Black Sea region has greatly changed over the past few years due to a shift in the interests of the major powers towards the Pontic-Caspian space, in Romania's proximity. Thus, the country has a potential to become a "gateway" to the economic and political regional bloc of the European Union and a transit space between the Central Asian producers and the large European consumers, taking into account

---

[22] "Commission to outline new energy action plan", http://www.euractiv.com/en/energy/commission-outline-new-energy-action-plan-news-418551

[23] Silviu Negut (coordinator), Aureliu Leca, "Directions regarding Romania's energy security", in *Strategy and Policy Studies (SPOS 2008)*, Study No. 2, European Institute of Romania, Bucharest, 2008, www.ier.ro

Romania's proximity to the deposits of energy resources and its considerable energy infrastructure (oil pipelines network, the largest refinery in the Black Sea basin – Midia Năvodari, the size of the petrochemical industry, the permissive topographic layout which doesn't pose particular economic and technical difficulties for the construction of new pipelines or pumping stations, the largest harbor in the Pontic Basin – Constanta, and so on)."

The study concluded with the following recommendations:

- A review of the energy strategy and policies is imperative and should be centered, especially, on the security of energy supply;
- The aim is to reduce Romania's dependence on imported energy resources; in the case of natural gas the aim is to reduce dependency on a single supplier and to diversify the sources of supply;
- The interconnection of the national electric power system should be with at least one of the other European countries in order to avoid the risk of power failures;
- Increased attention must be paid to the decision-making process in the national energy system taking into account its inertia (the time lag between the decision and the practical realization being approximately 4–20 years);
- To stimulate investments in order to update and streamline the energy sector (in all consumption sectors power losses are very high: 30–35%) and also to reduce the negative impact on the environment;
- To create a specific institutional frame for the energy sector (for example, a Ministry of Energy and Resources) because the dispersion of activities related to the energy sector among various departments in different ministries is no longer a viable option.

Regarding renewable sources of energy, the study put forward the following additional considerations:

- The share of renewable energy in Romania's balance of primary energy consumption is the largest in Central and Eastern Europe. Romania has a relatively important and economically usable potential for hydro energy and biomass (from agriculture and forestry, which have been practically ignored).
- The potential of other sources such as solar, wind, and geothermal energy may have been overrated. However, wind energy seems to receive increasing attention both in the EU and in Romania.
- With the exception of bio-fuel, few specific incentives have been provided in Romania for the development of renewable energy sources.
- By analyzing the structure of the national energy balance, the study produced the following findings:
- Romania is a country that is rich (in terms of a wide range) in poor (in terms of small amounts) and expensive (in terms of extraction costs) resources, accounting for a short span of assured production related to current consumption rates (14–15 years for oil, 15 years for natural gas, and 40–50 years for coal);

- Coal remains the main fuel for producing electric energy in thermo-power stations (with a contribution of at least 50% for a long period of time, the rest being covered by hydro, nuclear, and renewable energy sources). The fact that, in a medium term, the coal reserves will be depleted, corroborated with their poor quality, means that Romania will also have to import coal;
- Romania's balance of primary resources is based on hydrocarbons to the tune of 60–65% because the domestic resources are being rapidly depleted. The dependence on imported primary resources (mostly hydrocarbons) rose from 27.1% in 2000 to about 36% in 2005;
- The viability of the nuclear power plant at Cernavodă to function at its full capacity must be realistically reassessed (considering the specific risks associated with this type of power plant, the low quality, and poor reserves of domestic uranium resources but also the advantages to be gained from rendering two other reactors operational).

A 2020 Vision for the Black Sea Region – A Report by the Commission on the Black Sea[24] produced a comprehensive review of the regional issues and a set of conclusions and recommendations for the way forward. Conveying a sense of urgency about the need for a new conceptual approach to the Black Sea regionalism, the study stressed: "It is the Commission's conviction that it is realistic to envisage a cohesive, developed, integrated and stable region so long as we take action now."

Some of the evaluations and objectives contained in that study may also be relevant for the further work to prepare a renewed Energy Security Strategy for Romania:

- The regional actors must renounce the use of force in their political relations and respect each other's territorial integrity, the inviolability of their borders, international treaties, and the rule of law in their dealings.
- Interested outsiders must support efforts to secure good governance, the creation of interdependencies, and the regionalization of the Black Sea's politics and economy.
- The international community must encourage cooperative efforts and confidence building measures as well as action in favor of the peaceful resolution of disputes.

The Commission on the Black Sea put forth a set of specific policy recommendations with an emphasis on those economic issues that are likely to respond to the common challenges and real needs of the region:

- Promote the principles of sustainable development as the guiding philosophy of regional cooperation in the Black Sea area.

---

[24] *A 2020 Vision for the Black Sea Region* – A Report by the Commission on the Black Sea, www.blackseacom.eu

- In the context of the global financial and economic crisis that has severely affected most countries in the Black Sea region, it is essential to supplement the immediate mitigation measures with a realistic vision for the future.
- The importance to promote cross-border infrastructure and institutional linkages. These could include policy coordination and harmonization, cross-country regulation, enhanced information sharing in order to stimulate growth, and avoid overlapping activities.
- Creation of an early warning systems or reciprocal assistance mechanisms in order to reduce the vulnerability of countries to crises in the future.
- Policy measures to improve the business environment and to facilitate greater economic activity across borders.
- Regular policy dialogues between relevant officials concerned with finance, transport, energy, environment, fisheries, and so on.

ENERGY MIX: TRADITIONAL VERSUS RENEWABLE ENERGY SOURCES

Romania has a relatively well-balanced energy mix relying, in part, on domestic primary energy resources (coal, oil, natural gas, uranium, hydro, biomass). There is also a considerable potential for the development of renewable resources (wind, solar, geothermal).

Romania's fossil fuels reserves are limited; domestic production has been in steady, though gradual decline for years and few additional deposits have been found despite sustained exploration efforts. In early 2010, the oil reserves were estimated to be around 64 million tons (mt). Compared to the crude oil production of 14.7 mt in 1976, in 2006 Romania produced only 5 mt of its own oil. The estimation was based on two assumptions: first, the oil production would decrease at an annual rate of 2–4% due to the depletion of known reserves; second, the renewal of exploitable reserves would amount to only 15–20%. By the year of 2020 the estimated oil reserves would be about 28 mt.

The reserves of natural gas are estimated to be 141 bcm and actual domestic production has also been declining. Because of the further depletion of known reserves, the gas production is expected to decrease at an annual rate to 2–5%, and the renewal of exploitable reserves will cover only 15–30%. By the year of 2020, the estimated gas reserves will amount to 77 bcm.

The resources of coal are estimated at 2,245 million tons (mt), of which some 550 mt can be economically exploited in the existing concession areas.

The existing exploitable uranium reserves can cover the needs for two nuclear power units until 2017. Potential new discoveries of uranium deposits are not likely to significantly alter the existing situation.

Table 2. Statistic data for Romanian Energy Sector 1990-2007

	1990	1991	1992	1993	1994	1995	1996	1997	1998	1999	2000	2001	2002	2003	2004	2005	2006	2007
Production	41.40	36.85	33.99	33.64	31.94	32.60	33.09	31.77	29.33	28.12	28.80	27.87	29.53	29.66	27.72	28.17	28.18	29.90
Solid fuels	7.94	7.60	7.35	7.62	7.78	7.89	8.05	6.60	5.57	4.65	5.88	5.70	6.10	6.54	6.21	5.79	6.48	6.86
Oil	7.95	6.82	6.87	7.00	7.06	7.11	7.04	6.97	6.80	6.52	6.42	6.34	7.55	7.28	6.02	6.17	5.78	5.00
Gas	22.91	20.06	17.61	16.76	14.82	14.45	13.76	11.91	11.15	11.19	10.97	10.78	10.60	12.45	10.38	9.70	9.56	9.23
Nuclear							0.36	1.39	1.37	1.34	1.41	1.40	1.42	1.27	1.43	1.43	1.45	1.99
Renewables	2.61	2.37	2.15	2.26	2.27	2.80	3.85	4.87	4.64	4.40	4.04	3.42	3.75	4.06	4.59	4.98	4.83	4.72
Industrial Waste						0.36	0.04	0.03	0.03	0.03	0.09	0.22	0.11	0.09	0.09	0.09	0.08	0.11
Net Imports	22.61	15.59	13.94	12.50	11.30	14.53	14.94	14.80	11.85	7.97	8.13	9.51	9.16	10.24	12.00	10.84	11.90	12.82
Solid fuels	4.68	3.02	4.02	2.18	2.71	2.86	2.68	3.46	2.57	1.68	1.92	2.30	2.78	2.64	3.09	2.94	2.72	3.54
Oil	11.19	7.58	5.97	6.55	4.79	6.85	6.54	7.29	5.41	3.83	3.56	5.01	3.50	3.54	4.90	3.96	4.76	5.52
Gas	5.93	3.74	3.58	3.62	3.74	4.79	5.65	4.03	3.84	2.54	2.71	2.31	3.13	4.24	4.11	4.19	4.79	3.87
Electricity	0.81	0.61	0.36	0.16	0.06	0.03	0.07	0.02	0.04	-0.07	-0.06	-0.11	-0.25	-0.18	-0.10	-0.25	-0.37	-0.18
Renewables																		0.07
Gross Inland Consumption	63.74	52.87	46.92	46.25	43.62	47.10	48.21	45.45	41.52	36.91	37.13	36.90	38.49	40.23	39.60	39.25	40.73	40.08
Solid fuels	12.37	10.91	10.80	10.12	10.42	10.78	10.71	9.77	8.23	6.86	7.77	7.64	8.82	9.36	9.20	8.80	9.52	10.19
Oil	19.11	15.18	12.42	13.33	12.30	13.89	13.76	13.43	12.26	10.62	10.19	11.14	11.01	10.97	10.47	10.30	10.63	10.25
Gas	28.84	23.80	21.19	20.38	18.56	19.24	19.42	15.94	14.98	13.73	13.68	13.18	13.62	14.73	13.94	13.94	14.44	12.98
Nuclear							0.36	1.39	1.37	1.34	1.41	1.40	1.42	1.27	1.43	1.43	1.45	1.99
Renewables	2.61	2.37	2.15	2.26	2.27	2.80	3.85	4.87	4.64	4.40	4.04	3.42	3.75	4.00	4.57	4.94	4.78	4.75
Other	0.81	0.61	0.36	0.16	0.06	0.39	0.11	0.05	0.04	-0.04	0.04	0.11	-0.13	-0.09	-0.01	-0.16	-0.29	-0.07
Elec. Generation TWh	64.31	56.91	54.20	55.48	55.14	59.27	61.35	57.15	53.50	50.71	51.93	53.87	54.94	56.65	56.48	59.41	62.70	61.67
Coal TWh	22.54	19.95	19.00	10.09	19.63	20.59	20.47	16.86	14.49	14.68	18.93	19.69	20.31	23.34	21.47	21.92	25.15	25.10
Oil TWh	6.40	6.67	5.40	5.43	5.79	5.80	6.70	6.86	4.14	3.86	3.40	5.41	3.58	3.63	2.20	1.89	1.61	1.10
Gas TWH	21.38	18.92	18.02	18.11	16.68	16.18	17.04	10.50	10.67	8.68	9.38	9.40	9.48	11.50	10.75	9.83	11.95	11.77
Nuclear TWH							1.39	5.40	5.31	5.20	5.46	5.45	5.31	4.91	5.55	5.56	5.63	7.71
Renewables TWH	13.88	12.29	11.79	12.84	13.05	16.70	15.76	17.52	18.89	18.29	14.78	14.92	16.05	13.26	16.52	20.21	18.36	16.01
Other TWH	0.10	0.09																

Table 2. Statistic data for Romanian Energy Sector 1990-2007 - continued

	1990	1991	1992	1993	1994	1995	1996	1997	1998	1999	2000	2001	2002	2003	2004	2005	2006	2007
Final energy Consumption	37.05	30.12	24.09	22.38	25.08	26.79	29.66	28.74	26.16	22.43	22.52	23.04	23.09	24.20	25.50	24.68	24.77	24.02
by fuel/product																		
Solid fuels	3.22	2.61	2.03	1.89	1.55	1.70	1.74	1.81	1.53	1.08	1.05	1.04	1.23	1.55	1.58	1.65	1.57	1.48
Oil	8.03	7.55	6.24	4.95	5.55	5.56	6.69	7.07	6.51	4.89	5.30	6.17	6.38	6.00	7.39	6.59	6.34	6.87
Gas	20.50	15.48	5.77	6.11	9.08	10.25	9.73	8.38	6.99	6.76	6.89	7.22	7.40	8.11	7.88	7.72	8.25	7.04
Electricity	4.69	3.80	3.55	3.14	2.94	3.13	3.42	3.30	3.14	2.91	2.92	3.12	3.06	3.22	3.33	3.34	3.52	3.52
Renewables	4.69	3.80	3.55	3.14	2.94	3.13	3.42	3.30	3.14	2.91	2.92	3.12	3.06	3.22	3.33	3.34	3.52	3.52
Derrived Heat & Industrial Waste			5.92	5.45	4.83	4.88	5.61	4.89	5.01	3.97	3.63	3.37	2.76	2.48	2.26	2.20	2.04	1.83
by sector																		
Industry	25.30	18.79	11.04	10.07	13.54	15.03	14.88	12.93	10.71	8.94	9.12	9.70	10.36	10.36	10.17	9.99	9.56	9.13
Transport	4.41	3.77	3.91	3.19	3.24	3.06	4.05	4.15	3.89	3.14	3.38	4.07	4.34	4.34	5.18	4.20	4.34	4.66
Households	4.28	4.48	6.28	6.61	6.40	6.35	8.11	9.65	9.52	8.74	8.43	7.28	7.22	7.82	7.97	7.96	7.84	7.51
Agriculture	2.24	2.01	1.41	1.14	0.80	1.00	0.84	0.91	0.77	0.46	0.39	0.29	0.27	0.24	0.23	0.21	0.26	0.26
Services, etc	0.81	1.07	1.45	1.37	1.10	1.36	1.79	1.11	1.27	1.15	1.19	1.70	0.89	1.44	1.96	2.31	2.76	2.46
Non-Energy Uses	0.92	0.82	2.90	2.20	1.48	1.34	1.77	1.96	1.60	1.85	2.04	2.09	2.03	1.91	2.22	2.47	2.33	2.55
CO2 emissions (mt)	173	134	129	128	125	130	136	122	108	92	96	101	107	112	113	106	112	112
Energy Intensity	1320	1260	1225	1189	1079	1088	1071	1074	1031	928	913	869	853	847	768	731	703	652
CO2 Intensity	2.72	2.53	2.74	2.76	2.86	2.76	2.82	2.69	2.61	2.50	2.58	2.73	2.77	2.78	2.84	2.71	2.74	2.78
Import Dependency %	35.50	28.30	29.70	27.00	25.90	30.90	31.00	32.60	28.50	21.60	21.90	25.80	23.80	25.40	30.30	27.60	29.20	32.00
Energy per capita (kgoe/cap)	2747	2299	2059	2032	1919	2077	2131	2015	1845	1643	1654	1667	1766	1851	1826	1814	1887	1869
CO2 per capita (kg/cap)	464	5817	5648	5611	5483	5738	6018	5414	4812	4106	4268	4545	4892	5141	5191	4917	5172	5176

The hydro-power potential of Romania is 36,000 GWh/year, of which some 30,000 G Wh/year can actually be used in terms of economic efficiency according to the current market prices for energy.

Given the geographical configuration of Romania, the Energy Strategy for 2007–2013 identifies the following potential for renewable sources of energy by regions: Danube Delta for solar; Dobrogea for solar and wind; Moldova for micro-hidro, wind, and biomass; the Carpathian Mountains for biomass and micro-hidro; the Transylvanian plateau for micro-hidro; the West plain for geothermal; the Sub-Carpathian foothills for biomass and micro-hidro; the South flatlands for biomass, geothermal and solar energy.

The energy mix analysis reveals that, in addition to the domestic production of oil, gas and low-grade coal, Romania also has to import significant amounts of fossil fuels. The share of renewable resources in primary energy consumption and power generation is above the EU average. The contribution from nuclear energy is projected to grow. Coal is still the most important fuel for electric power generation. While the final energy consumption has decreased considerably since the early 1990s, the manufacturing industries remain the largest consumers of energy. In terms of energy intensity per unit of product and carbon dioxide emissions Romania still lags behind most of the EU countries.

According to Eurostat data for 2007, the relative shares of various sources in the total energy consumption in Romania were:

Coal and lignite (%)	Oil (%)	Gas (%)	Nuclear (%)	Renewables (%)	Industrial waste (%)	Imports-exports of electricity (%)	Total energy consumption (thousand TOE)
25.4	25.6	32.4	5.0	11.9	0.3	−0.4	40,083

The revised figures released by the Romanian Ministry of Economy in April 2009 indicate a somewhat different picture. Electric power generation amounted to 64.77 TWh in 2008, covering the domestic demand and allowing for exports of 4.43 TWh. Split by sources, electricity generation was provided to the tune of 54.3% by thermal power plants (66% coal fired, 30% burning natural gas, and 4% using heavy oil), 28.4% by hydropower stations, and 17.3% by nuclear facilities.

Energy import dependency is lower in Romania than the average for the EU-27. Forty percent of imported energy resources is represented by crude oil coming mainly from the Russian Federation and Kazakhstan. Romania also imports certain amounts of natural gas and solid fuels to meet domestic demand. Net imports have decreased by 41% since 1990.

TABLE 3. Romania energy profile [25]

Data	28 year history	2007				2008
		Romania	Europe	World	Rank	Romania
Petroleum (Thousand barrels per day)						
		114.15	5,421	84,392	48	115.25
Crude Oil Production		94.42	4,549	72,989	46	95.40
		229.00	16,135	86,142	52	225
Net Exports/Imports(−) Net Exports = Total Oil Production − Consumption. Negative numbers are Net Imports.		−114.85	−10,714	–	41	−110
Total Oil Exports to U.S.		1	917	13,469	52	NA
Refinery Capacity		517	16,832	85,355	33	517
Proved Reserves (Billion Barrels)		0.600	16	1,317	40	0.600
Production		441	11,195	103,977	36	0
Consumption		660	20,284	104,425	35	0
Net Exports/Imports		−219	−9,428	–	28	−170
Proved Reserves		3,550	200,745	6,124,016	51	2,225
Coal (Million Short Tons)						
		38.496	790	6,779	20	39.140
		44.410	1,050	6,737	19	45.460
Net Exports/Imports (Trillion Btu)		−90.256	−6,117	–	30	−139.211
Electricity (Billion Kilowatthours)						
Net Generation		59.28	3,554	18,015	39	NA
Net Consumption		48.43	3,297	16,379	41	NA
Installed Capacity (GWe)		19.747	810	4,012	32	19.22
Total Primary Energy (Quadrillion Btu)						
		1.171	48	469	51	NA
		1.678	86	472	39	NA
Energy Intensity (Btu per 2000 U.S. Dollars)		8,538	–	–	–	NA
Carbon Dioxide Emissions (Million Metric Tons of CO_2)						
Total from Consumption of Fossil Fuels		98.64		4,721	29,195	40

[25] ***, Romania Energy Profile, 2007, US Energy Information Administration, 2010, http://tonto.eia.doe. gov/country/country_energy_data.cfm?fips=RO

According to EU targets for 2020 on total energy consumption, Romania is expected to reach a target of 17.8%.[26]

ROMANIA'S ENERGY AGENDA AND THE ISSUE OF CLIMATE CHANGE

Romania has to submit its National Action Plan (NAP) for the implementation of the EU directives on the Climate Change and Energy Package to the European Commission. The financial crisis has so far deviated the Government's attention from other priorities, but the problem is not likely to go away.

Several zones in Romania are particularly vulnerable to natural disasters, which have grown in intensity and frequency as a result of global climate change. The most affected segments of the population are those who live in isolated areas with poor access to modern infrastructure. Huge areas are prone to flash floods, mudslides, severe drought and, more recently, tornados. Investments in warning systems and in logistical support have been reduced to a minimum in recent years. Although most climatologists predict rising climate instability and more frequent extreme weather events, concern about climate change still ranks rather low on the public agenda. Politicians are likely to favor a more down-to-earth approach, by prioritizing their worries about the economic and financial situation.

The Romanian officials in charge of environmental policies still seem to prefer to stick to the current Kyoto Protocol rules regarding the carry-over of the surplus emissions entitlements into a possible next Kyoto Protocol commitment period. Whatever the case may be, Romania should make the best use of the available EU and international tools, mechanism, funds and policies.

A relevant study on energy and the impact of climate change on the Romanian economy was published in 2009 by the European Institute of Romania.[27] Some of its conclusions and recommendations are of relevance for the substance of this paper.

(a) International Climate Change Policy and Its Impact on the Romanian Legislation

We live in "extraordinary times"[28] and face a large number of threats: social, economic, and environmental. However, the climate is the most unpredictable and dangerous one for the future of human society. This was the main conclusion of the groups of experts and scientists at the L'Aquila Summit of the Intergovernmental Panel on Climate Change.[29] The IPPC was awarded the Nobel Prize for its conclusive results, namely the four scenarios of possible outcomes in

[26] ***, Electricity Generation 2009–2010, Facts and figures, VGB PowerTech, www.vgb.org/daten_stromerzeugung.html?dfid=25744
[27] *Studii de strategie si politici (SPOS 2009): Sinteze*, coordinator Prof. Dr. Ing. Aurel Leca, Institutul European din România, Bucuresti, 2009.
[28] Political guidelines for the next Commission, http://ec.europa.eu/commission_2010-2014/president/pdf/press_20090903_en.pdf
[29] http://www.ipcc.ch/publications_and_data/publications_and_data.htm

the case of an increase over the 2°C threshold in global temperature. The measures taken under the United Nations Framework Convention on Climate Change may provide today a safer legacy for the generations to come.

The Gothenburg Summit in 2001 marked an important moment in the efforts to find a common global solution for fighting the effects of climate change. This was a result of earlier efforts to implement the Kyoto Protocol,[30] signed in 1997, which aimed to reduce the greenhouse gas emissions of the industrial countries and to provide assistance to the developing ones in order to improve their energy efficiency and reduce their GHG levels. At that time the action plan was perceived to have failed since the biggest global polluters, US and China (responsible for 1/3 of the total global GHG[31]) did not sign it.

Romania is one of the 7 EU member states that lobbied for the banking of Assigned Amount Units (AAUs) in the EU beyond the first Kyoto implementation period ending in 2012. Together with Bulgaria, Poland, Latvia, Lithuania, Hungary, and Slovakia, Romania signed the Kyoto Protocol, which required the reduction of emissions below a certain level relative to 1990 (1989 in the case of Romania). Actually, Romania reduced emissions far more than the level required by the Kyoto Protocol (by half compared to 1989) but this occurred as an automatic result of deindustrialisation.

Other, more developed countries have had difficulties in reducing their emissions because their energy sectors were much more technologically advanced than Romania's in the base-year 1990. The Kyoto Protocol allows those countries to add the reduction of emissions in third countries to their own by investing in more environmental friendly technologies there.

Since the Kyoto Protocol expires in 2012, it was expected that the follow-up procedures would be crystallized in the outcome of the FCCC Copenhagen Conference in December 2009.

An agreement to tackle climate change must deliver the cutbacks needed to get global GHG emissions on a pathway compatible with the Copenhagen Accord's objective to keep global warming below 2°C. Any unhampered carry-over of AAUs would make this target extremely implausible.

The accomplishment of an international legally binding agreement for the post-2012 period depends, among other things, on the participation (with determined guarantees) of the developing countries; this would be facilitated by attractive prices in the international carbon market, principally for the Clean Development Mechanism (CMD, a platform for trading emission allowances under the UNFCCC.

- The European Commission is currently carrying out an evaluation of the comparability of the targets taken by non-EU states in the Copenhagen Accord and also assessing the possibility for the EU to take a more ambitious pledge than the 20% target.

---

[30] http://www.kyotoprotocol.com/
[31] http://www.climateavenue.com/cl.ghg.sect.ctry.htm

- n 2009, EU member states closed AAU transactions amounting to 114.7 mt. However, possible AAU sales capacity of the EU is estimated at the level of 1,740 mt. The top EU trader is the Czech Republic (also the biggest global AAU seller), with 68.5 mt sold in 2009, out of 100 mt set-aside for sale by 2012. Other traders were Poland, Latvia, Hungary, Slovakia, with Lithuania, Bulgaria and Romania preparing to bring their AAUs to the marketplace.

In December 2008, the EU adopted through co-decision (now an ordinary legislative procedure) the Climate Change and Energy Package (CCEP). This revolutionary legislative package crowned a decade of European efforts to lay the basis for the EU climate change legislation; it was composed of four directives with a transposition deadline until 2013.[32]

Effort sharing: aiming at an increased exchange of expertise among the member states in the field of energy and climate change so as to reduce the burden of the less developed EU countries in achieving the 20/20/20 targets (by 2020 to reduce the GHG emissions by 20%, increase energy efficiency by 20%, and bring up renewable sources of energy to 20% of the EU energy mix).

Emission trading system (ETS): the creation of a trading platform for GHG emissions, "a cap and trade system" using 300 million allowances (1 allowance being equal to 1 ton of GHG). This directive had the role of creating a carbon trading market at the level of the EU interlinking it with the CDM.

Renewable sources of energy (RSE): support the member states to meet the 20% target of RSE within the energy mix and encourage future investments and research in this sector.

Carbon capture and storage: setting the legislative basis for the creation of underground facilities in order to store CHG emissions from industrial polluters (e.g. power plants, heavy industrial facilities). Romania has an important industrial sector that is dependent on coal and non-refined petroleum products.

This legislative package should represent the core element of the Romanian strategy to mitigate the effects of climate change.

The last important step for a global climate change policy had to be the Copenhagen Conference in December 2009, which was expected to bring a global agreement to set common binding targets for reducing the GHG emissions by at least 20%. Unfortunately, no definite agreement was reached in line with that target. Still, there was a general consensus on the notion that the developing countries needed to be helped. US$100 billion will be allocated until 2020 to support actions to fight the severe effects of climate change and to adjust to the new circumstances. This financing will have a fast-start allocation of $30 billion for the first 3 years.

As a successor to the Lisbon Strategy on Growth and Jobs, the new EU strategy "Europe 2020"[33] has as one of its flagship initiatives a "Resource Efficient Europe". It is an ambitious plan to support the 20/20/20 targets. If there is political consensus, the aim is to go beyond the 20% reduction target for

---

[32]http://ec.europa.eu/environment/climat/climate_action.htm
[33] http://ec.europa.eu/eu2020/index_en.htm

the GHG and to increase it up to 30%. The political momentum of the EC Communication can give a boost to international climate policy post-Copenhagen. The actions to reinforce global action on climate change[34] have to be enhanced through adequate national action plans (NAPs). In his political guidelines for the next Commission, President Barroso put climate change at the heart of his program to create a single market for GHG emissions trading, combined with a strong and competitive low-carbon, knowledge-based economy to be enhanced by a powerful research and development sector and a smart innovation policy. All these policies are strongly interdependent and have as a common denominator the development of an effective digital infrastructure (a digital agenda for a future single digital market) which will optimize the efficiency of managing the ETS markets and improve the GHG emissions monitoring system.

In this respect, Romania is still lagging behind the rest of the EU member states in terms of digital infrastructure. Only 33.4%[35] of the population has access to the internet. In order to implement a meaningful digital programme, Romania has to make the best of the €1 billion that is available to the EU member states in order to have 100% internet connectivity by 2013.

At the EU level, as well as at the global level, the economic and financial crisis has not slowed the race for increased competitiveness and assured supremacy in the international green, low-carbon markets. USA and China are already investing massively in the transformation of those specific industrial sectors. The value of the future financial carbon trading markets has been evaluated at €1.4 trillion.[36] This enormous amount of capital should provide the foundation for the future EU green economy. The European Economic Recovery Plan allocated €4 billion for supporting investment in wind farms and carbon capture and storage facilities. The Marguerite Fund,[37] financed by a consortium of EU countries with the support of the European Investment Bank, allocated €1.5 billion to support climate change programs. So far, Romania has not presented any project to this fund and may miss an important opportunity.

Fighting climate change would be a futile effort without a global consensus on meaningful action. According to the fourth IPCC scenario, after 2020 the effects would be irreversible. Some scientists are already arguing that even a minor rise in temperature is already causing serious effects around the planet: the icecaps are melting at a rapid pace and volcanic activity has increased. Volcanoes that used to be buried deep under the ice sheet are now coming back to life, as was the case in Iceland. Last but not least, there is a risk of freeing high concentrations of methane[38] (more than 300 trillion tons of CH_4) from the melting of the permafrost layer.

---

[34] http://ec.europa.eu/environment/climat/pdf/com_2010_86.pdf

[35] http://www.internetworldstats.com/stats.htm

[36] http://www.bloomberg.com/apps/news?pid=20601130&sid=aEMSUQUFvK6Q

[37] http://www.eib.org/about/press/2009/2009-242-europes-leading-public-financial-institutions-launch-marguerite-the-2020-european-fund-for-energy-climate-change-and-infrastructure.htm

[38] http://www.sciencedaily.com/releases/2007/08/070808213844.htm

(b) Climate Change Action in Romania

The Ministers responsible for European affairs of Bulgaria, Hungary, Poland, Romania, and Slovakia adopted in Warsaw, on 26 September 2008, a joint declaration which expresses concerns that the post-Kyoto deal would endanger economic growth and demand new sacrifices from those countries. The Warsaw declaration stated: "GHG reduction plans and the increased share of the renewable energy sources should be attained in a cost-effective manner to minimize the burden on the economies of the EU Member States, in particular those still undergoing the process of economic and social convergence [...].The specific situation of the less affluent should be taken into account [...]. The final allocation scheme of the emission reduction effort (including those under and outside the ETS) should also reflect the contribution to the common goals by those MS that have already undertaken significant reductions compared to their initial levels under the Kyoto Protocol's first commitment period."

The diplomatic language hides the disappointment at having to pay a higher price before catching up with the richer, more advanced Western countries. The difference between the mentalities of the two groups is clear. However, Romanian decision-makers stated that the country would have a flexible position during the negotiations, in contrast with the more inflexible stance of Poland and Hungary.

By supporting the rigid position of the group, Romania placed itself among those European states that support only minimal post-Kyoto commitments. According to the terms of the Protocol, Romania can effectively sell its surplus of GHG permits to developed countries, buying superior technology in exchange. Initially, the Romanian government did not want to sell its GHG surplus, thinking it might need some reserves for later. At the time, the country attracted more green-field investments compared to the European average. Later, the officials changed their minds and tried to sell a part of that surplus. Therefore, the commitment to further reduce emissions appeared to be questionable. One of the reasons was that the environmental movement in Romania was fragile and unable to exert serious pressure on the Government. Romania has not yet made any AAU transactions. It currently emits 43% less carbon than its limit set by the Kyoto Protocol, and is therefore looking to sell the 200 million AAUs it has in surplus. Considering that the signatory states of the Warsaw declaration would not be able to change the reference year, Romania have to follow a Plan B and negotiate potential compensations from the European Commission or other member states.

The Romanian government is looking to begin the sale of AAUs in the course of 2010. It plans to raise €1.2–1.5 billion by 2012 and up to €2 billion by 2015.[39] The sale of AAUs had been planned over the past 2 years but it was not undertaken in the absence of the appropriate legal framework (which Romania aimed to adopt in the first half of 2010). The money to be obtained by Romania from the sale of AAUs would be injected in green venture schemes, including, among others, the rehabilitation and de-sulphurisation of coal-fired power plants.

---

[39] http://english.hotnews.ro/stiri-top_news-7177857-romanian-government-approves-the-sale-co2-emission-certificates-romania-could-get-around-2-billion-euros.htm

For the time being, the new European consensus for the post-Kyoto policy did not go well for Romania. The decision to change the base year for reductions to the post-2000 period would mean that:

- it may lose the possibility to "sell" GHGs, since 2000 showed a minimum in energy consumption and, implicitly, emissions;
- it would be forced to reduce more than the nominal post-Kyoto commitment, and this would take place during a period of fast economic growth.

All the eastern EU member states are more or less in a similar position with regard to changing the reference year, since they all experienced an economic transition. A platform for cooperation has already been set.

At Copenhagen, Romania supported a reduction of 30% concerning the GHG.[40] The Romanian Government warned against one-sided measures which could put EU industry in jeopardy and was part of the group of six member states that contested the CO_2 quota set by the EC.

Romanian CO_2 emission had a descendent trend since 1989, as recent studies showed,[41] nowadays having reached a level which is 50% lower than the one imposed by the Kyoto Protocol. The Romanian industry has the chance to ensure substantial revenue from the use of the EC's emission trading scheme (ETC) using certificates that, in Romania, will be allocated on the basis of a National Action Plan (NAP). These revenues will be used to enhance Romania's energy security by supporting the use of renewable sources of energy and the innovation sector.

Like Belgium, Romania has to prepare for three different scenarios[42]:

- The virtual revenues are directly distributed to the national budget in order to compensate the extra costs of GHG emissions.
- The revenues are allocated to the energy infrastructure.
- The revenues are used in order to compensate the patronage taxes of social security. This scenario, like in the case of Belgium, is the most likely to be adopted due to the higher impact of the energy price on the lower-income social groups.
- The Romanian Strategy for adaptation to climate change identified four security risk zones[43]:
- The agricultural sector was affected during the past decade by intense drought and flooding.
- The forestry sector was equally affected by intensive logging and dry spells and its productivity will be severely affected after 2040. The National Weather Administration released a study demonstrating that, between

[40] http://www.greenpeace.fr/presse/climat/passage-de-l-UE-a-30.pdf

[41] Romanian Ministry of Environment http://www.mmediu.ro/departament_mediu/schimbari_climatice/1_Documentatie/SNSC_ro.pdf, accessed on 5 June 2009.

[42] Mouffe, Céline, "Paquet énergie-climat : quels impacts pour la Belgique?", Conseil central de l'économie in *Lettre mensuelle socio-économique*, No. 140, September 2008, p. 14.

[43] Romanian Ministry of Environment http://www.mmediu.ro/departament_mediu/schimbari_climatice/GASC.doc, accessed on 5 June 2010.

2001 and 2030 compared to the period 1961–1990, the temperature is likely to rise by 1.5°C in Romania in a business-as-usual scenario.[44]

Water reserves would go down because of higher temperatures. Natural and artificial dams would be at higher risk of destruction due to sudden flooding. This would also impact the hydro-power system which contributes 17%[45] of the energy mix. The energy deficit caused by lower hydro power output would make it necessary to resort more to conventional sources of energy which emit larger amounts of CO_2.

The most important security risk would be to the human settlements which are most vulnerable to storms that may disrupt power grids and affect various types of power plants.

The energy sector has an important part to play in fighting the effects of climate change in Romania. When the CCEP comes into force, conventional power plants will no longer receive free quotas of CO_2 emissions and will have to pay €21[46] for each ton of CO_2. This price is expected to rise to €40/certificate (1 ton of CO_2 emissions). Consequently, the price for the energy generated in those plants will be substantially higher, approaching price level for renewables. Experts in the energy sector envisage that 40% of the electric power production (29.7 TWh) would come from coal-fired plants at a cost of €1.2 billion (using the above higher cost per certificate).[47] This may lead to price instability. Therefore, Romania has to factor in the negative effects at the consumer end and devise strategies further to reduce the CO_2 emissions. €433 million were allocated by the EU and the national government over 2007–2013 for this purpose.[48] One solution could be to use carbon capture and storage (CCS) technologies for which the European Union allocated substantial funding, considering that a CCS facility can store 1 mt of CO_2 per year.

(c) Conclusions

Climate change policy has moved to the top of the global political agenda. Investment in research and innovative technologies is growing in support of this action. The EU has used this momentum to develop a low-carbon, knowledge-based economy and thus can keep its products globally competitive by supporting the European research centers to become beacons of excellence.

Romania must be part of that drive. That is why it is so important to match the actions to be taken in the energy sector with special programs for research, technological development, and innovation in order to enhance Romania's contribution to the worldwide efforts to prevent and mitigate the effects of climate change.

---

[44] Romanian scenarios of climate change for the period 2001–2030, Romanian Ministry of Environment http://www.mmediu.ro/departament_mediu/schimbari_climatice/1_Documentatie/SNSC_ro.pdf; accessed on 5 June 2009.

[45] Romanian Ministry of Environment, http://www.mmediu.ro/departament_mediu/schimbari_climatice/ GASC.doc , accessed on 5 June 2009.

[46] *Capital* newspaper, http://www.capital.ro/articol/planul-20-20-20-condamn-x103-rom-nia-la-energie-scump-x103-107514.html, accessed on 6 June 2010.

[47] Ibid

[48] *Capital,* http://www.capital.ro/articol/pachetul-schimbarilor-climatice-starneste-furtuna-la-bruxelles-113374.html, accessed on 6 June 2010.

THE NATIONAL SUSTAINABLE DEVELOPMENT STRATEGY
OF ROMANIA 2013–2020–2030

Sustainability and system resilience have become key concepts of modern development and they are certainly relevant to energy security. The origins of the modern notion of sustainable development can be traced about 40 years back to the early Club of Rome report on the *Limits of Growth* (1972). A subsequent series of international conferences under the aegis of the United Nations further refined the concept and eventually produced a universally accepted definition as the development path "that meets the needs of the present generation without compromising the ability of future generations to meet their own needs" in the G.H. Bruntdland report of the World Commission on Environment and Development (1987). This new conceptual frame proposed an integrated approach to policy design and executive decision-making and sought to restore and preserve a rational and enduring equilibrium between the needs of economic development and the integrity of natural environment in ways that society could understand, accept, and support. The idea of system resilience as a necessary complement to sustainability is of more recent vintage. It emerged as a result of the spectacular advances in mathematical modelling of the behavior and interaction of complex systems. It was no accident that the first area where this new conceptual frame was applied for operational purposes in a meaningful (and successful) way was energy security.

In the particular case of Romania, the awareness about the incongruity of the then prevailing development model (including the socio-political system) as related to the support capacity of natural capital gradually took root in the 1970s and 1980s of the past century. In the context of a hard-line communist regime those concerns were confined to certain narrow academic and intellectual circles and limited traction with political decision-makers.

After the revolutionary system change in late 1989, under the impact of the newly acquired freedom of information and public debate, the linkage between economic growth, environmental concerns, and social responsibility became official government policy. Within a space of months, numerous non-governmental organizations and even political parties with an environmental agenda came into being, mirroring (and in some cases mimicking) the initiatives that had long been in existence in most of Western Europe. New institutions were set up such as dedicated ministries and parliamentary committees on environment and sustainable development.

Basic elements of primary and secondary environmental legislation were enacted in line with the United Nations recommendations long before Romania became a credible candidate for EU membership. Romania was the first European country to ratify the Kyoto Protocol to the UN Framework Convention on Climate Change.[49] The link between energy policy and care for the state of the environment was thus firmly established in Romania as an official government doctrine

---

[49] Law 3/2001 of 2 February 2001.

since the early years of transition to pluralistic democracy and functional market economy.

With regard to the way in which Romania's energy policy gradually became more closely related to the principles and demands of sustainable development, two distinct phases can be discerned:

(i) After a protracted and often painful transformational period, in the late 1990s and the early years of the twenty-first century, Romania's strategic action initially focused on the diversification of the sources of supply for hydrocarbons. There were good reasons for such an approach, which now, in hindsight, appears to have been not very successful since it was limited to individual action at a national level. It is worth remembering that even during communist times Romania was paying by far the highest price of all COMECON (Soviet-dominated economic bloc) member states for imported oil and gas—and in hard currency. The accelerated depletion of national hydrocarbon resources compounded the problem. So, for several years during the transition, Romania spent a lot of time and effort to secure a favorable position as a pivotal country along the East-West strategic energy corridors around and across the Black Sea. Political action and diplomatic demarches were imbued with a sense of urgency about making sure that at least some of the main planned pipelines for oil and gas should reach the territory of Romania and pass through it. The presence of such vital elements of energy infrastructure was seen as a strategic asset for enhancing Romania's chances in its bid for NATO and EU membership. The main rationale was therefore political, while security of supply, though important, was considered a secondary consideration.

(ii) The second stage came with the start of serious preparations for, and the actual advent of Romania's accession to NATO (2004) and the European Union (2007). National strategies and executive actions were steered toward achieving full conformity with the principles, objectives, and institutional procedures of those two organizations, and public policies were adjusted accordingly. In line with Romania's commitments under the Treaty of Accession to the EU (signed on 25 April 2005, in effect as of 1 January 2007), a National Development Plan and a National Strategic Reference Framework for 2007–2013 were prepared and endorsed, along with a National Reform Programme and a Convergence Programme, plus a series of dedicated operational sectoral programs. At this stage, the strategic designs and policy orientations of NATO and the EU in the sphere of energy security are also Romania's own. Active involvement in shaping joint assessments and common energy policies has become the principal direction for government action to promote the specific interests of Romania.

In compliance with the requirements of EU membership, the National Sustainable Development Strategy of Romania 2013–2020–2030[50] was prepared and endorsed by the Government on 12 November 2008. The Strategy set specific objectives for moving, within a reasonable and realistic timeframe, toward a new model of development capable of generating high value added, motivated by

---

[50] Government Decision No. 1460 of 12 November 2008, in *Monitorul official* [Official Gazette] No. 824 / 8 December 2008.

knowledge and innovation and aimed at continued improvement of the quality of life and human relationships in harmony with the natural environment. The strategic goals envisaged three time horizons: by 2013 – to incorporate the principles and practices of sustainable development in all national programs and public policies; by 2020 – to reach the current average level of the EU countries; by 2030 – to match the average EU performance of that year (or come close to it) in terms of sustainable development indicators.

We have to make it clear that the current crisis and its fallout may cause an adjustment of priorities and intermediary targets in the short and medium terms, but it would not fundamentally alter the long-term strategic design – to implement a new model of sustainable development. Somewhat perversely, the crisis revealed the glaring fact that the old, wasteful, and uneconomic pattern of production and consumption could no longer be sustained and therefore had to be changed.

Following the logic of the renewed EU Sustainable Development Strategy, which was endorsed by the European Council on 9 June 2006, the Romanian National Strategy gave pride of place, as the first key challenge, to the chapter on "Climate change and clean energy".[51]

The document described the situation of the Romanian energy sector relying on the data available for 2007. After a decade of painful transition to a functional market economy, which was compounded by a difficult historical legacy, between 2001 and 2007 Romania's macroeconomic performance improved significantly. The average annual rate of GDP growth was in excess of 6%, one of the highest in the region. By 2007, the GDP totalled some €121.3 billion, three times over the figure for the year 2000, but still represented only about 41% of the EU average in terms of purchasing power parity. Structural adjustments in the economy and more efficient use of resources led to a sizeable reduction of primary energy intensity and electricity intensity per €1,000 of GDP, but still remained twice as high as the EU average. It was estimated that the national potential for energy saving, mainly through increased efficiency and reduced losses in the system could amount to 30–35% of primary resources used (20–25% in manufacturing industries, 40–50% in buildings, 35–49% in transport).

The dependence on imported primary resources went up in 2007 to about 30% for natural gas and 60% for crude oil because of the advancing depletion of national resources. This increasingly affected the balance of payments. The domestic production of coal, though still relatively abundant, was of low quality and uncompetitive in terms of cost. Internal reserves of nuclear fuel were also close to depletion.

Most of the functioning fossil fuel fired power generation plants and hydro-power stations were well past their designed lifetime, using technologies of the 1970s. The same applied to the technical obsolescence of the high-voltage trans-mission lines (50%), electricity substations (60%) and main gas pipelines (69%).

The aggregate energy consumption for all business activities represented 68.6% of the total (17.5% for manufacturing industries), while residential use accounted for 31.4%, compared to the EU average of 41%. The district heating

---

[51] Government of Romania, *National Sustainable Development Strategy 2013–2020–2030*, pp. 47–51.

network was relatively well developed, covering about 29% of the households mainly in urban areas. However, most of the large cogeneration units and distribution networks were equipped with badly outdated technology and registered very high losses in relation to fuel consumption (between 35% and 77%).

The National Sustainable Development Strategy of 2008 set the following specific targets for the year 2013, within the current EU financial planning exercise:

- To increase energy efficiency by reducing final energy consumption by 13.5% through legislative and regulatory measures, voluntary agreements, expanded energy-saving services, special financial instruments, and cooperative schemes.
- To upgrade the combined heat and power systems and to rehabilitate the thermal insulation of at least 25% of the multi-storied housing developments and office buildings.
- To reduce energy poverty and to revise the system of subsidies and support payments to the most vulnerable customers and to start introducing modern, eco-efficient heating systems in rural areas in order to provide energy services at a bearable cost.
- To render selective support to investments for the commissioning of new power production and co-generation facilities using clean energy techno- logies resulting in considerable reduction of emissions of greenhouse and other polluting gases, and in improved operational safety of the national energy system.
- To increase power production from renewable energy sources to 11.2% of the total.
- To complete the implementation of the Green Certificates scheme and the EU-wide Emission Trading System.

For the year 2020 the following additional targets were envisaged:

- To increase the share of renewable resources (wind, solar, biomass, geo- thermal, biogas, etc.) to 24% of the final energy consumption compared to the overall EU target of 20%.
- To further enhance energy efficiency resulting in a reduction of primary energy consumption by 20% and final energy consumption by 18%.
- To reduce primary and final energy intensity to a level close to the EU average of 2006.
- To rehabilitate an additional 35% of the multi-storied residential, administrative and commercial buildings.
- To commission new power generation units and to connect them to the national grid in order to meet the projected rise in demand, including two new reactors at the Cernavoda nuclear plant and several hydropower stations.
- To upgrade the national power grid and pipeline system and to develop new interconnectors with the neighbor countries in order to achieve full integration with the EU energy networks.

For reasons of prudence and professional realism, no specific targets were set for the year 2030. Romania would continue to contribute effectively, in keeping with the guidelines to be agreed at international and EU level, toward the shaping and implementation of common objectives on energy security and climate change.

A renewed version of the National Sustainable Development Strategy of Romania is due for completion and submission to the European Commission by June 2011. It will have to take into account the lessons learned from the preparation of the Strategy of 2008 and its early implementation, to introduce some necessary corrections in light of the consequences of the current financial and economic crisis, to consider carefully the realistic potential of the recent breakthroughs in energy-related technologies, and to adjust to the new realities of the energy market and to the demands for the mitigation of climate change in Europe and worldwide.

The conceptual framework for this endeavor is provided in the new EU strategic document *Europe 2020*[52] which sets precise, mandatory targets for the member states and the Community as a whole. The very title of that document suggests the new quality of growth to be pursued by the EU in the post-crisis period: it should be *smart* by relying on knowledge and innovation; it should be *sustainable* by promoting a more effective use of resources, enhancing ecological performance and improving competitiveness; and it should be *inclusive* by improving employment and strengthening social and territorial cohesion in the European Union.

It is important to note that three of the five headline targets and five of the seven flagship initiatives that are detailed in the new EU Strategy have a direct bearing on energy security and related policies.

Critical Infrastructure Protection – Energy Security

THE CRITICAL INFRASTRUCTURE CONCEPT

Infrastructure is essential for economic prosperity, national security, and the quality of life in any country. The proper functioning of infrastructure is thus the means whereby modern society aims to secure its present and develop its future.

Infrastructures can be grouped into three large categories, depending on their location, role and importance for the stable functioning of the economic and societal structures as well as for the safety and security of systems:

- Ordinary infrastructures;
- Special infrastructures;
- Critical infrastructures.

Ordinary infrastructures represent the frame which enables the development and functioning of a system. These infrastructures do not display special qualities

---

[52] Communication from the European Commission, *Europe 2020: A Strategy for Smart, Sustainable and Inclusive Growth*, COM (2010) 2020, Brussels, 3 March 2010.

beside those that justify their existence and presence within the frame of systems and processes.

A country, for example, will always have roads, railways, towns, schools, libraries, etc. As time goes by, some of these may become special, or even critical, depending on the new role they may have, on the dynamic of their importance and other criteria. For example, the cities that have airports, powerful communication centers, nuclear plants, railway nodes, etc. can be part of special infrastructures and, under certain circumstances, even part of the critical ones.

Special infrastructures play a particular role in the functioning of systems and processes, ensuring their enhanced effectiveness, quality, comfort, and perform-ance. Generally, special infrastructures are performance infrastructures. Some of those, especially the ones that undergo extension or transformation (modernization), can have an important role in the stability and security of systems, can also be critical infrastructures.

Critical infrastructures are generally those on which the stability, safety, and security of systems and processes depend. They can be part of the special infra-structure category. However, it is not mandatory for all infrastructures that are or can become critical at some point to be part of the category of special infrastructures.

Depending on the situation, other elements can also intervene and even some of the ordinary infrastructures – as for example country roads, irrigation systems, etc. – become critical infrastructures. This leads to the conclusion that there is a flexibility criterion in the identification and evaluation of such structures.

The European Programme for Critical Infrastructure Protection in 2006 proposed this definition: "Critical infrastructures consist of those physical and information technology facilities, networks, services and assets which, if disrupted or destroyed, would have a serious impact on the health, safety, security or economic well-being of citizens or the effective functioning of governments in the member states."

Infrastructures are accounted to be critical for several reasons:

- Singularity within the frame of infrastructures of a system or process;
- Vital importance as a material or virtual (net-like) support in the functioning of systems and the unfolding of processes – economic, social, political, informational, military etc.;
- Important, non-replaceable role that they play in the stability, reliability, safety, functionality and especially in the security of systems;
- Increased vulnerability to direct threats, as well as to threats targeting the systems these infrastructures are a part of;
- Special sensitivity in case of variation of the conditions and especially in case of sudden changes of the situation.

The importance of critical infrastructures also results from the fact that they can be defined as being those industrial capabilities, services, and facilities which, in case of interruption of their normal functioning, can affect human life and, moreover, can harm or destroy human life. The protection of life and of the lifestyle of people envisages especially the preservation of the continuous functioning of these services and facilities.

Identifying, optimizing, and securing critical infrastructures is an undisputed concern both for the managers of systems and processes as well as for those aiming to attack, destabilize, or destroy the targeted systems and processes.

Critical infrastructures are not and do not become critical only because of attacks, but also due to other causes, human, as well as technical, some of them difficult to be identified and analyzed. Especially after the terrorist attacks of 11 September 2001 on the World Trade Center and the Pentagon, it is considered that infrastructures are or can become critical due to terrorist attacks or other threats, especially asymmetrical ones.

- The criteria used for such analysis are variable, even if their area of coverage could be the same. The predominant criteria for analysis, as mentioned in specialized literature, are the following:
- Physical criterion, regarding the positioning within other infrastructures (size, spread, endurance, reliability etc.);
- Functional criterion, regarding the infrastructure's role (what exactly does it "do");
- Security criterion (what is its role for the overall safety and security of the system);
- Flexibility criterion (reflecting the dynamic and flexibility in defining infrastructures as critical; some of the ordinary infrastructures become under certain circumstances critical ones and vice-versa);
- Unpredictability criterion (considering that some of the ordinary infra-structures can become suddenly critical infrastructures);
- Critical infrastructures have at least three components of critical phases:

Internal component, defined through the increase (either direct or induced) of infrastructure vulnerabilities with an important role in the functioning and security of the system;

External component, referring to the exterior stability and functioning in relationship to the systems the infrastructure is integrated in or associated to;

Interface component, defined through the multitude of neighboring infra-structures, which do not belong to the system as such, but influence its stability, functioning, and security.

THE GEOPOLITICS OF CRITICAL INFRASTRUCTURES

Besides the classic civil protection concept, forms of "physical protection" of citizens and protection against imminent threats, new vulnerabilities have evolved in the modern, increasingly globalized societies.

Terrorist attacks on air, rail, underground, and road means of transport or on key information systems have had an impact on government-level debates and on political and military decision-making, as well as on the documents and decisions issued after 11 September 2001.

There are three angles to approach critical infrastructure protection:

(a) Many of the currently operational critical infrastructure systems are the consequence of the Cold War both in the West and in the East (especially the legacy of the communist system).

(b) The prospect of new types of vulnerabilities, the imperative to keep key critical infrastructures operational, and the need for modernization require considerable financial support. Critical infrastructures in the spatial dimension had not suffered major events, until the Tamil Tigers attacked a satellite.

A 3-week wave of massive cyber-attacks on the small Baltic country of Estonia, the first known incidence of such an assault on a state, caused alarm across the European Union and NATO alliance; both structures had to examine the offensive, its implications and further required measures.

Since the beginning of 2008, the under the sea level dimension has suffered several optical fiber cable cut-offs, thus affecting the Internet information transfer across the continents, with damages that have not been well quantified yet. The attacks on those cables highlighted the enormous amount of Internet traffic that uses the undersea cable system, which carries many times more traffic than the satellite system does.[53]

(c) The evolution towards a multi-polar world, involving issues of energy security and information security will have effects that can hardly be estimated at present. Both the United States and Russia have therefore taken due action, but the "play" has grown more complex due to the experience and importance of new actors coming from the Asian region.

Following these new dimensions of risk and vulnerabilities, we can speak about critical infrastructure geopolitics mainly in the sector of energy security, but new sectors of critical infrastructures from water supply to food supply security could be added in the following years. Given the circumstances, new developments may occur, each calling for a complex assessment of certain systems, or systems of systems, as well as for specific measures.

First, there is a need for developing the ability to forecast and interpret specific events that are going to take place or have already started. Natural hazards, man-made disasters, technical accidents, the intervention of an external criminal hand, of an "enemy within", or a possible terrorist attack are rather hard to identify and determine in the early stage of a major event.

Second, increasingly complex systems and the interdependence existing among various categories of critical infrastructures call for expert interdisciplinary training that includes both international experience and concrete aspects deriving from "lessons learned" from previous events.

Within this context, the top management of public and private enterprises must provide more time and resources (financial, human, and material) to the people responsible for the smooth functioning of the critical infrastructures existing on their premises, as well as of those connected to national and trans-national networks.

---

[53] Prof. Dr. Adrian Gheorghe, Dr. Liviu Muresan, "Critical Infrastructures Protection, Resiliency and Maritime Security - MIPS" , Organized by the 5th US Fleet, Bahrain, 2008.

At the same time, specific *"security culture"* action is needed to ensure a flow of correct, complete, and timely information not only for own employees but also for the general public and especially for the local communities where the institutions operate.

Third, given the security environment dynamics, it is necessary to be aware of the new threats to date and of the respective vulnerabilities for critical infrastructures. New vulnerabilities can thus appear for any critical infrastructure but also for the "nodal points" where they are interconnected with local, national, and international networks.

Fourth, based on objective prioritization, authorities should build a list of critical infrastructures that require protection measures. Due to objective reasons, limited budgets, lack of qualified personnel, lack of specific protection technology, and lack of time to find solutions in complex situations, authorities cannot take extensive measures for the protection of all critical infrastructures at the same level.

A choice among ordinary infrastructures and specific infrastructures can be made starting from the "history of events" reported at national level, international experience, specific classified information and alerts received from special services, etc.

Fifth, the decision to take protection measures in specific cases is accounted for not only by technological considerations but also by political, social, economic, and cultural implications such as business continuity, quality of life, and others. Smooth functioning or, conversely, malfunctioning in some critical infrastructures that directly affect the quality of life and the functioning of modern society (electricity, water supply, heating, transportation, health care, waste disposal, etc.), having a direct impact on citizens, can bear a high political cost. Therefore, political decision-makers will carefully monitor the reactions if such situations occur, particularly of those "charged with electoral potential".

Sixth, the authorities and the leadership of both the public and private sector must employ staff and provide all the resources in time, when it comes to highly specialized critical infrastructures such as the ones in the information systems sector or in case of cyber threats to energy security. The training and the stability of highly qualified information specialists are challenges for all sectors with workforce mobility in a competitive environment.

In this area, specialists are "headhunted" by institutions in the national security sector, by the IT sector, by finance and banking institutions, by the national and multinational private sector, or by international organizations or by non-state actors like organized crime and terrorist groups. The outsourcing of IT services to overseas companies can provide well-paid jobs with delocalization at distances of hundreds or thousands of kilometers.

Seventh, the central and local authorities must be aware of the need for the continuous assessment of the state of infrastructures, particularly of critical ones, and to set specific standards and clear responsibilities for their protection. Suitable legislation is needed to set responsibilities for the authorities, key institutions, to integrate their activities into a comprehensive civil protection concept, to train the personnel having special responsibilities, to integrate all into a coherent, flexible system, etc.

Eighth, the public-private partnership is mandatory for critical infrastructures, considering that the majority percentage is owned by the private sector in this domain.

Legislation must provide acknowledgement and regulations with respect to mutual rights and responsibilities. There is a need for a continuous dialog on critical infrastructure issues between the authorities and the private sector. The dialog must start during periods of normal conditions so that cooperation can function smoothly right from the beginning of an emergency or critical situation.

In this respect, authorities must identify incentives for a functional partnership and at the same times apply sanctions when the usage of specific infrastructures is obstructed.

A special case is that of foreign companies or institutions operating on the national territory, either independent firms or multinationals.

Ninth, the responsibility for providing information regarding the location, physical state, and legal nature of key critical infrastructures rests with the national and local authorities. Even though critical infrastructure protection necessarily requires restrictive information circulation, a national database must be designed and managed according to the national legislation, as well as the regulations and obligations that are incumbent on Romania as an EU and NATO member.

The present EU legislation does not reflect the new environment in the field of energy after the EU enlargement, the new challenges to the supply security, renewable energies climate change policy, as well as the notification of new investment projects. The European Parliament allowed the Commission that every 4 years it requires the member states should collect and notify data on energy infrastructure investment projects concerning production, storage and transport of oil, gas, coal, renewable energy, electric power, as well as major projects for district heating and cooling and carbon dioxide capture, transport and storage, planned under construction in their territory, including interconnections with third countries.

Energy companies would be obliged to provide the data to their own Member State on: capacities, location, timetables, technologies used in the interest of supply, carbon capture systems of retrofitting mechanisms, and comments on delays and obstacles on the implementation.[54]

Tenth, critical infrastructures and the need for their protection are key issues that must be included in strategies of national security, energy security, information security, food supply security, health security, transportation security, etc.

A coherent national strategy on critical infrastructures, integrated into the network of the above-mentioned strategies, is a determining factor for a nation's resilience capability.

Recent international security evolutions have shown a period of relatively high instability, probably followed by a period of "stabilization of instability".

In this context, we consider several issues such as critical infrastructures, their protection and resilience could contribute to security and stability. If we compare

---

[54] Adina Ioana Valean (ALDE, RO), Industry Committee – European Parliament, "*Investment in energy infrastructure: more data to help improve planning*", a report adopted with 551 votes in favor, 24 against and 25 abstentions, February 2010.

a critical infrastructure system with an articulated concrete block (used for river bank stabilization), several articulated concrete blocks can be seen as a system of systems that could break the current "instability waves".

In this context, the Next Generation Infrastructures Research Programme, launched in 2004 in the Netherlands by Prof. Dr. Margot Weijnen, connects the world of technology, policy, and behavior and develops methods and technologies that measure up to the complexity of the world making infrastructures better prepared for the future. They take into consideration that the world of infrastructure managers, administrators, and decision-makers is swiftly changing, with government retreating, markets being liberalized, a growing world population, rapidly developing economies, technological advances, the environmental problem, the financial crisis, and the increasing scarcity of fossil fuels.

CRITICAL INFRASTRUCTURE PROTECTION: EUROPEAN AND EURO-ATLANTIC PERSPECTIVES

As natural disasters increase in amplitude and frequency and as terrorist events acquire unprecedented scopes, critical infrastructures require enhanced protection from threats and risks. Because of that, governments worldwide have shown special concern for ensuring the security of the population and of the state authority.

The European Union member states have generally included in the category of critical objectives: telecommunications, water and energy sources, distribution networks, production and distribution of food, health institutions, transport systems, financial and banking systems, defense and public order institutions (army, gendarmerie, and police). In this sense, a critical infrastructure represents a material good or a complex objective that is vital for the overall functioning of the economy and society and is usually interconnected to other infrastructures.

In the analysis of this topic two axioms are accepted:

- it is practically impossible to ensure 100% protection of a critical infrastructure;
- there are no single or universal solutions for solving this problem.

During times of crisis or war, adversaries can try to intimidate or block the freedom of action of national political leaders by attacking the critical infrastructures and the basic functions of the economy or by eroding public trust in governance or information systems.

Cyber attacks on the information networks of any country can have serious consequences, such as the interruption of the functioning of key components, causing losses of material and intellectual property or even of human lives.

EUROPEAN CRITICAL INFRASTRUCTURES

The actions mentioned earlier lead to the fact that a process was started at the European Commission level to develop normative proposals in the field of critical infrastructure protection. These projects were finalized and presented to the

European Parliament, some of them started in 2005 and the rest of the documents in December 2006.

Up to the present moment, several counties – Austria, France, Germany, Great Britain, Italy, Norway, Sweden, Switzerland, and Spain – have created specific organisms, developed methodologies, and allocated substantial funds for the protection of the infrastructures they defined as critical.

The European Council, at its June 2004 meeting, asked the European Commission and the High Representative to develop a global strategy regarding the consolidation of critical infrastructures and their protection.

Especially after the dramatic events of 11 September 2001 in the Unites States and 11 March 2004 in Madrid, but also on 7 July 2005 in London, the risks associated with terrorist attacks on European infrastructures rose. The consequences of such attacks are considered to be variable.

It is estimated that a cyber-attack would make few or no human victims as a direct consequence, but it could lead to the interruption of the functioning of vital infrastructures. For example, a cyber-attack against the transmission networks would lead to the interruption of telephone conversations, data transmissions, television, and radio. Until the damage is repaired, serious consequences can occur as a result the chain-like propagation of unpredictable events due to the social impact caused especially through the psychological effect on the population and the major disruptions of governance at local and state level.

There is, however, also another perspective regarding the attacks on critical infrastructures. An attack on the command-and-control systems of chemical installations or of the transport and distribution networks for electrical energy or gas and oil products could cause many victims and significant material damage. Even more, due to the interdependence of interconnected systems, the effects could multiply and unfold in a chain reaction.

An attack on the electricity networks could have very big effects in terms of the functioning of industrial installations, computer networks, banking sector, communication networks, etc. Where there are no adequate electric energy sources, there can be also serious effects on the vital medical equipment used for patients undergoing surgery or under monitored control. Long lasting electricity interruptions in large areas of North America and Europe pointed out once again that infrastructures in the field of energy are especially critical and vulnerable.

According to the definition mentioned in the European Council Directive 2008/114/EC of 8 December 2008 on the identification and designation of European critical infrastructures and the assessment of the need to improve their protection, critical infrastructure represents "an asset, system or part thereof located in Member States which is essential for the maintenance of vital societal functions, health, safety, security, economic or social well-being of people, and the disruption or destruction of which would have a significant impact in a Member State as a result of the failure to maintain those functions."

The same document defined "European critical infrastructure" (ECI) as the critical infrastructure located in Member States the disruption or destruction of which would have a significant impact on at least two Member States. The significance of the impact shall be assessed in terms of cross-cutting criteria. This includes effects resulting from cross-sector dependencies on other types of infrastructure.

The European Commission suggests three essential criteria for the identification of potentially critical infrastructures:

Extent or surface. The deterioration of critical infrastructure is evaluated depending on the geographical region which would suffer consequences; the international, national, regional/territorial, or local dimension;

The degree of seriousness. The incidence or degradation can be null, minimal, moderate, or high. The main criteria for the evaluation of the degree of seriousness are: economic incidence, incidence on the public, incidence on the environment, dependence, political incidence;

Effect in time. This criterion shows the moment when the degradation of the infrastructure can have a major incidence or a serious effect – immediately, after 24–48 h, in a week or within a longer period of time.

It is the duty of every state to identify, through its governmental structures, the critical infrastructures on its territory. However, the European states do not stand alone; they are not isolated but operate in extremely tightly knit, complex relationships. The absolute independence concept has disappeared a long time ago. Europe becomes more and more interdependent and responsible for everything that is going on, not only in international relations but also in the territory of each state.

In other words, the responsibility for identifying, evaluating, protecting, and securing critical infrastructures becomes, in the context of increased interdependency and proliferation of threats, a vital aspect for the good functioning of human society.

This is another important conclusion for the management of critical infrastructure security.

The international dimension of this responsibility resides in the following reality:

- Most of critical infrastructures, or those that can become critical, reach out beyond the geographical area of one state;
- The increased vulnerabilities of the critical infrastructures of one state determine, one way or another, the raising of vulnerabilities of all infrastructures in the area and/or network;
- The network configuration and philosophy accentuate the interdependence and equally raise the vulnerabilities of all participating structures but also their capacity and resistance to perturbations and threats.

Obviously, it is not possible to protect all critical infrastructures completely and always.

In this context, the management of security is defined by the European Commission as a "deliberate process which envisages the evaluation of risk and the implementation of the actions aimed at bringing the risk at a determined and acceptable level, at an acceptable cost". This requires:

Identifying the risk associated to the system and process vulnerabilities of the critical infrastructures, the dangers and threats these face;

Analyzing and evaluating the risk;

Controlling the dynamics of the risk;

Risk control within desirable limits.

Due to the complexity of the earlier mentioned aspects, the Programme of the European Commission envisages only trans-national critical infrastructures, protection of the national ones remaining the responsibility of the EU member states EU within a common framework.

In this sense, there are already numerous directives and regulations which impose means and procedures for releasing information on accidents, establishing intervention plans in cooperation with the civil protection, the administration, the emergency services, etc. There are, for example, action and reaction programmes in civil and military emergencies, such as nuclear, industrial, chemical, environmental, oil-related accidents, natural disasters, etc.

The relevant Communication of the European Commission, which involves all the analyses and sector measures, constitutes the basis of a European Programme for Critical Infrastructure Protection (EPCIP) and aims to find solutions for their security. The objectives of the program are:

Identifying, through the governments of the member states, all the critical infrastructures of each state and adding them to a central inventory, according to the priorities established through EPCIP;

The collaboration of enterprises and companies in the respective sectors along with the governments for the dissemination of relevant information and reducing the risk of incidents susceptible of creating extended or durable disturbances to critical infrastructure;

The common approach to the issue of critical infrastructure security thanks to the collaboration of private and public actors.

The European Programme targeted, among others, the reunion of every structure specialized in protecting the critical infrastructure of the member states into a network. This could lead to the development of an early warning network for critical situations: Critical Infrastructure Warning Information Network – CIWIN.

The European Programme for Critical Infrastructure Protection (EPCIP) has the following main objectives:

Identifying and designating the European critical infrastructure and the measures that are required to protect and improve them. The proposed directive sets the procedures for the identification and designation of the European Critical Infrastructure (ECI) and the accepted measures for improving the protection of these infrastructures;

The promotion of specific measures that facilitate the implementation of the European Programme for Critical Infrastructure Protection (EPCIP), including a Plan of Action for the Protection of European Critical Infrastructures and Warning Information Network (CIWIN), involving a group of experts at European Union level, a system for permanent exchange of useful information, as well as identifying and analyzing the interdependence of critical systems;

Support for the member states regarding the National Critical Infrastructures (NCI), which can optionally be used by other countries;

Financial support for some of the agreed measures and particularly for the EU Program regarding "Prevention, Protection and Removal of Terrorist Action Effects and Other Security Risks", for 2007–2013, providing financial support for the protection of critical infrastructures which have the potential to be transferred and applied at EU level.

The European Programme for Critical Infrastructure Protection focuses its attention on 33 vital sectors and services connected to them.

Sector	Service or Product
I. Energy	1. Production of oil and gas, refinery, treatment and storage, including pipelines
	2. Production of electric energy
	3. Energy, gas and oil transport
	4. Energy, gas and oil distribution
II. Information and communication technology	5. Information and network systems
	6. Command, automation and instrumentation systems
	7. Mobile and land telecommunication services
	8. Navigation and radio communication services
	9. Satellite communication services
	10. Broadcasting services
III. Water supply	11. Drinking water supply
	12. Water quality control
	13. Dam building and water quantity control
IV. Food supply	14. Food supply, food safety, security and protection
V. Health	15. Medical support and hospital services
	16. Drugs, serums, vaccines , and pharmaceutical products
	17. Bio laboratories and bio agents
VI. Finance	18. Payment services/related structures
	19. Governmental financial systems
VII. Defence, public order, national security	20. Country defense, public order and national security
	21. Integrated management of borders
VIII. Administration	22. Government
	23. Armed forces
	24. Administration and services
	25. Emergency services
IX. Transport	26. Road Transport
	27. Railways
	28. Sea, river and ocean transport
	29. Air transport
X. Chemicals and nuclear energy	30. Production, processing, and storage of chemical and nuclear substances
	31. Dangerous chemical substances pipes
	32. Air traffic
XI. Space	33. Outer Space (Proposal made at ESRIF Workshop, September 2007, by Prof. Dr. Adrian Gheorghe, Dr. Liviu Muresan and Dr. Dumitru Prunariu, Astronaut – EURISC Foundation)

ENERGY FOR EUROPE WITH SECURE SUPPLY

On 19 September 2007, in Brussels, a new momentum was given to an energy policy for Europe. The European Commission adopted the third package of legislative proposals to ensure an opportunity for real and effective choice of supplier and accruing benefits to every single EU citizen. The Commission's proposals put consumer choice, fairer prices, cleaner energy, and security of supply at the center of its approach.

The legislative package consists of:

A Regulation establishing the EU Agency for the Cooperation of National Energy Regulators;

An Electricity Directive amending and completing the existing Electricity Directive 2003/54;

A Gas Directive amending and completing the existing Gas Directive 2003/55;

An Electricity Regulation amending and completing the existing Electricity Regulation 1228/03;

A Gas Regulation amending and completing the existing Gas Regulation 1775/05.

The above-mentioned package promotes sustainability by stimulating energy efficiency and guaranteeing that even smaller companies, for instance those that invest in renewable energy, have access to the energy market.

Discussing the European agenda for infrastructure, Susanne Nies considers two future important factors: the development of LNG and new technologies.[55]

To make the internal market work for all consumers, whether large or small, and to help the EU achieve more secure, competitive and sustainable energy, the Commission proposed a number of measures to complement the existing rules.

Separation of Production and Supply from Transmission Networks: Network ownership and operation should be "unbundled". This refers to the separation of the networks operating in the area of electricity and gas transmission from supply and generation activities. The proposals make it clear that the Commission's preferred option in this respect is ownership unbundling – in other words, a single company can no longer both own transmission facilities and be occupied in energy production or supply activities. In addition, the Commission proposes a second option, the "independent system operator" which makes it possible for existing vertically integrated companies to retain network ownership, provided that the assets are actually operated by a company or body completely independent from it. Either one of these options will create new incentives for companies to invest in new infrastructure, inter-connection capacity, and new generation capacity thereby avoiding blackouts and unwarranted price surges.

---

[55] Susanne Nies, "Gaz et Petrole vers l'Europe – Perspectives pour les infrastructures ", *Les Etudes IFRI*, Paris, 2008.

Facilitating Cross-Border Energy Trade: The Commission decided to establish an Agency for the Cooperation of National Energy Regulators, with binding decision-making powers, to complement the work of national regulators. This Agency is charged with ensuring the proper handling of cross-border cases, thus enabling the EU to develop a real European network working as one single grid, promoting diversity and security of supply.

More Effective National Regulators: The Commission proposes measures to strengthen and guarantee the independence of national regulators in member states.

Promoting Cross-Border Collaboration and Investment: The Commission proposes a new European Network for Transmission System Operators. EU grid operators would cooperate and develop common commercial and technical codes and security standards, as well as plan and coordinate the investments needed at EU level. This would also ease cross-border trade and create a level playing field for operators.

Greater Transparency: Steps to improve market transparency in network operation and supply will guarantee equal access to information, make pricing more transparent, increase trust in the market and help avoid market manipulation.

Increased Solidarity: By bringing national markets closer together, the Commission foresees an increased potential for member states to assist one another in the face of energy supply threats.

Customers would also benefit from a new Energy Customers' Charter, which was launched in 2008. This includes measures to address fuel poverty, information for customers to choose a supplier and supply options, actions to reduce red tape when changing energy suppliers and to protect citizens from unfair selling practices. A separate information campaign would inform customers about their rights.

The Commission's proposals for the internal energy market are an integral part of the Lisbon Strategy and the EU Energy Strategy and will be further discussed at the level of Heads of State and Government at their regular Summits.

On 23 December 2008, the European Union published in the *Official Journal* the new "Directive on the Identification and Designation of European Critical Infrastructures (ECI) and the assessment of the need to improve their protection" (Council Directive 2008/114/EC).

This Directive establishes a procedure for the identification and designation of European critical infrastructures (ECI), and a common approach to the assessment of the need to improve the protection of such infrastructures in order to contribute to the protection of citizens.

In terms of the energy sector operation and in particular with regard to the methods of electricity generation and transmission (electricity supply), it is understood that, wherever it is deemed appropriate, electricity generation may include the electricity transmission segments of nuclear power plants, but would exclude specifically the nuclear elements covered by relevant nuclear legislation, including treaties and Community law.

PROTECTION OF CRITICAL INFRASTRUCTURES IN ROMANIA

Romanian infrastructures are almost in their entirety of a critical nature from at least a few essential viewpoints:

They used to be related to giant industrial facilities, they are inflexible and hardly adaptable to the requirements of market economy;

Romanian society and its economy are in different stages of transition and still need to be synchronised;

Environmental degradation resulting from human action, massive deforestation, deficient land use, and the lack of a coherent and effective agricultural, ecological, and environmental safety policy generate dangers for critical infrastructure;

The participation of Romania in the anti-terrorist coalition and other peace-keeping, crisis management, and conflict prevention missions can generate new types of threats to human lives or living conditions and to vital economic, social, and information infrastructures.

Of course, the potential dangers and threats are much more numerous. Some responses were provided through legislative initiatives, others were included in the National Security Strategy, the energy security programs and other important documents, yet not all of them are fully supervised, managed and controlled.

PROMOTING THE CONCEPT

Notable steps were taken at institutional level by the Romanian Presidency, Parliament (Defence, Safety and Public Order Commission of the Chamber of Deputies), the Romanian Intelligence Service (supported by the Autonomous Company RASIROM and the Information Centre for Security Culture – CICS) and in the framework of the Ministry of Economy (through the security structure of the General Directorate for Energy Policy and the National Power Grid Company TRANSELECTRICA S.A.).

In the private sector, the EURISC Foundation, the Romanian Power Grid Company TRANSELECTRICA S.A., the UTI Group, the RASIROM Company and more recently the Romanian Association for Security Technology (ARTS) had a major role in promoting the concept, organizing series of round tables, national, and international conferences or workshops, and running research projects in partnership with other institutions. The Critical Infrastructure Problematic is now included in the curricula of National Defence University "Carol I", National Defence College, Academy of Economic Studies, University of Bucharest, National Intelligence Academy, Police Academy, etc. In the framework of the Academy of Scientists of Romania ("Academia Oamenilor de Stiinta") a section for critical infrastructure was created.

The EURISC Foundation organized several events on the topic of critical infrastructure protection at an early stage, when this new concept had just started to be promoted in US and EU:

The Ministry of Defence, General Staff, hosted a presentation by the EURISC Foundation on the "Clinton Report" regarding critical infrastructure (1997);

The International Seminar on Risk Governance and Critical Infrastructure was organized by the EURISC Foundation under the auspices of the President of Romania and with the participation of experts from the US and Switzerland as well as 200 Romanian civilian and military specialists from the Supreme Council of National Defence (CSAT) (2001);

The International Seminar on Critical Infrastructure Protection organized by the National Company TRANSELECTRICA S.A. and USAID (Bucharest, 2003);

The International Seminar on Critical Infrastructure Protection organized by the EURISC Foundation under the auspices of the President of Romania with the participation of experts from the US and Switzerland, and 150 civilian and military specialists from the Supreme Council of National Defence (CSAT) (2004);

An International Critical Infrastructure Protection (C.I.P.) Seminar organized at the Palace of Parliament in Bucharest by the Defence, Safety and Public Order Commission of the Chamber of Deputies in partnership with the RASIROM Company, National Company TRANSELECTRICA S.A. and the Romanian Energy Regulatory Authority (ANRE), with the support of the EURISC Foundation, the UTI Group and the Information Centre for Security Culture (CICS) of the Romanian Intelligence Service (2005);

The International C.I.P. Seminar organized by the National Company TRANSELECTRICA S.A. together with the EURISC Foundation, the RASIROM Company and the UTI Group in partnership with "Reseaux Transport Electricité" (EDF) (Sibiu, 2006);

The Roundtable on Critical Infrastructure of the Ministry of Economy and Trade, organized by the National Company TRANSELECTRICA S.A. and the EURISC Foundation (2006);

The Roundtable on Critical Infrastructure Protection organized by the EURISC Foundation and attended by representatives of the Romanian Presidency, the National Company TRANSELECTRICA S.A., the RASIROM Company, the UTI Group and others (Bucharest, 2007);

The International Seminar on Critical Infrastructures Protection under the aegis of the World Security Forum was organized by the EURISC Foundation with the participation of representatives of the Romanian Presidency, the Government, the Ministry of Economy and Finances, the Ministry of Foreign Affairs, the National Company TRANSELECTRICA S.A., the RASIROM Company, ambassadors and other diplomats, experts from the US, Norway, Greece and Romania in risk analysis and protection of energy critical infrastructure focusing on energy security and the role of transit countries. The participants proposed to draft a regional security strategy, to organize an association of transit countries and to launch a new concept of good energy governance (Bucharest, 2007).

With an aim to promote the implementation of the European Programme for Critical Infrastructure Protection in Romania, from 2007 to 2010 the EURISC Foundation together with some of the above-mentioned partners organized in Romania and abroad five seminars, four conferences, six workshops, eight round tables, two NATO and EU expert meeting, four Public-Private Partnership

National Forums for the Security Sector on with topics regarding urban security and critical infrastructure resilience.

An important step toward promoting the critical infrastructure protection concept was made on 13 November 2009 at the International Conference on Energy and Environment (CIEM'09), which was organized in partnership with the Academy of Romanian Scientist and the Polytechnic University of Bucharest, Electrical Engineering College. It gave an opportunity to members of academia, specialists from various sectors, and students to present their research, opinions, and proposals concerning theoretical and practical aspects of critical infrastructure protection.

It is also to be noted that Romanian experts from the Ministry of Economy, the National Power Grid Company TRANSELECTRICA S.A., the Ministry of Interior and Administration, and the EURISC Foundation have been included in European and Euro-Atlantic committees that promote the concept of critical infrastructure.

Since July 2009, EURISC Foundation has cooperated with the Department for National Security of the Presidency of Romania and the National Security Structure of the cabinet of the Prime Minister of Romania for the creation of a working group dedicated to the improvement of national legislation for critical infrastructure protection.

Pending further elaboration of new concepts, strategies, legislation, and specific norms, specialized structures, dedicated logistics, and allocation of financial resources for the functioning, endowment, training, simulations, exercises, security culture, and information for the public, the management of different types of emergency situations involving critical infrastructure is regulated by special laws (on national security, emergency situations, civil protection, etc) and provided by permanent or temporary structures that operate at the level of ministries, departments and specific sectors.

Most of the interventions that are beyond the duties and possibilities of the sectoral structures are carried out by the Inspectorate General for Emergency Situations (IGSU), which is subordinated to the Ministry of Interior and Administration and co-ordinates the Inter-ministerial Group for the Protection of Critical Infrastructure (2007).

The Working Group at the level of the Ministry of Economy and Commerce (2006) provided the general framework for debates and decisions on the development of a coherent Energy Security Strategy.

The Consultative Expert Group on Energy Security, established (2007) at the initiative of the Prime Minister of Romania, developed valuable insights regarding the critical infrastructure protection issues in the energy sector, which were accepted and included in the Energy Strategy of Romania 2007–2020.

The participation of Romanian research units in European projects under the Framework Programme on Science and Technology (FP7) also provides a good opportunity to establish contacts with experienced partners in other EU member states and to promote transfer of expertise in security-related areas.

The Ministry of Education and Research of Romania, with the support of the Romanian Space Agency (ROSA), organizes national competitions for the financing of projects on security research, some which being dedicated to topics on energy security and critical infrastructure protection.

In the framework of security research, several public-private partnerships were established in Romania over the past few years to debate numerous subjects related to energy security and critical infrastructure protection (the EURISC Foundation, the National Power Grid Company TRANSELECTRICA S.A., the RASIROM Company, the UTI Group, etc.).

The ESRIF Report, considered the only large–scale high-level initiative of its kind in Europe was an important source of valuable research knowledge for the ROSTREC Project.[56]

FROM "WAR GAMES" IN MILITARY TO "SERIOUS GAMING" IN ENERGY

"War games" are probably as old as human conflicts. For thousands of years, the military used manned versions of the war games using maps, symbols for personnel and equipment, simulating conflicts, tactics, and strategies. Now the situation is different. The military are using sophisticated equipment and software to play the war games of the twenty-first century.

Due to the complexity of energy issues in the new security environment, not only at a national level but also at an international level, it is high time to move from the classical maps, statistics analysis, reports, etc. in energy to a new tool for enhanced simulation and decision-making for energy suggestively titled "serious gaming".

It is almost impossible, at the level of companies both state owned and private, of national institutions in charge of coordination, regulatory framework development or monitoring the energy players etc., of international organisations, institutions, associations or groups of interests to "digest" the huge amount of data, to manage, propose and take decisions in real time with the lessons learned from the past to the alternative scenarios for the future.

"Serious gaming" for energy is now helping overcome challenges in studying poorly understood dynamic processes, as well as the poorly understood stake-holder institutional interactions by opening to the participations to many different perspectives and special competencies. On a country basis and, over time, it makes it possible to prepare the players in the domain of energy for operational responsibility, analysis research, etc. (see Brewer, G. 2007. Inventing the future: scenarios, imagination, mastery and control" *Sustainability Sciences* 2(2) 159-177)

---

[56] The European Security Research and Innovation Forum (ESRIF) based in a joint initiative of the European Commission and the 27 EU Member States was established in September 2007. More than 600 experts and a plenary of 65 members from 32 countries including independent representatives from industry, public and private end-users, research establishments, universities, non-governmental organizations as well as EU bodies, worked on a European Security Research and Innovation Agenda, with a perspective for the next 20 years. Dr. Liviu Muresan, president of EURISC Foundation participated to the project as a member of the ESRIF plenary. More information on *www.esrif.eu*

In the framework of this project, a ROSTREC scenario gaming solution was developed for actual energy security, by Prof Dr. Adrian Gheorghe and was launched on 26 May 2010 in Brussels.[57]

CONCLUSIONS FOR FURTHER WORK

Action at EU Level

The European institutions recognize that the proper functioning of Europe's critical infrastructures is essential to safe energy supply, but their protection is insufficiently addressed at all levels of governance. Therefore, based on new methods and technologies, as well as an extended dialogue with EU officials, operators, and users, the European Organization for Security (EOS)[58] urged the European Commission (EC) to:

1. Develop an overarching EU policy for the protection and resilience of energy infrastructures unambiguously linked to the EU security (of supply) policy.

2. Develop a consistent EU-wide approach supported by an EU Energy Infrastructure Protection & Resilience Programme in order to promote a sufficient and equal level of protection across borders, thus enhancing the resilience of the entire EU energy network.

3. Address the financial as well as operational obstacles that prevent operators from implementing security measures for their energy infrastructures by setting up adequate incentives and defining an acceptable–risk-based liability model.

EOS considered that: "The present EU energy strategy, connecting different sectoral energy activities in a way that allows for the harmonious and simultaneous pursuit of all of its energy objectives, is not including systematically the protection of energy infrastructures and enhancing the resilience of the networks. This lack of global approach and governance entails an ineffective use of existing funding and fragmented initiatives."

To support and foster the EU's global energy governance, and create an EU Energy Security Programme, EOS put forward a set of actions for consideration. These proposals were recommended to the European Commission and can possibly serve as food for thought for a Romanian future energy strategy.

Short term measures at EU level (2010–2012)

- Develop an EU policy for protection and resilience of energy infrastructures;

---

[57] "Is Europe's Energy Security Policy a Reality or an Ambition?" Policymaker's dinner Bibliotheque Solvay, By Security Defence Agenda, Friends of Europe and EURISC Foundation, 26 May, 2010.

[58] EOS White Paper, *A Global European Approach for Energy Infrastructure Protection & Resilience,* EOS Energy Infrastructure Protection & Resilience Working Group, http://www.eos-eu.com/

- Create a Pan-European Forum of Energy Infrastructure Protection for Public-Private Dialogue (EC, insurance and energy industry organisations and industry suppliers) to discuss common grounds, issues, alternative ideas, good practices, existing methodologies and techniques, etc., to support the definition of a global EU energy policy;
- Develop minimum requirements for products and services delivered by the security industry in cooperation with all relevant stakeholders;
- Provide improved coordination and support to security R&D programs, pilot projects, and experimentations dealing with the protection of energy infrastructures;
- Develop a framework for a risk and threat assessment common methodology and identification of vulnerable points;
- Adopt a liability model that will manage security risks and improve resilience most effectively;
- Broaden the scope of the directive on ECIs so as to encompass also national critical infrastructures;
- Provide elements to the new European Commission, the European Parliament, and Member States to justify the creation of an EU Energy Infrastructure Protection & Resilience Programme and consequent financial support to be envisaged in the 2014–2020 EU financial perspectives.

Medium term measures at EU level (2013–2016)

- Elaborate consistent implementation strategies and harmonies for the content of the OSPs through the development, organization, and sharing of operational good practices and procedures;
- Provide financial and/or operational incentives for CIEOs to encourage the implementation of security measures;
- Support the deployment of the ECI Directive, including the necessary implementation and monitoring mechanisms to ensure compliance;
- Create an EC funded body, a "European Security of Energy Operations Council" to be the custodian of the developed common requirements and good practices for ongoing management and development;
- Promote pilot projects to validate solutions and procedures for risk management (possibly as proposed by EURACOM);
- Create an EU Energy Infrastructure Protection & Resilience Programme.

Long term measures at EU level (2016–2020)

- Deploy the previously established EU policies and foresee the necessary harmonized implementation across EU MS and monitoring mechanisms to ensure compliance;
- Adopt and use a specific and consistent budget for the deployment of security solutions within operators' infrastructures compliant with the ECI Directive and, when needed, specific regulations.

Envisaged Action in Romania

(a) The current legislation, norms, and rules in force as well as other regulations for specific sectors that were previously issued, adopted, harmonized, and published progressively in the Romanian Official Gazette starting with the year 2003 offer a substantial legal database which will have to be taken into consideration by ministries and central administration authorities in order to promote a single legislative initiative defining the critical infrastructure concept in Romania.

(b) Further efforts will have to be undertaken in order to design working procedures, norms, and standards for risk evaluation and analysis, information exchange between specialized structures, qualifications and responsibilities regarding guidance, control and coordination at central, departmental and local or sectoral levels, logistical endowment (material and financial), and the assignation of forces for necessary interventions.

(c) Romania's strategic options have in view of the rapid development of effective, specialized infrastructure networks that should be compatible with the European and Euro-Atlantic ones and be capable to sustain the development and accelerated modernization of the economy and the strengthening of national security. For this purpose, comprehensive programs of national investments will have to be launched in cooperation and partnership with other member states of the European Union and NATO.

(d) The Government should take appropriate steps toward the improvement of national and international critical infrastructure protection through the development of a standardized methodology regarding the identification and classification of threats, provision of adequate communication, co-ordination, effective cooperation, and coherent implementation of the established protection measures, and setting clear responsibilities for all subordinated structures with competences in this field, while paying special attention to terrorist threats.

(e) Special attention will have to be paid to critical infrastructures in the sectors of energy and transport, which are considered as particularly sensitive and having a huge impact on national security, economic activity, and the personal safety of citizens. Failure to do so would have a negative psychological effect on the population and would cause a loss of credibility for the Government and the state administrative structures.

(f) The facilities in the field of energy infrastructure, even though they are operational, are not sufficient to cover all the needs of the domestic market. Further steps must be taken to reinforce Romania's position as a reliable transit country that has important possibilities and responsibilities for the security of the current and future energy routes to Central and Western Europe.

(g) Romania has the potential to become more actively involved in the current debates in the framework of the European Union institutions, European Parliament, and other structures, including NATO, aimed at developing a common energy strategy and designing rules that are acceptable to all partners for regulating the relationship between producer, consumer and transit countries.

(h) It is in Romania's interest to have agreed standards of good governance in the field of energy in keeping with the rules of the free market and providing for a flexible regulatory mechanism for the trade and transit of energy, exchange of relevant information, promotion of public/private partnerships, and involvement of local communities and civil society in sharing the responsibilities and benefits of a cooperative approach.

(i) The new international requirements in the areas of critical infrastructure protection, energy supply, transport security, and stronger cyber defense at the level of both NATO and the European Union level are mandatory for Romanian national authorities and impose specific obligations for the development and implementation of a coherent set of effective policies in this sensitive field.

(j) The expertise acquired by the EU and NATO in setting international standards and the relevant legislation developed by the more advanced member states represent the main sources for the Romanian authorities to design a new, long-term strategic vision for identifying, analyzing, assessing, and managing the national critical infrastructure, for ensuring effective risk management and for improving national capacity to respond to threats.

(k) Given the abiding interest of Romania to promote security and stability in the wider Black Sea region, an important role can be assigned to further initiatives addressing mainly the issues of critical infrastructure and improved regional cooperation to ensure the normal functioning of the interconnected systems, while attaching priority importance to the transit systems for electricity, gas and oil, as well as to air, maritime, railway, and road transport networks.

(l) In accordance with the commitments undertaken by Romania as regional factor of stability in a geographic area that is subject to asymmetric risks, Romanian official representatives and experts will continue to take active part in the work of the specialized structures of the EU and NATO Senior Civil Emergency Planning Committee (SCEPC). Other projects aimed at promoting the improvement of civil protection include the signing of memoranda of understanding with all neighboring countries and bilateral arrangements with other European countries and with the United States Federal Emergency Management Agency (FEMA).

(m) Following the EU and NATO integration of Romania, several Romanian experts won international recognition as system designers or consultants in the field of civil protection and critical infrastructure protection. Some of them came up with very important contributions in such areas as energy infrastructure protection, resilience and mitigation, or the promotion of a new conceptual approach to on risk assessment and mitigation for interdependent systems of systems.

(n) The NATO Summit that was held in Bucharest, in April 2008, provided a good opportunity for specific Romanian contributions to an in-depth examination of issues related to critical infrastructure protection with particular emphasis on energy security as important components of the national security of all member states and of the North Atlantic Alliance as a whole. The Bucharest Summit Declaration, which was issued by the Heads of State and Government on 3 April 2008, highlighted a new NATO approach to the subject of critical infrastructure protection and emphasized NATO's role in energy security as one of the sectors that had a strong impact on the defense capabilities of the allied nations.

The Norway Model – A Regional Framework for Cooperation

THE NATIONAL FRAMEWORK FOR ENERGY SECTOR IN NORWAY

The Storting (parliament) sets the political framework for the energy sector and water resource management in Norway. The Ministry of Petroleum and Energy (MPE) has overall administrative responsibility for these sectors. It is the ministry's responsibility to ensure that the management follows the guidelines set by the Storting.

The Ministry of Petroleum and Energy: The Ministry of Petroleum and Energy has the overall responsibility for an integrated energy policy based on efficient utilization of energy resources.

It comprises four departments:

- The Energy and Water Resources Department,
- The Oil and Gas Department,
- The Technology and Industry Department and
- The Department for Economic and Administrative Affairs (as shown in Figure 1.)

The Oil and Gas Department is responsible for administering oil and gas activities on the Norwegian continental shelf.

The Technology and Industry Department is responsible for research and development activities, research programs within the energy and petroleum sectors, and grants to the development of a Norwegian based international competitive energy industry. The Department also handles the ministry's work on national climate policy and follows up international environmental issues, including international climate negotiations, and regional climate cooperation work. The Department manages the governmental control of Gassnova.

The Department for Economic and Administrative Affairs supervises the government's owner interests in petroleum activities. The Department also handles the ministry's administrative tasks and general services. This includes organizational matters, personnel management, budgeting, and economic affairs.

The responsibilities of **the Energy and Water Resources Department** are the subject of this publication. The Department's main objective is to ensure sound management, in both economic and environmental terms, of water and hydropower resources and other domestic energy sources. The Department acts for the government as owner of the Statnett and Enova state enterprises.

On 1 January 2002, responsibility for exercising the government's ownership function for Statkraft SF was transferred from the Ministry of Petroleum and Energy to the Ministry of Trade and Industry.

The Energy and Water Resources Department consists of the following sections:

- *Water Resources and Area Planning* The Section's working area includes water resource management (issues linked to the use and protection of watercourses and management of the licensing of small power stations), land use planning for energy plants, emergency planning, and water course safety. Administrative responsibility for the NVE also lies here.

- *Hydro Power- and Energy Law and EEA* The Section's main responsibilities are legal issues related to administration of the energy sector. This includes watercourse regulation and hydropower development, licenses for acquisition of waterfalls, reversion, and license management of electricity plants, power lines, and district heating in accordance with the Energy Act. The Section also is responsible for EEA issues linked to watercourse and energy management and to Nordic energy cooperation.

- *Electricity Market* The Section's main working areas are: issues linked to the power market in Norway, power trading with foreign countries, owner follow up of Statnett SF, and the follow up of Statkraft's contracts with power intensive industries. Regulation of grid activities and issues relating to electricity grid tariffs are also a part of the section's work. The Section also works with power industry financial issues, including taxes and fees, and is responsible for power supply emergency planning.

- *Energy Policy* The Section's main responsibilities are general energy policy issues and analyses relating to the energy and power balance. The Section is responsible for the use of the means available for restructuring energy usage and production. Owner follow up of the state owned company Enova, other international energy issues and administrative issues are also a part of the Section's working area.

The Norwegian Water Resources and Energy Directorate. The NVE is a subordinate agency of the Ministry of Petroleum and Energy responsible for administration of Norway's water and energy resources. Its job is to ensure coherent and environmentally-sound management of river systems and to promote efficient electricity trading, cost-effective energy systems, and effective energy use. It also plays a central role in emergency response to flooding and dam failure, and heads contingency planning for power supply. Other duties relate to research and development work and to international cooperation within its sphere of responsibility. The directorate serves as Norway's national hydrological institution.

The Norwegian Petroleum Directorate. This is a subordinate agency of the Ministry of Petroleum and Energy. From 1 January 2004, the Directorate was split into two independent agencies, the NPD and the Petroleum Safety Authority Norway (PSA) – see below.

The NPD's most important duties are:

to exercise the administrative and financial control required to ensure that exploration for and production of petroleum is in accordance with statutes, regulations, decisions, license terms, and so forth;

to ensure that exploration for and production of petroleum resources is in accordance with the guidelines laid down by the Ministry;

to advise the Ministry on issues relating to exploration for or production of submarine natural resources.

The Petroleum Safety Authority Norway. The PSA was established on 1 January 2004 through a division of the NPD. This regulator is responsible for safety, emergency response, and the working environment in the petroleum business and is a subordinate agency of the Ministry of Labor and Social Inclusion.

The Statnett SF. Statnett SF was founded in 1992. The Ministry of Petroleum and Energy acts as its owner on behalf of the government, as specified in the Act of 30 August 1991 relating to state enterprises (Figure 2). Statnett SF is responsible for construction and operation of the central grid. It owns about 87% of the central grid, and operates the entire system. As the transmission system operator (TSO) in Norway, it is also responsible for short- and long-term system coordination. This means that it coordinates the operation of the entire Norwegian electricity supply system and ensures that the amount of electricity generated is always equal to the amount consumed. Statnett's revenues are regulated by the NVE as part of its regulation of monopoly operations.

The Enova SF. Enova SF was founded on 22 June 2001. Based in Trondheim, it is subordinate to the Ministry of Petroleum and Energy. On 1 January 2002, Enova became responsible for government efforts to restructure energy production and use. This work had previously been split between the NVE and the electricity distribution utilities. Enova's activities are financed via an Energy Fund. The fund receives income from a supplement added to the grid tariff of NOK 0.01 per kWh. Its tasks are to promote more efficient energy use, the production of new renewable forms of energy, and environment-friendly uses of natural gas. Quantitative goals have been set for Enova's activities.

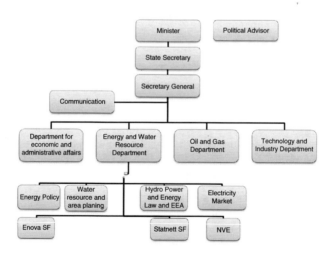

Figure 2. Organisation of the Ministry of Petroleum and Energy.

The Gassnova. Gassnova was established on 1 January 2005 to manage the authority's focus on the development of a gas fired power station with CO_2 processing. Gassnova's goal is to develop cost effective and future orientated technologies for a power station with CO_2 processing, through providing financial support for test and demonstration projects. Gassnova, through its work, is to contribute to the realization of the government's goals relating to the development and use of CO_2 processing technologies. The activity's main finance source is yields from the NOK 2 billion Gas Technology Fund. This was almost NOK 92 million in 2006.

THE MAIN STRATEGIC FRAMEWORK – ENERGI21

The Energi21 initiative was launched by Minister of Petroleum and Energy Odd Roger Enoksen in the winter of 2007 with the aim of designing a broad-based, collective R&D strategy for the energy sector. The mandate for the initiative identified value creation potential, environmental considerations, supply security, and efficient utilization of energy resources as essential criteria for such a strategy. In addition, importance was to be attached to further developing existing research and industrial expertise.

Energi21 unifies energy stakeholders, for the first time, behind a shared vision and R&D strategy that forms the basis for a collective boost. Although the strategy is built on industry priorities, it also stresses closer collaboration between the authorities, trade and industry, and other players in the research arena.

The Energi21 Process

Norway's Minister of Oil and Energy, Odd Roger Enoksen, initiated a process with a decisive impact on Norwegian energy research for decades. The process was called *Energi21*, and its purpose was to establish a broad yet unifying R&D strategy in the energy sector.

The strategy document was presented by Mr. Enoksen at the beginning of 2008 to the interested partners and institutions. Mr. Enoksen appointed a strategic committee to be in charge of this effort. Sverre Gotaas of Statkraft chaired the committee and Hans Otto Haaland from the Research Council of Norway headed its secretariat.

The mandate of the committee was established a broad strategy for the energy sector that brought together a wide-range of R&D goals and communities. The objective of the strategy was to provide a secure platform for the growth of sustainable economic activity and supply-side security in the energy sector by promoting and coordinating a commitment to research, development, demonstration, and commercialization of new technology.

The Entire Chain of Innovation

Energi21 addressed topics relevant to stationary production of energy, energy transport, and energy use. Moreover, Energi21 encompasses the entire chain of innovation with the exception of independent basic research – in other words, everything from strategic energy research to the introduction of a new technology in the market. Relevant social science research was also included.

An inclusive process:

Work on this strategy has taken place in the context of an inclusive process which encompassed the entire energy industry (energy companies, suppliers and research communities) in an R&D strategy to which most of them were able to commit.

Work on the strategy was taken status of R&D in this sector in order to establish a road map of the industry and identify the most important technological challenges and knowledge deficits. This has provided a basis for defining a number of strategic target areas – an essential step in deciding the strategy's structure and focus. Individual strategy plans has been formulated for these target areas at a later stage of the process.

Specific objectives:

One aspect of the strategy effort has been to identify specific R&D objectives and formulate recommendations on what needs to be done to achieve these objectives.

What encompasses Energi21?

Energi21 is open for all technology relevant to:

stationary energy production

energy distribution (transport)

energy use

The focus is on technology to be used in Norway with potential for international application.

Dialog

The strategy group invited interested parties to take part in a dialog and offered suggestions/proposals through a range of channels. The group urged interested parties to contribute ideas through their own organizations directly to group members, through special gatherings that were held, and through a website, www.energi21.no. The intention of an interactive blog on this website was to ensure access for the fullest possible scope of good ideas and suggestions/proposals.

The Findings and Strategic Directions

The strategic committee was convinced that this was an ideal time to introduce such a strategy; however, it should be noted that there were three essential elements that must be in place if a collective effort of this nature is to succeed:

- The industry's investments in R&D are based on the belief that they will lead to future profits. Stable, competitive framework conditions are therefore critical.
- Public R&D allocations as well as financial incentives directly aimed at the industry are necessary to increase the pace of R&D activity. The SkatteFUNN tax deduction scheme is one example, and others must be considered.
- Activities must be given priority under a cohesive strategy – which is the core message of this paper.

A key point of the Energi21 collective strategy was to reduce this hold-up and to increase contributions by the industry. It is difficult to quantify the need for investment in research and development in the energy sector. Given the attention being focused on renewable and environmentally friendly energy, the strategic committee feels that ambitions should seek to be both high and realistic.

The strategic committee suggests doubling the Ministry's R&D contributions in 2009, while the needs from 2010 onward will be much higher. These include public allocations of NOK 400 million per year (excluding funding for CO2 management and nuclear power), which over time is expected to attract private investment of at least NOK 2.4 billion annually. It was not easy to estimate the need for funding for demonstration and commercialization projects. Experience shows that these types of projects require substantial amounts of funding. An increase in the budgetary framework for R&D therefore carries with it a need for more funding for demonstration projects to ensure the commercialization of R&D results.

Based on the efforts carried out, the strategic committee has established the following vision for Energi21.

Norway: Europe's energy and environment-conscious nation – from a national energy balance to green energy exports Norway has the natural resources, community of experts, and social framework to become Europe's leading energy and environment-conscious nation, which will in turn make Norway:

- A society with low greenhouse emissions and high energy efficiency;
- A major supplier of environmentally friendly power to Europe;
- A nation whose R&D strategy and industrial policy will attract the world's foremost energy and technology companies.

To achieve this vision, the strategic committee recommended that:

- An R&D initiative to be established that focuses on five priority areas, as well as investment across the board in education and basic and applied research.

- Energy research to be strengthened through a plan of escalating funding that includes doubling today's allocations over a 2-year period.
- Energi21 to be organized as a permanent activity with its own Board, to ensure that the stakeholders remain involved and committed to realizing the Energi21 vision.
- The industry and the authorities must cooperate closely on the objectives for and organization of energy research.

The strategic committee proposed also to concentrate on five priority areas:

- Research on efficient use of energy in buildings, households, and industry will be a key component of Energi21. Research in this area will comprise not only technological research and development, but also social science research in areas such as consumer behaviour.
- Given the natural advantages of Norway and the technological and industrial areas in which it excels, initiatives targeting more climate-friendly power from hydropower, wind power, and solar energy are recommended.
- Investment in further strengthening the communities involved in solar cell technology was also recommended.
- R&D initiatives relating to CO_2-neutral heating, including bio-resources, utilization of heat from surroundings, and waste heat.
- Within bio-energy, initiatives should encompass access to resources, production processes, distribution, and market systems.

This strategy was an instrument applied in Norway's efforts to find solutions to the challenges being posed by global climate change. The Energi21 process has found that there is great willingness on the part of the stakeholders to commit and contribute to a collective strategy.

NORWAY ENERGY SECTOR IN THE EUROPEAN AND REGIONAL FRAMEWORK

Participation in EU Energy Programmes

Norway has participated since 1996 in the EU's Save and Altener programs on energy efficiency and renewable energy sources. On 26 June 2003, the EU resolved to establish a new overarching energy program for 2003–2006 called Intelligent Energy Europe. This extends Save and Altener and introduces Steer, a program aimed at the transport sector, and the Coopener program on cooperation with developing countries over energy issues. Intelligent Energy Europe was incorporated in the EEA agreement in November 2003 and, somewhat later, Coopener was incorporated in EEA in November 2004. The Commission established an Intelligent Energy Executive Agency that is responsible for the operational aspects of the program, while the Commission will continue to deal with the policy-related issues together with the member countries in EU/EEA. Planning the next phase of the Intelligent Energy Europe from 2007 has already started by the integration of the energy program in a larger, sector overreaching program, the 'Competitiveness and Innovation Programme' (CIP), that will run from 2007 to 2013. Norwegian participation is anticipated in the program. Norway contributes funding to the programs and Norwegian interests can apply for project support from the Intelligent Energy Europe program. Such applications must be made jointly with one or more partners from within the EU.

The Nordic Cooperation

The Nordic countries have a long tradition of cooperation in the energy field. At government level, collaboration has been established under the Nordic Council of Ministers. In addition, there is extensive collaboration between the system operator networks in each country, and collaboration between the Nordic regulators in NordREG.

The Nordic energy ministers meet once a year. Between these meetings, energy collaboration is headed by a committee of senior civil servants. The collaboration in the energy sector is concentrated in the following areas: Electricity, sustainable energy (climate, energy efficiency implementation, renewable energy and energy technologies, and regional cooperation with neighboring areas. The energy ministers have agreed on an Action Plan for the Nordic political energy collaboration from 2006 to 2009. Norway held the chair in 2006. In the Akureyri Declaration issued after the Ministers Meeting on 2 September 2004 and the Greenland Meeting in August 2005, the Nordic countries agree to further develop the Nordic electricity market and to collaborate on supply safety issues in Scandinavia. The Scandinavian countries agree to focus in particular on the possibilities for coordination of system operator responsibility and common approaches to investments in the central network. Nordel, the organization for system operator networks in Scandinavia, will submit a report on further harmonization of the power market in Scandinavia.

The Baltic Cooperation

The Bergen Declaration on a sustainable energy supply around the Baltic was issued by the Nordic prime ministers in 1997. It forms the basis for energy cooperation in this region, and has been followed up subsequently by the energy ministers. Following the energy minister meetings in Stavanger in 1998 and Helsinki in 1999, a more permanent regional energy collaboration was established as the Baltic Sea Region Energy Cooperation (BASREC). BASREC is organized as part of the collaboration within the Baltic Council which embraces 11 countries – Russia, Germany, Poland, Estonia, Latvia, Lithuania, Sweden, Finland, Denmark, Iceland, and Norway – and the European Commission with work headed by a group of senior energy officials (GSEO). BASREC's Energy Minister Meeting in Reykjavik 2005 issued a new mandate for BASREC operations which will apply for the period from 2006 to 2009. The power companies in the Baltic region have established their own collaboration, known as Baltrel, to help create a single market for the area. Baltrel cooperates with Baltic Gas, a similar organization for the gas companies.

The Economic Commission for Europe (ECE)

The Economic Commission for Europe is one of the UN's five regional commissions. It was established in 1947, and has a committee for sustainable energy in which Norway participates. This committee provides a meeting place for 55 nations, including the USA, Canada, European countries, and most of the former Soviet republics in Central Asia. It has working groups for energy efficiency, gas, and coal. In addition to discussing key energy policy issues of mutual interest, these groups focus on information dissemination and knowledge transfer between the member countries, with a particular emphasis on measures for energy efficiency in central Europe.

The European Energy Charter

The European Energy Charter forms the political framework for pan-European energy cooperation, including former Soviet Union republics and East European nations, as well as Japan and Australia. Signed in December 1991, its objective is to promote long-term energy cooperation based on the principles of the market economy and non-discrimination. The European Energy Charter treaty was signed in Lisbon in 1994. Fifty-one countries have signed both the treaty and a protocol on energy efficiency. After 30 countries had ratified both the treaty and the energy efficiency protocol, these two agreements came into force in the spring of 1998. Norway signed the concluding document of the Conference and signed the treaty in 1995, but has yet to ratify it.

The Cooperation with Russia and the Barents Area

Norway signed an energy efficiency agreement with Russia in 1996. The objective of the agreement is to facilitate projects on energy efficiency and utilization of new renewable energy sources in north-western Russia. Six energy efficiency centers have been established under this agreement in the Russian part of the Barents region. Collaboration between the two countries in this area has continued after the agreement expired in 2002. Expertise transfer, demonstration projects, development of funding models, and information dissemination are important elements in this bilateral collaboration. The Barents Euro-Arctic Council adopted an action plan in 1998 for improving energy supply in the Russian part of the Barents region. It also decided to appoint an Energy Working Group (EWG) to pursue the objectives of the plan. The EWG includes representatives from various sectors and regions in Norway, Finland, Sweden, and Russia. Denmark and Iceland participate sporadically, and the EU has observer status. Work in the EWG has concentrated on establishing networks and disseminating information. Particular attention has been focused on energy efficiency and the use of renewable energy sources. The energy efficiency centers backed by Norway in north-western Russia have gained official status through the EWG as Barents Energy Focal Points. A special group of experts on bio-energy was appointed in 2002 and submitted its final report in the spring of 2004. Norway held the EWG presidency in the period 2001–2004. After the regions were included more actively in the group in 2004, the Chair was assumed jointly by Finnmark County Municipality and Russia.

The International Energy Agency (IEA)

The International Energy Agency embraces 26 of the 30 members of the Organization for Economic Cooperation and Development (OECD). The EU Commission also participates in its work. The IEA was established in response to the 1973–1974 oil crisis as an independent organization associated with the OECD, and has subsequently developed into an important element in the political and scientific cooperation on energy pursued between the OECD countries. Norway is associated with the IEA through a separate agreement which provides that Norway cannot be made subject to the same obligations as other members in the event of an oil supply crisis. The country otherwise participates on equal terms with other countries in this collaboration, on its board and in its sub-committees. Issues related to electricity generation and supply, energy use. and energy efficiency are mainly discussed in the committee for long-term cooperation (SLT). Analyses are also carried out of output/production and demand for various energy carriers, such as electricity, gas, coal, and nuclear power. The IEA's activities also embrace energy research and development. The member countries' energy policies were the subject of regular detailed reviews led by representatives from the other member countries. The IEA defines minimum storage volumes in crude oil and oil products for all members and operates an important crisis sharing system. Norway was last reviewed in 2005. Energy and Environmental issues have gained a more prominent place in the IEA's energy policy agenda and the agency has become an important contributor to various international fora.

Conclusions and Recommendations for Good Energy Governance

There is no single source of new energy that can replace oil and gas during the next 20 years.

There are a number of sources that can contribute, but lead times are long and one cannot assume that the invisible hand of the market will provide timely solution.

The visible hand of the government intervention may compensate through subsidies, investment support, taxation, etc., but there is no free lunch, and as the economic crisis continues, focus will turn to fiscal discipline rather that fiscal stimulus.

Romania is a relatively good position thanks to its development of nuclear power, and with a significant production of oil and gas.

To reach climate targets and improve competitiveness, it is important to find areas with the lowest mitigation cost (cost per ton of CO_2 removal): improving energy efficiency within the projected energy mix that meets the target established by 2020 plan.

Probable landing of one of the major European projects may offer exciting opportunities for Romania.

Strategic targets for Romania should be:

- Improved energy security, definition of vulnerability/resilience levels;
- Reduced CO2 emissions;
- Reduces import requirements for oil and gas;
- Increased economic activity and optimal use of EU financial support, mobilization of private and foreign capital.

I. Deregulation challenges

> **Review of the national energy strategies**
>
> **Cope with negative impacts on other sectors**
>
> **Social protection instruments for energy consumption**

1. Review of the existing national and energy strategies, promotion of a public debate among the stakeholders concerning a common strategy to cope with negative effects of the transition phase in the sector.

2. Reducing the financial and economic crisis impact by promoting decisions and steps to manage the transition to a liberalized energy sector/market.

3. Reducing the negative impact on the human resource involved in the energy sector within the context of restructuring process, especially the nuclear sector, but also the other energy sectors which need highly qualified labor force.

4. The process of deregulation should be implemented taking into consideration the possible negative impact over other connected sectors from the national economy, especial industrial activity.

5. Given the fact that the primary energy source will remain the coal energy sector, Romania should develop a long term and coherent vision of the coal energy use within the European recommendations concerning the standards of coal energy production.

6. Taking into consideration that subsidies and social protection instruments for energy consumption were reduced, there is a need for finding a solution in order to provide a protection alternative. For example, the use of cohesion EU funds or possible revenues can provide substantial social benefits both country and consumers.

II. Regulation mechanisms and energy security priorities

7. Despite the fact that it was modified a few times, the primary normative framework should be improved.

8. The main goal of an integrative energy strategy is providing and ensuring energy at competitive prices.

9. Enhancing the role of ANRE in the energy market as an active player at the national and as an interface with the European framework.

ANRE should address the issues of planned energy sectors, competition, integration of the national energy market into regional, European and international market, improving the quality of the energy sector.

10. A coherent and comprehensive strategy to involve private actors in the energy sector according to European provisions in the competition area.

11. Improving the regulatory framework in respect of new energy sources and in accordance with the EU recommendations (solar, wind, bio-fuel, etc.).

Improvement of the primary normative framework

Ensuring energy at competitive prices

Enhancing the role of the Romanian Energy Regulatory Authority – ANRE

Promote new energy sources and energy efficiency

European Policy Synergies towards the Wider Black Sea Area

---

12. To review the existing legal framework concerning the Romanian strategy toward non EU energy actors.

13. A better use of the existing instruments outside the EU market (Black Sea Synergy, Black Sea Economic Cooperation Organization, Regional Bilateral, and trilateral strategic partnerships.

14. It is in Romania's interest to have agreed standards of good governance in the field of energy in keeping with the rules of the free market and providing for a flexible regulatory mechanism for the trade and transit of energy, exchange of relevant information, promotion of public/private partnerships, and involvement of local communities and civil society in sharing the responsibilities and benefits of a cooperative approach.

15. The below programs have to be revised and updated in light of the recent international changes and their format will have to comply with the transposition of the EU's legislation. Further, the below initiatives have to receive a broader scope in order to be able to cover all the EU's Climate Change Action Plans.

→ *The National Strategy on Environment Protection*, Climate Change chapter has to be revised in order to integrate the EU new actions.

→ *Strategy and Plan of Action on Atmosphere Protection*, contains outdated information and it has to be refreshed with new data from the European Environment Agency.

→ National Strategy on Climate Change for Romania (SNSC) which summarizes the Romanian strategy for meeting its commitments to the EU and

UNFCCC has to be revised in line with the outcomes of the international negotiations.

→ The National Plan of Action for Climate Change (PNASC) structured by:

→ The industrial policy of Romania.

→ The energetic sector network , Romanian Government's Decision HG no. 890/2003.

→ The National Strategy related to the Energy Efficiency (HG no. 163/ 2004).

→ The strategy on renewable energy sources (HG no. 1535/2003).

→ The Romanian Medium Term National Strategy of Economic Development.

16. Romania has to take appropriate measures to mitigate the risk of becoming the subject of having opened an EU's infringement procedure; hence the Ministry of the Environment and Sustainable Development has to carefully check that all the EU's legislation is respected and correctly transposed.

III. Integration into the European energy community

Meet the 2020 energy efficiency targets

Contributions to European energy legislative initiatives

Access European Programs including funds for smart grids carbon capture and storage

Waste to energy projects

17. Romania, as a member of EU, should follow the six key areas highlighted by the European Commission in order to create policies amenable to its energy security.

- The EU looks at its energy mix and continues attempts to meet its 2020 goals: a 20% reduction in greenhouse gas emissions from 1990 levels, a 20% increase in the share of renewable energy sources in final energy consumption, and a 20% increase in energy efficiency by 2020. In order to meet these goals, Romania should be part of the EU efforts to develop new technologies.
- Sooner or later, EU will address, and Romania should be prepared for, an EU debate concerning the legally binding framework for energy efficiency.
- The second area is the strength of the internal market. In the view of the past Ukrainian crisis, this objective is an important stability element.
- Third, the strength of any EU member state's internal market is based on the strength of its infrastructure. To respond to this, the EU Commission will put forth an outline in November 2010 as to what our energy infrastructure policy should be. The current energy recovery program,

which has allotted one billion Euros to energy infrastructure development, will be followed up in the 2014–2020 financial perspective.

- Fourth, Romania should take into consideration the EU proposal for the EU to be better prepared for upcoming crises through emergency preparedness.
- The fifth point is diversifying Europe's energy supply and especially the southern corridor which includes among others, projects as Nabucco (a strategic priority for Romania). The EU is interested in the diversification of three elements: supply countries, supply companies, and supply routes.
- The last element is the process of slowly replace of previous inter-governmental agreements with the new EU framework.

18. Active involvement of Romania in the preparation of the EU energy-related legislative and regulatory framework which may eventually shape (or have an impact on) agreed international standards in areas such as:

- Smart grids and smart metering;
- Standardization of required infrastructure for electric vehicles, including car-charging, battery exchange facilities and interfaces with interoperable plugs;
- Operational requirements for carbon capture and storage facilities;
- Optimal incentives for the development of power production from renewable and alternative sources;
- Definition of energy poverty and agreed ways to cope with it.

19. Improving and identifying new national instruments to increase the absorption capacity of European funds concerning the energy sector, e.g. smart grids and carbon capture storage projects. For the private sector, Romania should encourage investments in small energy units. At the European level, energy production diverges from scale economies.

20. Waste to energy projects should be supported and promoted and regarded as a high priority for the Romania Government, not only as a win-win situation but also using the local community potential for solving stringent problems of waste, energy, and workplaces.

21. Romania should support EOS proposal for the creation of an EC funded body, a "European Security of Energy Operations Council" to be the custodian of the developed common requirements and good practices for ongoing management and development.

22. At the European level Romania should support the building of a framework for a risk and threat assessment common methodology and identification of vulnerable points.

23. Romania has to comply with the position of the other EU MS and to accept the changes brought by the EU's decision to modify the base year for the GHG gas reductions to the post-2000 period. In compensation for its allowances lost it should receive funds to support the transition to a low carbon industry and infrastructure.

IV. Medium and Long term strategic vision

24. Promote Romania's position towards a unified European strategy and a common energy market. Access to market is one of the strategic factors in ensuring energy security for all public and private stakeholders.

25. Support the existing and the future energy projects aiming to diversify sources and reduce the energy dependency.

26. Continue to develop the energy infrastructure, with the aim of interconnection at the regional and European level at all energy sectors. The implementation of smart grids projects and transformation of electric networks into active networks, will follow one of the main European priorities.

Diversify sources and reduce energy dependency

Contribution and participation in the European Energy Infrastructure

Infrastructure protection including new defense against cyber attacks

PROCSI - new tools for education in the field of resilience

Public private partnerships

Energy diplomacy

Serious Gaming: ROSTREC Online applications for policy formulation

27. One of the existing challenges is the technological gap among EU members. Romania should find solutions in order reduce this gap, to assure the transfer of know-how and new technologies by privatizations, joint ventures, direct and indirect foreign investments, etc.

28. Romania's strategic options have in view the rapid development of effective, specialized infrastructure networks that should be compatible with the European and Euro-Atlantic ones and capable to sustain the development and accelerated modernization of the economy and the strengthening of national security. For this purpose comprehensive programmes of national investments will have to be launched in cooperation and partnership with other member states of the European Union and NATO.

29. The Government should to take appropriate steps toward the improvement of national and international critical infrastructure protection through the development of a standardized methodology regarding the identification and classification of threats, provision of adequate communication, co-ordination, effective cooperation, and coherent implementation of the established protection measures and set clear

responsibilities for all subordinated structures with competences in this field, while paying special attention to terrorist threats.

30. The facilities in the field of energy infrastructure, even though they are operational, are not sufficient to cover all the needs of the domestic market. Further steps must be taken to reinforce Romania's position as a reliable transit country that has important possibilities and responsibilities for the security of the current and future energy routes to Central and Western Europe.

31. Romania should develop two-way interconnection capacities to neighboring countries and the European gas grid.

32. Given the abiding interest of Romania to promote security and stability in the wider Black Sea region, an important role can be assigned to further initiatives within the existing regional structures addressing mainly the issues of critical infrastructure and improved regional cooperation to ensure the normal functioning of the interconnected systems, while attaching priority importance to the transit systems for electricity, gas, and oil as well as to air, maritime, railway, and road transport networks.

33. In Romania, there is a need for building a new leadership culture in the energy sector, given the generation change and the reforms that should be continued. There is a need for long term approach in training, energy security will be in demand for intensive, and strategic courses oriented towards EU developments. Educational programs will have a positive impact only if they benefit from a European integrated vision with a strong intensive simulation dimension such as CAE[59]-type applications for crisis management and benefiting from new technologies including virtual networks and e-learning tools. PROCSI – the Regional Center for Training and Research on Critical Services and Infrastructures Protection – and the EURISC Foundation Project – could, under the above mentioned factors, develop a pilot-educational project in the field of the protection and resiliency of the critical services and infrastructures, at national and international levels for energy security.

34. Special attention should be paid to strategic investors. There is a need to improve the existing legal and economic framework for public-private partnership in the energy sector. The public-private partnership is represented by the trilateral relationship: state sector, the private Romanian sector, and the foreign private sector (there are several foreign energy players from Austria, Azerbaijan, Canada, Czech Republic, Germany, France, Hungary, Italy, Russia, Spain, Kazakhstan, USA, etc.).

35. For a coherent and integrated energy strategy, Romania should adapt to the new trends of using, developing, and promoting such concepts as energy diplomacy as a part of foreign policy. This integrative concept covers the political, economic, security and societal dimensions of foreign policy.

36. Promote EU solidarity and crisis sharing, IEA system for oil and oil products could be a model. Define minimum storage volumes and key pipeline two-way connections.

---

[59] Computer Assisted Exercises.

37. Accelerate the possibility of utilizing depleted Romanian gas fields as strategic gas storage.

38. In the field of R&D, Romania should establish projects with maximum EU-funding under the SET (Strategic Technologies Plan) programme.

39. One of the Romanian cities can become, under the EU energy efficiency: Smart Cities Initiative by 2020 the initiative should put 25–30 European cities at the forefront to a law carbon future: smart networks, a new generation of buildings and low carbon transport solutions.

40. Romania has to actively be involved in the SET Plan: the sustainable nuclear fission initiative. The European target is to have in place and operate by 2020 the first generation IV Prototype. The first co-generation reactor could also appear as demonstration projects to test the technology for coupling with industrial process.

41. As a solution proposed by ROSTREC project is the promotion and use of "Serious Gaming: ROSTREC Online" type of applications by public and private companies, state institutions in Romania, as well as promoting this decision aiding solution to international institutions and organisations such as the EU, NATO, World Bank, IMF etc.

AUTHOR INDEX

A. Gheorghe and L. Muresan (eds.), *Energy Security: International and Local Issues,*   413
Theoretical Perspectives, and Critical Energy Infrastructures,
DOI 10.1007/978-94-007-0719-1, © Springer Science+Business Media B.V. 2011

Printed by Books on Demand, Germany